Solutions Manual for

Algebra $\frac{1}{2}$

An Incremental Development

Third Edition

John H. Saxon, Jr.

SAXON PUBLISHERS, INC.

Saxon Publishers gratefully acknowledges the contributions of the following individuals in the completion of this project:

Author: John Saxon

Editorial: Brian E. Rice

Editorial Support Services: Christopher Davey, Jack Day, Shelley Turner

Production: Adriana Maxwell, Brenda Lopez

Printed in the United States of America

ISBN: 978-1-56577-131-4

ISBN: 1-56577-131-1

26 0928 21

4500819413

Preface

This manual contains solutions to every problem set in the *Algebra $\frac{1}{2}$*, Third Edition, textbook. Early solutions of problems of a particular type contain every step. Later solutions omit obvious steps. We have attempted to stay as close as possible to the methods and procedures outlined in the textbook. These solutions are representative of a student's work, but many problems have alternate solutions. For ease of grading, the final answer of each problem is set in boldface type.

The following individuals were instrumental in the development of this solutions manual, and we gratefully acknowledge their contributions: Clint Keele and Matt Maloney for writing and revising the solutions; Karen Bottoms for working the problems and checking the answers; Eric Atkins and Jane Claunch for typesetting the manual; Sariah Adams, Jeremy Eiken, and Chad Morris for proofreading the manual; Wendy Chitwood, David LeBlanc, Tonea Morrow, Dan O'Connor, Jason Purswell, and Nancy Rimassa for creating the graphics; Chris Davey for copyediting the manual; and Carrie Brown and Brian Rice for supervising the project.

PROBLEM SET 1

1. (a) This 6 is in the ten-thousands' place, so it has a value of $6 \times 10,000$, or **60,000.**

   (b) This 9 is in the hundreds' place, so it has a value of 9×100, or **900.**

   (c) This 3 is in the units' place, so it has a value of 3×1, or **3.**

2. **334,333**

3. **3,666,766**

4. **39,959,992**

5. **41,000,200,520**

6. **507,640,090,042**

7. **407,000,090,742,072**

8. **980,000,470**

9. 517,236,428 is **five hundred seventeen million, two hundred thirty-six thousand, four hundred twenty-eight**

10. 90,807,060 is **ninety million, eight hundred seven thousand, sixty**

11. 32,000,000,652 is **thirty-two billion, six hundred fifty-two**

12. 3,250,009,111 is **three billion, two hundred fifty million, nine thousand, one hundred eleven**

13. 6,040,000 is **six million, forty thousand**

14. 99,019,900 is **ninety-nine million, nineteen thousand, nine hundred**

15. $(3 \times 100,000) + (4 \times 1000) + (2 \times 10)$
    $= 300,000 + 4000 + 20 = \mathbf{304,020}$

16. $(7 \times 10,000) + (8 \times 100) + (6 \times 10)$
    $= 70,000 + 800 + 60 = \mathbf{70,860}$

17. $(9 \times 1000) + (4 \times 100) + (5 \times 1)$
    $= 9000 + 400 + 5 = \mathbf{9405}$

18. $(7 \times 1,000,000) + (2 \times 10,000) + (6 \times 1000)$
    $= 7,000,000 + 20,000 + 6000 = \mathbf{7,026,000}$

19. $5280 = 5000 + 200 + 80$
    $= \mathbf{(5 \times 1000) + (2 \times 100) + (8 \times 10)}$

20. $408 = 400 + 8 = \mathbf{(4 \times 100) + (8 \times 1)}$

21. $70,600 = 70,000 + 600$
    $= \mathbf{(7 \times 10,000) + (6 \times 100)}$

22. $21,000 = 20,000 + 1000$
    $= \mathbf{(2 \times 10,000) + (1 \times 1000)}$

23. $4005 = 4000 + 5 = \mathbf{(4 \times 1000) + (5 \times 1)}$

24. $9080 = 9000 + 80 = \mathbf{(9 \times 1000) + (8 \times 10)}$

25.
$$
\begin{array}{r}
43 \\
76 \\
84 \\
+\ 91 \\
\hline
\mathbf{294}
\end{array}
$$

26.
$$
\begin{array}{r}
4628 \\
5734 \\
+\ 8416 \\
\hline
\mathbf{18,778}
\end{array}
$$

27.
$$
\begin{array}{r}
\$53.58 \\
+\ \$52.78 \\
\hline
\mathbf{\$106.36}
\end{array}
$$

28.
$$
\begin{array}{r}
9056 \\
4708 \\
+\ 9076 \\
\hline
\mathbf{22,840}
\end{array}
$$

29.
$$
\begin{array}{r}
432 \\
846 \\
943 \\
+\ 721 \\
\hline
\mathbf{2942}
\end{array}
$$

30.
$$
\begin{array}{r}
\$3.64 \\
\$0.52 \\
+\ \$9.00 \\
\hline
\mathbf{\$13.16}
\end{array}
$$

PROBLEM SET 2

1.

2. **−415, 145, 154, 451, 514**

3. **249, 294, 429, 924, 942**

4.
$$4,185,\textcircled{2}70$$
The rounded number is **4,185,300.**

5.

$$83,72\textcircled{1},525$$

The rounded number is **83,722,000.**

6.

$$415,\textcircled{2}37,842$$

The rounded number is **415,200,000.**

7. **777,727,757**

8. **3,634,733**

9. **107,047,020**

10. **93,462,000,047**

11. 731,284,006 is **seven hundred thirty-one million, two hundred eighty-four thousand, six**

12. 903,721,625 is **nine hundred three million, seven hundred twenty-one thousand, six hundred twenty-five**

13. 9,003,001,256 is **nine billion, three million, one thousand, two hundred fifty-six**

14. 7,234,000,052 is **seven billion, two hundred thirty-four million, fifty-two**

15. $(7 \times 10,000) + (6 \times 100) + (5 \times 10)$
$+ (4 \times 1)$
$= 70,000 + 600 + 50 + 4 = \mathbf{70,654}$

16. $(3 \times 100,000) + (9 \times 1000) + (7 \times 100)$
$+ (6 \times 10) + (3 \times 1)$
$= 300,000 + 9000 + 700 + 60 + 3$
$= \mathbf{309,763}$

17. $(9 \times 1000) + (6 \times 100) + (9 \times 1)$
$= 9000 + 600 + 9 = \mathbf{9609}$

18. $109,326 = 100,000 + 9000 + 300 + 20 + 6$
$= (1 \times 100,000) + (9 \times 1000) + (3 \times 100)$
$+ (2 \times 10) + (6 \times 1)$

19. $68,312 = 60,000 + 8000 + 300 + 10 + 2$
$= (6 \times 10,000) + (8 \times 1000) + (3 \times 100)$
$+ (1 \times 10) + (2 \times 1)$

20. $903,162 = 900,000 + 3000 + 100 + 60 + 2$
$= (9 \times 100,000) + (3 \times 1000) + (1 \times 100)$
$+ (6 \times 10) + (2 \times 1)$

21.
$$\begin{array}{r} \$93.17 \\ \$45.26 \\ + \ \$90.15 \\ \hline \mathbf{\$228.58} \end{array}$$

22.
$$\begin{array}{r} 7316 \\ 4582 \\ + \ 9143 \\ \hline \mathbf{21,041} \end{array}$$

23.
$$\begin{array}{r} 88,871 \\ 40,012 \\ + \ 90,375 \\ \hline \mathbf{219,258} \end{array}$$

24.
$$\begin{array}{r} 78,524 \\ 91,325 \\ 70,026 \\ + \ 91,358 \\ \hline \mathbf{331,233} \end{array}$$

25.
$$\begin{array}{r} 42,715 \\ 90,826 \\ 41,222 \\ + \ 39,057 \\ \hline \mathbf{213,820} \end{array}$$

26. $37,251 + 81,432 + 90,256 + 21,312$
$= \mathbf{230,251}$

27. $14 + 32 + 16 + 21 + 932 + 21 = \mathbf{1036}$

28. $1 + 2 + 21 + 12 + 122 + 1222 = \mathbf{1380}$

29. $33¢ + \$3 + \$13.99 = \mathbf{\$17.32}$

30. $4 + 314 + 134 + 13,245 = \mathbf{13,697}$

PROBLEM SET 3

1. **225,223**

2. **70,777**

3. **4,144,444**

4. **14,705,052**

5. **500,000,465,182**

6. $64,030 = 60,000 + 4000 + 30$
$= (6 \times 10,000) + (4 \times 1000) + (3 \times 10)$

7. $79,003 = 70,000 + 9000 + 3$
$= (7 \times 10,000) + (9 \times 1000) + (3 \times 1)$

8. $123{,}419 = 100{,}000 + 20{,}000 + 3000 + 400$
$\quad\quad + 10 + 9$
$\quad = (1 \times 100{,}000) + (2 \times 10{,}000) + (3 \times 1000)$
$\quad\quad + (4 \times 100) + (1 \times 10) + (9 \times 1)$

9.
$$\begin{array}{r} 551 \\ -\ 174 \\ \hline \mathbf{377} \end{array}$$

10.
$$\begin{array}{r} 853 \\ -\ 284 \\ \hline \mathbf{569} \end{array}$$

11.
$$\begin{array}{r} 936 \\ +\ 474 \\ \hline \mathbf{1410} \end{array}$$

12.
$$\begin{array}{r} 839 \\ +\ 472 \\ \hline \mathbf{1311} \end{array}$$

13.
$$\begin{array}{r} \$60.00 \\ -\ \$49.49 \\ \hline \mathbf{\$10.51} \end{array}$$

14.
$$\begin{array}{r} 4017 \\ -\ 3952 \\ \hline \mathbf{65} \end{array}$$

15.
$$\begin{array}{r} X \\ -\ 245 \\ \hline 276 \end{array} \quad \begin{array}{r} 276 \\ +\ 245 \\ \hline \mathbf{521} \end{array} \quad \text{Check:} \begin{array}{r} 521 \\ -\ 245 \\ \hline 276 \end{array}$$

16.
$$\begin{array}{r} 800 \\ -\ M \\ \hline 436 \end{array} \quad \begin{array}{r} 800 \\ -\ 436 \\ \hline \mathbf{364} \end{array} \quad \text{Check:} \begin{array}{r} 800 \\ -\ 364 \\ \hline 436 \end{array}$$

17.
$$\begin{array}{r} 735 \\ +\ A \\ \hline 1211 \end{array} \quad \begin{array}{r} 1211 \\ -\ 735 \\ \hline \mathbf{476} \end{array} \quad \text{Check:} \begin{array}{r} 735 \\ +\ 476 \\ \hline 1211 \end{array}$$

18.
$$\begin{array}{r} 925 \\ +\ F \\ \hline 1111 \end{array} \quad \begin{array}{r} 1111 \\ -\ 925 \\ \hline \mathbf{186} \end{array} \quad \text{Check:} \begin{array}{r} 925 \\ +\ 186 \\ \hline 1111 \end{array}$$

19. $(8 \times 10{,}000) + (5 \times 1000) + (3 \times 100)$
$\quad + (2 \times 10) + (5 \times 1)$
$\quad = 80{,}000 + 5000 + 300 + 20 + 5 = \mathbf{85{,}325}$

20. $(6 \times 1000) + (6 \times 100) + (6 \times 1)$
$\quad = 6000 + 600 + 6 = \mathbf{6606}$

21. $(3 \times 100{,}000) + (2 \times 10{,}000) + (9 \times 100)$
$\quad + (7 \times 1)$
$\quad = 300{,}000 + 20{,}000 + 900 + 7 = \mathbf{320{,}907}$

22. 5,803,125,702 is **five billion, eight hundred three million, one hundred twenty-five thousand, seven hundred two**

23. $295 + 486 + 588 + 714 = \mathbf{2083}$

24. $\$2 + \$9.37 + 86¢ = \mathbf{\$12.23}$

25.
$$\begin{array}{r} 90{,}125 \\ 40{,}061 \\ 30{,}627 \\ +\ 95{,}132 \\ \hline \mathbf{255{,}945} \end{array}$$

26.
$$\begin{array}{r} 1125 \\ 986 \\ 139 \\ +\ 2364 \\ \hline \mathbf{4614} \end{array}$$

27.
$\downarrow$
$7\,①\,6{,}487{,}250$
The rounded number is **720,000,000.**

28.
$\downarrow$
$716{,}4\,⑧\,7{,}250$
The rounded number is **716,490,000.**

29.
```
 ←—•—+—•—+—•—+—+—+—+—•—→
  -4-3-2-1 0 1 2 3 4 5 6
```

30. $-213,\ 123,\ 132,\ 231,\ 321$

PROBLEM SET 4

1. **7333**

2. **3,339,933**

3. **47,000,014**

4. **14,000,042,755**

5.
$\downarrow$
$71\,⑥\,{,}487{,}250$
The rounded number is **716,000,000.**

6. $-176, -76, 167, 617, 671$

7. $7650 = (7 \times 1000) + (6 \times 100) + (5 \times 10)$

8. $(5 \times 1000) + (6 \times 10) + (7 \times 1) = \mathbf{5067}$

9. $\dfrac{\$25.11}{9}$

$$\begin{array}{r} \mathbf{\$2.79} \\ 9{\overline{\smash{\big)}\,\$25.11}} \\ \underline{18} \\ 7\,1 \\ \underline{6\,3} \\ 81 \\ \underline{81} \\ 0 \end{array}$$

10. $2800 \div 50$

$$\begin{array}{r} \mathbf{56} \\ 50{\overline{\smash{\big)}\,2800}} \\ \underline{250} \\ 300 \\ \underline{300} \\ 0 \end{array}$$

11. $$\begin{array}{r} \mathbf{1115\ r\ 42} \\ 45{\overline{\smash{\big)}\,50{,}217}} \\ \underline{45} \\ 5\,2 \\ \underline{4\,5} \\ 71 \\ \underline{45} \\ 267 \\ \underline{225} \\ 42 \end{array}$$

12. $\dfrac{9114}{7}$

$$\begin{array}{r} \mathbf{1302} \\ 7{\overline{\smash{\big)}\,9114}} \\ \underline{7} \\ 21 \\ \underline{21} \\ 01 \\ \underline{0} \\ 14 \\ \underline{14} \\ 0 \end{array}$$

13. $4165 \div 40$

$$\begin{array}{r} \mathbf{104\ r\ 5} \\ 40{\overline{\smash{\big)}\,4165}} \\ \underline{40} \\ 16 \\ \underline{0} \\ 165 \\ \underline{160} \\ 5 \end{array}$$

14. $$\begin{array}{r} \mathbf{1438\ r\ 17} \\ 21{\overline{\smash{\big)}\,30{,}215}} \\ \underline{21} \\ 9\,2 \\ \underline{8\,4} \\ 81 \\ \underline{63} \\ 185 \\ \underline{168} \\ 17 \end{array}$$

15. $$\begin{array}{r} 285 \\ \times\ 321 \\ \hline 285 \\ 570 \\ 855 \\ \hline \mathbf{91{,}485} \end{array}$$

16. $$\begin{array}{r} \$5.06 \\ \times\ \ \ 75 \\ \hline 2530 \\ 3542 \\ \hline \mathbf{\$379.50} \end{array}$$

17. $$\begin{array}{r} 512 \\ \times\ 320 \\ \hline 10240 \\ 1536 \\ \hline \mathbf{163{,}840} \end{array}$$

18. $25 \times 40 \times 100$

$$\begin{array}{r} 25 \\ \times\ 40 \\ \hline 1000 \end{array} \qquad \begin{array}{r} 1000 \\ \times\ \ 100 \\ \hline \mathbf{100{,}000} \end{array}$$

19. 500×420

$$\begin{array}{r} 500 \\ \times\ 420 \\ \hline 10000 \\ 2000 \\ \hline \mathbf{210{,}000} \end{array}$$

20. $6 \times 12 \times 24$

$$\begin{array}{r} 12 \\ \times\ 6 \\ \hline 72 \end{array} \qquad \begin{array}{r} 72 \\ \times\ 24 \\ \hline 288 \\ 144 \\ \hline \mathbf{1728} \end{array}$$

21. $$\begin{array}{r} 943 \\ -\ \ X \\ \hline 274 \end{array} \qquad \begin{array}{r} 943 \\ -\ 274 \\ \hline \mathbf{669} \end{array} \qquad \text{Check:} \begin{array}{r} 943 \\ -\ 669 \\ \hline 274 \end{array}$$

22.
$$\begin{array}{r} 605 \\ +\ M \\ \hline 927 \end{array} \qquad \begin{array}{r} 927 \\ -\ 605 \\ \hline \mathbf{322} \end{array} \qquad \text{Check:} \quad \begin{array}{r} 605 \\ +\ 322 \\ \hline 927 \end{array}$$

23.
$$\begin{array}{r} K \\ -\ 2257 \\ \hline 925 \end{array} \qquad \begin{array}{r} 2257 \\ +\ 925 \\ \hline \mathbf{3182} \end{array} \qquad \text{Check:} \quad \begin{array}{r} 3182 \\ -\ 2257 \\ \hline 925 \end{array}$$

24. $25 \times N = 400$

$$\begin{array}{r} \mathbf{16} \\ 25)\overline{400} \\ \underline{25} \\ 150 \\ \underline{150} \\ 0 \end{array} \qquad \text{Check:} \quad \begin{array}{r} 25 \\ \times\ 16 \\ \hline 150 \\ \underline{25} \\ 400 \end{array}$$

25. $625 \div W = 25$

$$\begin{array}{r} \mathbf{25} \\ 25)\overline{625} \\ \underline{50} \\ 125 \\ \underline{125} \\ 0 \end{array} \qquad \text{Check: Our solution provides its own check.}$$

26. $\dfrac{X}{54} = 7$

$$\begin{array}{r} 54 \\ \times\ 7 \\ \hline \mathbf{378} \end{array} \qquad \text{Check:} \quad \begin{array}{r} 7 \\ 54)\overline{378} \\ \underline{378} \\ 0 \end{array}$$

27.
$$\begin{array}{r} 408,627 \\ 915,634 \\ 589,062 \\ +\ 113,093 \\ \hline \mathbf{2,026,416} \end{array}$$

28.
$$\begin{array}{r} 957,125 \\ 826,015 \\ 902,121 \\ +\ 313,947 \\ \hline \mathbf{2,999,208} \end{array}$$

29. $73 + 816 + 92 + 47 + 321 + 5432 = \mathbf{6781}$

30. $92¢ + \$31.82 + \$21 = \mathbf{\$53.74}$

Problem Set 5

1.
$$\begin{array}{r} 6190 \\ -\ 5320 \\ \hline \mathbf{870}\ \textbf{people} \end{array}$$

2.
$$\begin{array}{r} 530 \\ +\ N \\ \hline 778 \end{array} \qquad \begin{array}{r} 778 \\ -\ 530 \\ \hline \mathbf{248}\ \textbf{dancers} \end{array} \qquad \text{Check:} \quad \begin{array}{r} 530 \\ +\ 248 \\ \hline 778 \end{array}$$

3.
$$\begin{array}{r} 725,000,000 \\ -\ 550,000,000 \\ \hline \mathbf{175,000,000}\ \textbf{people} \end{array}$$

4.
$$\begin{array}{r} \$25.17 \\ -\ \$\ 8.56 \\ \hline \mathbf{\$16.61} \end{array}$$

5. **86,838,887**

6. **1,555,155**

7. $\dfrac{9300}{7}$

$$\begin{array}{r} \mathbf{1328\ r\ 4} \\ 7)\overline{9300} \\ \underline{7} \\ 23 \\ \underline{21} \\ 20 \\ \underline{14} \\ 60 \\ \underline{56} \\ 4 \end{array}$$

8. $\$41.63 \div 23$

$$\begin{array}{r} \mathbf{\$1.81} \\ 23)\overline{\$41.63} \\ \underline{23} \\ 18\ 6 \\ \underline{18\ 4} \\ 23 \\ \underline{23} \\ 0 \end{array}$$

9.
$$\begin{array}{r} \mathbf{2607\ r\ 13} \\ 16)\overline{41,725} \\ \underline{32} \\ 9\ 7 \\ \underline{9\ 6} \\ 12 \\ \underline{0} \\ 125 \\ \underline{112} \\ 13 \end{array}$$

10.
$$\begin{array}{r} \$7.89 \\ \times\ \ \ \ 9 \\ \hline \mathbf{\$71.01} \end{array}$$

11.
$$\begin{array}{r} 506 \\ \times\ \ 27 \\ \hline 3542 \\ \underline{1012} \\ \mathbf{13,662} \end{array}$$

12.
```
    512
  × 632
   1024
   1536
   3072
 323,584
```

13.
```
    A      526    Check:   963
  − 526  + 437          − 526
   437    963             437
```

14.
```
   743    743    Check:   743
  −  N   − 188          − 555
   188    555             188
```

15.
```
    F     123    Check:   412
  − 123  + 289          − 123
   289    412             289
```

16.
```
    K     1001   Check:   676
  + 325  − 325          + 325
   1001   676            1001
```

17.
```
   643    1112   Check:   643
  +  X   − 643          + 469
   1112   469            1112
```

18.
```
    S     915    Check:   188
  + 727  − 727          + 727
   915    188             915
```

19.
```
    32      7    Check:   32
  ×  W  32)224         ×  7
   224    224            224
            0
```

20. $D \div 26 = 7$
```
     26                      7
   ×  7    Check:  26)182
    182                    182
                             0
```

21. $\dfrac{990}{P} = 45$
```
       22                    45
  45)990    Check:  22)990
     90                      88
     90                     110
     90                     110
      0                       0
```

22. $(7 \times 10{,}000) + (4 \times 100) + (3 \times 10)$
$= \textbf{70{,}430}$

23. $2109 = (2 \times 1000) + (1 \times 100) + (9 \times 1)$

24. **47,000,014**

25. **14,000,042,755**

26. **−691, −619, 169, 196, 916, 961**

27. 5021 is **five thousand, twenty-one**

28.
```
              ↓
  716,48⑦,250
```
The rounded number is **716,487,000.**

29.
```
    408,627
  + 915,634
  1,324,261
```

30. $\$9.99 + 33¢ + \$1.07 = \textbf{\$11.39}$

PROBLEM SET 6

1.
```
   37,918,500
 − 19,099,009
   18,819,491
```

2.
```
    725,000,000
  + 475,000,000
  1,200,000,000 people
```

3.
```
   10,000
 −  5,420
   4,580 runners
```

4. **6,264,666**

5. **372,333**

6.
```
  14.0300   Check:   14.0168
 − 0.0132          + 0.0132
  14.0168           14.0300
```

7.
```
  941.20    Check:   926.97
 − 14.23           + 14.23
  926.97            941.20
```

8.
```
  3624
  ────
   23

       157 r 13
  23)3624
     23
    ───
     132
     115
    ───
     174
     161
    ───
      13
```

9. $12.75 ÷ 17

```
    $0.75
17)$12.75
    11 9
      85
      85
       0
```

10.
```
     811 r 1
51)41,362
   40 8
      56
      51
      52
      51
       1
```

11.
```
     81 r 2
27)2189
   216
    29
    27
     2
```

12. 2546 ÷ 41
```
     62 r 4
41)2546
   246
    86
    82
     4
```

13. $\dfrac{92,438}{51}$
```
      1812 r 26
51)92,438
   51
   41 4
   40 8
     63
     51
    128
    102
     26
```

14.
$$\downarrow$$
0.02⓪532

The rounded number is **0.021**.

15.
$$\downarrow$$
14.1193①627

The rounded number is **14.11932**.

16. 3178.0285 is **three thousand, one hundred seventy-eight and two hundred eighty-five ten-thousandths**

17. 504,327.001510512 is **five hundred four thousand, three hundred twenty-seven and one million, five hundred ten thousand, five hundred twelve billionths**

18. **63,000.0214**

19. **0.000029**

20.
```
  625     913    Check:   625
+  Z    - 625           + 288
  913     288             913
```

21.
```
  921     921    Check:   921
-  Y    - 199           - 722
  199     722             199
```

22.
```
   X      763    Check:   952
- 763   + 189           - 763
  189     952             189
```

23.
```
    W
  × 8
$50.00
```
```
     $6.25    Check:  $6.25
8)$50.00            ×     8
   48               $50.00
    2 0
    1 6
      40
      40
       0
```

24. $N ÷ 18 = 15$
```
    18                       15
  × 15      Check: 18)270
    90                       18
    18                       90
   270                       90
                              0
```

25. $\dfrac{945}{V} = 45$
```
      21                      45
45)945       Check: 21)945
   90                        84
   45                       105
   45                       105
    0                         0
```

26.
```
      0.005
     21.620
      9.035
 +  5165.200
   5195.860
```

27.
```
    70.0200
     0.0013
     9.0620
 +   0.1420
    79.2253
```

28. $3917 = (3 \times 1000) + (9 \times 100) + (1 \times 10) + (7 \times 1)$

29. $(9 \times 10,000) + (4 \times 1000) + (5 \times 100) + (7 \times 10) + (9 \times 1) = $ **94,579**

30.
```
     703
  × 579
    6327
   4921
   3515
 407,037
```

PROBLEM SET 7

1.
```
   1569
 + 1237
   2806 heroic acts
```

2.
```
   35,264
 − 17,927
   17,337 fans
```

3.
```
   72,000
 − 39,400
   32,600 delegates
```

4. **747,777,677,277**

5.
```
    1.4160      Check:   1.3992
 − 0.0168              + 0.0168
    1.3992                1.4160
```

6.
```
   23.410      Check:   20.744
 −  2.666              +  2.666
   20.744                23.410
```

7.
```
   38.04       Check:   36.44
 −  1.60              +  1.60
   36.44                38.04
```

8.
```
 Estimate:    10      Evaluate:    14.13
            × 20                 × 21.6
             200                  8478
                                  1413
                                 2826
                               305.208
```

9.
```
 Estimate:   900      Evaluate:   914.23
           × 0.01               × 0.0132
            9.00                 182846
                                 274269
                                  91423
                              12.067836
```

10.
```
 Estimate:  0.004     Evaluate:   0.00413
           ×  0.3               ×  0.312
           0.0012                  826
                                   413
                                  1239
                             0.00128856
```

11. $\$90.16 \div 23$
```
         $3.92
   23)$90.16
      69
      21 1
      20 7
         46
         46
          0
```

12.
```
        1812.585
  41)74,316.000
     41
     33 3
     32 8
        51
        41
       106
        82
        24 0
        20 5
         3 50
         3 28
          220
          205
           15
```

The rounded number is **1812.59.**

13.
$$\frac{90.327}{0.08}$$

$$\begin{array}{r} 1129.087 \\ 008.\overline{)9032.700} \\ 8 \\ \overline{10} \\ 8 \\ \overline{23} \\ 16 \\ \overline{72} \\ 72 \\ \overline{07} \\ 0 \\ \overline{70} \\ 64 \\ \overline{60} \\ 56 \\ \overline{4} \end{array}$$

The rounded number is **1129.09.**

14. $42 \div 0.009$

$$\begin{array}{r} 4666.666 \\ 0009.\overline{)42000.000} \\ 36 \\ \overline{60} \\ 54 \\ \overline{60} \\ 54 \\ \overline{60} \\ 54 \\ \overline{60} \\ 54 \\ \overline{60} \\ 54 \\ \overline{60} \\ 54 \\ \overline{6} \end{array}$$

The rounded number is **4666.67.**

15.
$$91.648\,\textcircled{5}\,73$$
The rounded number is **91.6486.**

16.
$$0.84165\,\textcircled{8}\,39$$
The rounded number is **0.841658.**

17. $-13, -10, -2, 0, 6, 12$

18. **4007.09742**

19. **702.00942**

20. 14,372.015264 is **fourteen thousand, three hundred seventy-two and fifteen thousand, two hundred sixty-four millionths**

21. 9001.321 is **nine thousand one and three hundred twenty-one thousandths**

22.
$$\begin{array}{ll} X & 493 \\ -\,493 & +\,409 \\ \overline{409} & \overline{\textbf{902}} \end{array}$$
Check:
$$\begin{array}{r} 902 \\ -\,493 \\ \overline{409} \end{array}$$

23.
$$\begin{array}{ll} S & 1151 \\ +\,473 & -\,473 \\ \overline{1151} & \overline{\textbf{678}} \end{array}$$
Check:
$$\begin{array}{r} 678 \\ +\,473 \\ \overline{1151} \end{array}$$

24.
$$\begin{array}{ll} 126 & 1152 \\ +\,A & -\,126 \\ \overline{1152} & \overline{\textbf{1026}} \end{array}$$
Check:
$$\begin{array}{r} 126 \\ +\,1026 \\ \overline{1152} \end{array}$$

25. $276 \div X = 12$
$$\begin{array}{r} 23 \\ 12\overline{)276} \\ 24 \\ \overline{36} \\ 36 \\ \overline{0} \end{array}$$
Check:
$$\begin{array}{r} 12 \\ 23\overline{)276} \\ 23 \\ \overline{46} \\ 46 \\ \overline{0} \end{array}$$

26. $\frac{N}{7} = 27$
$$\begin{array}{r} 27 \\ \times\,7 \\ \overline{\textbf{189}} \end{array}$$
Check:
$$\begin{array}{r} 27 \\ 7\overline{)189} \\ 14 \\ \overline{49} \\ 49 \\ \overline{0} \end{array}$$

27. $13 \times P = 117$
$$\begin{array}{r} 9 \\ 13\overline{)117} \\ 117 \\ \overline{0} \end{array}$$
Check:
$$\begin{array}{r} 13 \\ \times\,9 \\ \overline{117} \end{array}$$

28. $15 + 54¢ + 31.62 = \textbf{47.16}$

29.
$$\begin{array}{r} 3.164 \\ 75.236 \\ +\,4328.914 \\ \overline{\textbf{4407.314}} \end{array}$$

30.
$$\begin{array}{r} 3.0624 \\ +\,9053.2160 \\ \overline{\textbf{9056.2784}} \end{array}$$

PROBLEM SET 8

1.
$$32,015,032$$
$$-\ 14,000,642$$
18,014,390 light-years

2.
$$1743$$
$$+\ 1234$$
2977 orations

3. $\dfrac{41,362.68}{100}$ = **413.6268**

4. 305.2165×100 = **30,521.65**

5. $9315.21 \div 1000$ = **9.31521**

6. $32.1652 \times 10,000$ = **321,652**

7.
$$0.00526$$
$$\times \qquad 3.14$$
$$2104$$
$$526$$
$$1578$$
0.0165164

8.
$$2.315$$
$$\times \qquad 413$$
$$6945$$
$$2315$$
$$9260$$
956.095

9.
$$0.00312$$
$$\times \qquad 0.642$$
$$624$$
$$1248$$
$$1872$$
0.00200304

10.
$$313.65$$
$$\times \qquad 0.9$$
282.285

11.
$$392.163 \qquad \text{Check:} \qquad 388.086$$
$$-\ \ 4.077 \qquad\qquad\qquad +\ \ 4.077$$
$$388.086 \qquad\qquad\qquad\qquad 392.163$$

12.
$$3.2421 \qquad \text{Check:} \qquad 1.8791$$
$$-\ 1.3630 \qquad\qquad\qquad +\ 1.3630$$
$$1.8791 \qquad\qquad\qquad\qquad 3.2421$$

13.
$$A \qquad\ 28.9 \qquad \text{Check:} \qquad 15.5$$
$$+\ 13.4 \qquad -\ 13.4 \qquad\qquad +\ 13.4$$
$$28.9 \qquad\ \ 15.5 \qquad\qquad\qquad 28.9$$

14.
$$116.04 \qquad\ 116.04 \qquad \text{Check:} \qquad 116.04$$
$$-\qquad X \qquad -\ 107.06 \qquad\qquad\qquad -\quad 8.98$$
$$107.06 \qquad\quad 8.98 \qquad\qquad\qquad\qquad 107.06$$

15. $\dfrac{Q}{12.2}$ = 6

$$12.2 \qquad\qquad\qquad\qquad\quad 6$$
$$\times \qquad 6 \qquad \text{Check: } 122.\overline{)732.}$$
$$73.2 \qquad\qquad\qquad\qquad 732$$
$$\qquad\qquad\qquad\qquad\qquad\quad 0$$

16. $9 \times R = 587.7$

$$\qquad 65.3 \qquad \text{Check:} \qquad 65.3$$
$$9\overline{)587.7} \qquad\qquad\qquad \times \qquad 9$$
$$\underline{54} \qquad\qquad\qquad\qquad\quad 587.7$$
$$47$$
$$\underline{45}$$
$$2\ 7$$
$$\underline{2\ 7}$$
$$0$$

17. $58.8 \div M = 4.9$

$$\qquad 12 \qquad\qquad\qquad\qquad 4.9$$
$$49.\overline{)588.} \qquad \text{Check: } 12\overline{)58.8}$$
$$\underline{49} \qquad\qquad\qquad\qquad \underline{48}$$
$$98 \qquad\qquad\qquad\qquad 10\ 8$$
$$\underline{98} \qquad\qquad\qquad\qquad \underline{10\ 8}$$
$$0 \qquad\qquad\qquad\qquad\quad 0$$

18. $\dfrac{59.329}{0.4}$

$$\qquad\ 148.322$$
$$04.\overline{)593.290}$$
$$\underline{4}$$
$$19$$
$$\underline{16}$$
$$33$$
$$\underline{32}$$
$$1\ 2$$
$$\underline{1\ 2}$$
$$09$$
$$\underline{8}$$
$$10$$
$$\underline{8}$$
$$2$$

The rounded number is **148.32.**

19. $52 \div 7$

$$
\begin{array}{r}
7.428 \\
7\overline{)52.000} \\
49 \\
\overline{30} \\
28 \\
\overline{20} \\
14 \\
\overline{60} \\
56 \\
\overline{4}
\end{array}
$$

The rounded number is **7.43.**

20. $\dfrac{321.4}{0.071}$

$$
\begin{array}{r}
4526.760 \\
0071.\overline{)321400.000} \\
284 \\
\overline{374} \\
355 \\
\overline{190} \\
142 \\
\overline{480} \\
426 \\
\overline{54\,0} \\
49\,7 \\
\overline{4\,30} \\
4\,26 \\
\overline{40} \\
0 \\
\overline{40}
\end{array}
$$

The rounded number is **4526.76.**

21. $3.22 \div 0.0022$

$$
\begin{array}{r}
1463.636 \\
00022.\overline{)32200.000} \\
22 \\
\overline{102} \\
88 \\
\overline{140} \\
132 \\
\overline{80} \\
66 \\
\overline{14\,0} \\
13\,2 \\
\overline{80} \\
66 \\
\overline{140} \\
132 \\
\overline{8}
\end{array}
$$

The rounded number is **1463.64.**

22.

42.12345⑥78

The rounded number is **42.123457.**

23.

31.6③72052

The rounded number is **31.64.**

24. **47,000,067,000.00417**

25. **1003.04742**

26. 0.00006184 is **six thousand, one hundred eighty-four hundred-millionths**

27. 4,000,062.013 is **four million, sixty-two and thirteen thousandths**

28. **0.0164, 0.0426, 0.0461, 0.0614**

29.
$$
\begin{array}{r}
7186.132 \\
+\ \ 185.620 \\
\hline
\mathbf{7371.752}
\end{array}
$$

30.
$$
\begin{array}{r}
42.1600 \\
0.0032 \\
+\ \ 3.1650 \\
\hline
\mathbf{45.3282}
\end{array}
$$

PROBLEM SET 9

1.
$$
\begin{array}{r}
35,000 \\
-\ 19,763 \\
\hline
\mathbf{15,237}\ \textbf{musicians}
\end{array}
$$

2.
$$
\begin{array}{r}
23,215 \\
-\ 16,219 \\
\hline
\mathbf{6,996}\ \textbf{canvases}
\end{array}
$$

3. $\dfrac{3164.215}{100} = \mathbf{31.64215}$

4. $3164.215 \times 100 = \mathbf{316{,}421.5}$

5. $\dfrac{417,365.20}{1000} = \mathbf{417.36520}$

6. $2.15 \times 10,000 = \mathbf{21{,}500}$

7.
$$
\begin{array}{r}
0.0316 \\
\times\ \ \ \ 2.4 \\
\hline
1264 \\
632 \\
\hline
\mathbf{0.07584}
\end{array}
$$

8.
$$
\begin{array}{r}
2.862 \\
\times\ \ \ 61 \\
\hline
2862 \\
17172 \\
\hline
\mathbf{174.582}
\end{array}
$$

9.
$$
\begin{array}{r}
8.123 \\
\times\ \ \ 0.9 \\
\hline
\mathbf{7.3107}
\end{array}
$$

10.
$$
\begin{array}{r}
43.240000 \\
-\ \ 0.000613 \\
\hline
\mathbf{43.239387}
\end{array}
\qquad
\begin{array}{r}
\text{Check:}\quad 43.239387 \\
+\ \ 0.000613 \\
\hline
43.240000
\end{array}
$$

11.
$$
\begin{array}{r}
3.065 \\
-\ 0.210 \\
\hline
\mathbf{2.855}
\end{array}
\qquad
\begin{array}{r}
\text{Check:}\quad 2.855 \\
+\ 0.210 \\
\hline
3.065
\end{array}
$$

12. $PR = PQ + QR$

$3.065 = PQ + 1.423$

$$
\begin{array}{r}
3.065 \\
-\ 1.423 \\
\hline
\mathbf{1.642}\ \textbf{units}
\end{array}
\qquad
\begin{array}{r}
\text{Check:}\quad 1.642 \\
+\ 1.423 \\
\hline
3.065
\end{array}
$$

13. $(6 + 10 + 20 + 10 + 26 + 20)$ in. = **92 in.**

14. $(40 + 7 + 34 + 15 + 34 + 10 + 40 + 32)$ in.
= **212 in.**

15. $0.002215 \div 0.042$

$$
\begin{array}{r}
0.052 \\
0042.\overline{)0002.215} \\
2\ 10 \\
\hline
115 \\
84 \\
\hline
31
\end{array}
$$

The rounded number is **0.05.**

16. $\dfrac{16.032}{0.024}$

$$
\begin{array}{r}
668 \\
0024.\overline{)16032.} \\
144 \\
\hline
163 \\
144 \\
\hline
192 \\
192 \\
\hline
0
\end{array}
$$

17. $\dfrac{416.5}{0.07}$

$$
\begin{array}{r}
5950. \\
007.\overline{)41650.} \\
35 \\
\hline
66 \\
63 \\
\hline
35 \\
35 \\
\hline
0
\end{array}
$$

18.

$61.373737\,\textcircled{8}\,42$

The rounded number is **61.3737378.**

19.

$433.6851\,\textcircled{4}\,72$

The rounded number is **433.68515.**

20. **742,000,537.010948**

21. **0.0001748**

22. 0.00128647 is **one hundred twenty-eight thousand, six hundred forty-seven hundred-millionths**

23. 27,000,316.081 is **twenty-seven million, three hundred sixteen and eighty-one thousandths**

24. **2.00321, 2.01465, 2.0155, 2.04285**

25.
$$
\begin{array}{r}
904.682 \\
513.976 \\
+\ 214.685 \\
\hline
\mathbf{1633.343}
\end{array}
$$

26.
$$
\begin{array}{r}
2172.062 \\
+\ 5091.799 \\
\hline
\mathbf{7263.861}
\end{array}
$$

27.
$$
\begin{array}{r}
204.63 \\
-\quad A \\
\hline
39.67
\end{array}
\qquad
\begin{array}{r}
204.63 \\
-\ 39.67 \\
\hline
\mathbf{164.96}
\end{array}
\qquad
\begin{array}{r}
\text{Check:}\quad 204.63 \\
-\ 164.96 \\
\hline
39.67
\end{array}
$$

28.
$$
\begin{array}{r}
88.762 \\
+\quad L \\
\hline
89.800
\end{array}
\qquad
\begin{array}{r}
89.800 \\
-\ 88.762 \\
\hline
\mathbf{1.038}
\end{array}
\qquad
\begin{array}{r}
\text{Check:}\quad 88.762 \\
+\ 1.038 \\
\hline
89.800
\end{array}
$$

29. $\dfrac{Z}{9} = 0.52$

$$
\begin{array}{r}
0.52 \\
\times \quad 9 \\
\hline
\mathbf{4.68}
\end{array}
$$

Check: $9\overline{)4.68}$
$$
\begin{array}{r}
0.52 \\
\hline
4\,5 \\
\hline
18 \\
18 \\
\hline
0
\end{array}
$$

30. $0.8 \times R = 1.44$

$$
08.\overline{)14.4}
$$
$$
\begin{array}{r}
1.8 \\
8 \\
\hline
6\,4 \\
6\,4 \\
\hline
0
\end{array}
$$

Check:
$$
\begin{array}{r}
1.8 \\
\times \ 0.8 \\
\hline
1.44
\end{array}
$$

PROBLEM SET 10

1.
$$
\begin{array}{r}
2{,}300{,}619 \\
- 1{,}219{,}312 \\
\hline
\mathbf{1{,}081{,}307 \ people}
\end{array}
$$

2.
$$
\begin{array}{r}
2941 \ m \\
- 1627 \ m \\
\hline
\mathbf{1314 \ m}
\end{array}
$$

3. **762,000,442.12792**

4. **0.014702**

5. $(34 + 16 + 29 + 9 + 5 + 25)\,m = \mathbf{118\ m}$

6. $(12 + 23 + 13 + 8 + 25 + 31)\,m = \mathbf{112\ m}$

7.
$$
\begin{array}{r}
0.0352 \\
\times \quad 2.24 \\
\hline
1408 \\
704 \\
704 \\
\hline
\mathbf{0.078848}
\end{array}
$$

8.
$$
\begin{array}{r}
305 \\
\times \ 2.42 \\
\hline
610 \\
1220 \\
610 \\
\hline
\mathbf{738.10}
\end{array}
$$

9.
$$
\begin{array}{r}
3.062 \\
\times \quad 410 \\
\hline
30620 \\
12248 \\
\hline
\mathbf{1255.420}
\end{array}
$$

10.
$$
\begin{array}{r}
70 \\
\times \ 0.9 \\
\hline
\mathbf{63.0}
\end{array}
$$

11.
$$
\begin{array}{r}
4.016 \\
- 3.217 \\
\hline
\mathbf{0.799}
\end{array}
$$
Check:
$$
\begin{array}{r}
0.799 \\
+ 3.217 \\
\hline
4.016
\end{array}
$$

12.
$$
\begin{array}{r}
23.2100 \\
- \ 0.0034 \\
\hline
\mathbf{23.2066}
\end{array}
$$
Check:
$$
\begin{array}{r}
23.2066 \\
+ \ 0.0034 \\
\hline
23.2100
\end{array}
$$

13.
$$
\begin{array}{r}
2.049 \\
- \quad N \\
\hline
0.684
\end{array}
\qquad
\begin{array}{r}
2.049 \\
- 0.684 \\
\hline
\mathbf{1.365}
\end{array}
$$
Check:
$$
\begin{array}{r}
2.049 \\
- 1.365 \\
\hline
0.684
\end{array}
$$

14.
$$
\begin{array}{r}
F \\
+ 963.09 \\
\hline
1750.24
\end{array}
\qquad
\begin{array}{r}
1750.24 \\
- 963.09 \\
\hline
\mathbf{787.15}
\end{array}
$$
Check:
$$
\begin{array}{r}
787.15 \\
+ 963.09 \\
\hline
1750.24
\end{array}
$$

15. $\dfrac{14.28}{X} = 0.51$

$$
051.\overline{)1428.}
$$
$$
\begin{array}{r}
28. \\
102 \\
\hline
408 \\
408 \\
\hline
0
\end{array}
$$
Check: $28\overline{)14.28}$
$$
\begin{array}{r}
0.51 \\
14\,0 \\
\hline
28 \\
28 \\
\hline
0
\end{array}
$$

16. $M \cdot 5 = 8.5$

$$
5\overline{)8.5}
$$
$$
\begin{array}{r}
1.7 \\
5 \\
\hline
3\,5 \\
3\,5 \\
\hline
0
\end{array}
$$
Check:
$$
\begin{array}{r}
1.7 \\
\times \quad 5 \\
\hline
8.5
\end{array}
$$

17. $0.0030 \div 0.031$

$$
0031.\overline{)0003.000}
$$
$$
\begin{array}{r}
0.096 \\
2\,79 \\
\hline
210 \\
186 \\
\hline
24
\end{array}
$$

The rounded number is **0.10**.

18. $\dfrac{18.034}{0.04}$

$$
\begin{array}{r}
450.85 \\
004.\overline{)1803.40} \\
16 \\
\hline
20 \\
20 \\
\hline
03 \\
0 \\
\hline
3\,4 \\
3\,2 \\
\hline
20 \\
20 \\
\hline
0
\end{array}
$$

19. $2.77 \div 0.0055$

$$
\begin{array}{r}
503.636 \\
00055.\overline{)27700.000} \\
275 \\
\hline
200 \\
165 \\
\hline
35\,0 \\
33\,0 \\
\hline
2\,00 \\
1\,65 \\
\hline
350 \\
330 \\
\hline
20
\end{array}
$$

The rounded number is **503.64.**

20. $\dfrac{50.93}{9}$

$$
\begin{array}{r}
5.658 \\
9\overline{)50.930} \\
45 \\
\hline
5\,9 \\
5\,4 \\
\hline
53 \\
45 \\
\hline
80 \\
72 \\
\hline
8
\end{array}
$$

The rounded number is **5.66.**

21. (a) The following numbers are divisible by 2, since the last digit of each is even: **300, 4888, 9132,** and **72,654.**

    (b) The following numbers are divisible by 3, since the sum of the digits of each is divisible by 3: **300, 9132,** and **72,654.**

    (c) The following numbers are divisible by 5, since the last digit of each is either 5 or 0: **235** and **300.**

    (d) The following number is divisible by 10, since its last digit is 0: **300.**

22.
$$
\downarrow
$$
$$4\,\textcircled{2}\,83.52162$$
The rounded number is **4300.**

23.
$$
\downarrow
$$
$$478.64\,\textcircled{3}\,85$$
The rounded number is **478.644.**

24. $47{,}123 \div 1000 = $ **47.123**

25. $40.265 \times 1000 = $ **40,265**

26. $\dfrac{0.00143}{100} = $ **0.0000143**

27. 472.058 is **four hundred seventy-two and fifty-eight thousandths**

28. **0.341, 0.417, 0.471, 0.704, 0.714**

29. $MR = MN + NR$

    $11 = 7.43 + NR$

$$
\begin{array}{r}
11.00 \\
-\ 7.43 \\
\hline
\textbf{3.57 cm}
\end{array}
\qquad
\text{Check:}
\begin{array}{r}
7.43 \\
+\ 3.57 \\
\hline
\textbf{11.00 cm}
\end{array}
$$

30.
$$
\begin{array}{r}
416.520 \\
3.006 \\
+\ 215.006 \\
\hline
\textbf{634.532}
\end{array}
$$

PROBLEM SET 11

1.
$$
\begin{array}{r}
50 \\
\times\ 15 \\
\hline
250 \\
50 \\
\hline
\textbf{750 players}
\end{array}
$$

2. $15 \times N = 660$

$$
\begin{array}{r}
\textbf{44 students} \\
15\overline{)660} \\
60 \\
\hline
60 \\
60 \\
\hline
0
\end{array}
$$

3.   19.0863 in.
   + 24.4506 in.
   43.5369 in.

4.   43,219
   − 26,314
   16,905 delegates

5.   (31 + 14 + 25 + 11 + 6 + 25) in. = **112 in.**

6.   (31 + 17 + 12 + 8 + 19 + 25) in. = **112 in.**

7.   132,116 ÷ 10,000 = **13.2116**

8.   136.13 × 1000 = **136,130**

9. $\dfrac{123.6}{1000}$ = **0.1236**

10.   162
     × 2.25
       810
       324
       324
     364.50

11.   1.811
     ×   20.1
        1811
        3622
      36.4011

12.   19
     × 0.91
        19
       171
      17.29

13.   61.8100   Check: 61.8088
    − 0.0012   + 0.0012
     61.8088   61.8100

14.   129.631   Check: 127.151
    −   2.480   +   2.480
     127.151   129.631

15.   2.110   Check: 1.079
    − 1.031   + 1.031
     1.079   2.110

16.   (a) The following numbers are divisible by 2, since the last digit of each is even: **6132, 9130,** and **6130.**

(b) The following numbers are divisible by 3, since the sum of the digits of each is divisible by 3: **6132** and **6111.**

(c) The following numbers are divisible by 5, since the last digit of each is either 5 or 0: **6325, 9130,** and **6130.**

(d) The following numbers are divisible by 10, since the last digit of each is 0: **9130** and **6130.**

17. $\dfrac{411.23}{61}$

```
       6.741
   61) 411.230
       366
        45 2
        42 7
         2 53
         2 44
           90
           61
           29
```

The rounded number is **6.74.**

18.   0.0016 ÷ 0.011

```
         0.145
   0011.) 0001.600
             1 1
             50
             44
              60
              55
               5
```

The rounded number is **0.15.**

19. $\dfrac{11.031}{3.1}$

```
        3.558
   31.) 110.310
        93
        17 3
        15 5
         1 81
         1 55
           260
           248
            12
```

The rounded number is **3.56.**

20.

223.0⑨2870

The rounded number is **223.09.**

21.

$$1 \textcircled{6} 21.32161$$

The rounded number is **1600**.

22. **1,625,000,250,025.123**

23. 123.9621 is **one hundred twenty-three and nine thousand, six hundred twenty-one ten-thousandths**

24. 223,092,870 is **two hundred twenty-three million, ninety-two thousand, eight hundred seventy**

25.
$$\begin{array}{r} 1135.62 \\ + \quad 32.61 \\ \hline \mathbf{1168.23} \end{array}$$

26. **An acute angle is an angle less than a right angle (90° angle).**

27. $DF = DE + EF$
$6.32 = 3.12 + EF$

$$\begin{array}{r} 6.32 \\ - \ 3.12 \\ \hline \mathbf{3.20 \ cm} \end{array} \qquad \text{Check:} \quad \begin{array}{r} 3.20 \\ + \ 3.12 \\ \hline 6.32 \ cm \end{array}$$

28.
$$\begin{array}{r} A \\ - 22.49 \\ \hline 68.73 \end{array} \qquad \begin{array}{r} 68.73 \\ + 22.49 \\ \hline \mathbf{91.22} \end{array} \qquad \text{Check:} \quad \begin{array}{r} 91.22 \\ - 22.49 \\ \hline 68.73 \end{array}$$

29. $X \cdot 0.3 = 116.91$

$$\begin{array}{r} \mathbf{389.7} \\ 03.\overline{)1169.1} \\ \underline{9} \\ 26 \\ \underline{24} \\ 29 \\ \underline{27} \\ 2\,1 \\ \underline{2\,1} \\ 0 \end{array} \qquad \text{Check:} \quad \begin{array}{r} 389.7 \\ \times \quad 0.3 \\ \hline 116.91 \end{array}$$

30. $\dfrac{0.0768}{D} = 0.48$

$$\begin{array}{r} \mathbf{0.16} \\ 048.\overline{)007.68} \\ \underline{4\,8} \\ 2\,88 \\ \underline{2\,88} \\ 0 \end{array} \qquad \text{Check:} \quad \begin{array}{r} 0.48 \\ 016.\overline{)007.68} \\ \underline{6\,4} \\ 1\,28 \\ \underline{1\,28} \\ 0 \end{array}$$

PROBLEM SET 12

1.
$$\begin{array}{r} 12,000.04100 \ \text{in.} \\ - \quad 1,021.00002 \ \text{in.} \\ \hline \mathbf{10,979.04098 \ in.} \end{array}$$

2.
$$\begin{array}{r} 14,000,762.0075 \\ + \quad 842,015.0070 \\ \hline \mathbf{14,842,777.0145 \ units} \end{array}$$

3. $5 \times N = 740$

$$\begin{array}{r} \mathbf{148 \ containers} \\ 5\overline{)740} \\ \underline{5} \\ 24 \\ \underline{20} \\ 40 \\ \underline{40} \\ 0 \end{array}$$

4. $25 \times N = 602$

$$\begin{array}{r} 24 \\ 25\overline{)602} \\ \underline{50} \\ 102 \\ \underline{100} \\ 2 \end{array}$$

2 buses will have 25 fanatic fans; 23 buses will have 24 fanatic fans.

5. (a) The following numbers are divisible by 2, since the last digit of each is even: **302, 9172, 3132,** and **62,120.**

(b) The following number is divisible by 3, since the sum of its digits is divisible by 3: **3132.**

(c) The following numbers are divisible by 5, since the last digit of each is either 5 or 0: **625** and **62,120.**

(d) The following number is divisible by 10, since its last digit is 0: **62,120.**

6. $(52 + 23 + 42 + 7 + 10 + 30) \ \text{ft} = \mathbf{164 \ ft}$

7. $(31 + 10 + 16 + 15 + 15 + 25) \ \text{ft} = \mathbf{112 \ ft}$

8. $91,865 \div 100 = \mathbf{918.65}$

9. $36.8211 \times 1000 = \mathbf{36,821.1}$

10. $5 + 7 + 11 + 13 + 17 + 19 + 23 = \mathbf{95}$

11. $2 + 3 + 5 + 7 + 11 + 13 + 17 + 19 = \mathbf{77}$

12.
```
      913
   × 0.19
     8217
      913
   173.47
```

13.
```
    0.0316
   ×   8.9
     2844
     2528
   0.28124
```

14. (a) $\dfrac{95}{5} = 19$

$95 = 5 \cdot 19$

(b) $\dfrac{720}{5} = 144; \quad \dfrac{144}{3} = 48; \quad \dfrac{48}{3} = 16;$

$\dfrac{16}{2} = 8; \quad \dfrac{8}{2} = 4; \quad \dfrac{4}{2} = 2$

$720 = 2 \cdot 2 \cdot 2 \cdot 2 \cdot 3 \cdot 3 \cdot 5$

(c) $\dfrac{2862}{2} = 1431; \quad \dfrac{1431}{3} = 477; \quad \dfrac{477}{3} = 159;$

$\dfrac{159}{3} = 53$

$2862 = 2 \cdot 3 \cdot 3 \cdot 3 \cdot 53$

15.
```
  162.1330    Check:   162.1207
-   0.0123           +   0.0123
  162.1207            162.1330
```

16.
```
  1.329    Check:   0.330
- 0.999           + 0.999
  0.330            1.329
```

17.
```
       A        $ 9.93    Check:   $51.22
  - $41.29     + $41.29          - $41.29
    $9.93        $51.22            $9.93
```

18. $25 \times D = 1953$
```
        78.12    Check:    78.12
  25)1953.00            ×     25
     175                    39060
     203                    15624
     200                  1953.00
       30
       25
        50
        50
         0
```

19. $\dfrac{N}{3.1} = 200.32$
```
       200.32                        200.32
   ×      3.1       Check:   31)6209.92
      20032                   62
      60096                   09
    620.992                    0
                               99
                               93
                               62
                               62
                                0
```

20. $\dfrac{3012.3}{12}$
```
       251.025
  12)3012.300
     24
     61
     60
     12
     12
      0 30
        24
        60
        60
         0
```

The rounded number is **251.03.**

21. $\dfrac{32.631}{0.03}$
```
     1087.7
  3)3263.1
    3
    026
     24
     23
     21
      2 1
      2 1
        0
```

22. $18.621 \div 6.1$
```
      3.052
  61)186.210
     183
      3 21
      3 05
        160
        122
         38
```

The rounded number is **3.05.**

23.

$\downarrow$

4⑥92.83215

The rounded number is **4700**.

24.

$\downarrow$

4113.6②185

The rounded number is **4113.62**.

25. **961,313,000,025**

26. 0.001621 is **one thousand, six hundred twenty-one millionths**

27. 16.0562 is **sixteen and five hundred sixty-two ten-thousandths**

28. **An obtuse angle is an angle greater than a right angle (90°) but less than a straight angle (180°).**

29.
$$\begin{array}{r} 931.62 \\ 621.73 \\ + \ 631.81 \\ \hline \mathbf{2185.16} \end{array}$$

30.
$$\begin{array}{r} 0.7210 \\ + \ 4.3906 \\ \hline \mathbf{5.1116} \end{array}$$

PROBLEM SET 13

1.
$$\begin{array}{r} 79{,}864 \\ \times \qquad 26 \\ \hline 479184 \\ 159728 \\ \hline \mathbf{2{,}076{,}464 \ points} \end{array}$$

2.
$$\begin{array}{r} 2.53 \\ \times \qquad 38 \\ \hline 2024 \\ 759 \\ \hline \mathbf{96.14 \ percent} \end{array}$$

3. $100 \times N = 26{,}000$

$26{,}000 \div 100 = \mathbf{260 \ bags}$

4.
$$\begin{array}{r} 14{,}742{,}000.000170 \\ - \qquad 800{,}000.000042 \\ \hline \mathbf{13{,}942{,}000.000128 \ hectares} \end{array}$$

5. $13 + 17 + 19 + 23 + 29 + 31 + 37 + 41$
$= \mathbf{210}$

6. (a) The following numbers are divisible by 2, since the last digit of each is even: **1020, 130, 1332,** and **132.**

(b) The following numbers are divisible by 3, since the sum of the digits of each is divisible by 3: **1020, 1332,** and **132.**

(c) The following numbers are divisible by 5, since the last digit of each is either 5 or 0: **1020, 125, 130,** and **185.**

(d) The following numbers are divisible by 10, since the last digit of each is 0: **1020** and **130.**

7. $(31 + 20 + 25 + 5 + 6 + 25) \ \text{cm} = \mathbf{112 \ cm}$

8. $(31 + 25 + 31 + 25) \ \text{cm} = \mathbf{112 \ cm}$

9. $31{,}621 \div 1000 = \mathbf{31.621}$

10. $311.836152 \times 10{,}000 = \mathbf{3{,}118{,}361.52}$

11.
$$\begin{array}{r} 621 \\ \times \ 8.11 \\ \hline 621 \\ 621 \\ 4968 \\ \hline \mathbf{5036.31} \end{array}$$

12.
$$\begin{array}{r} 2.28 \\ \times \ 22.4 \\ \hline 912 \\ 456 \\ 456 \\ \hline \mathbf{51.072} \end{array}$$

13. (a) $\dfrac{360}{5} = 72;\quad \dfrac{72}{3} = 24;\quad \dfrac{24}{3} = 8;\quad \dfrac{8}{2} = 4;$

$\dfrac{4}{2} = 2$

$360 = \mathbf{2 \cdot 2 \cdot 2 \cdot 3 \cdot 3 \cdot 5}$

(b) $\dfrac{720}{5} = 144;\quad \dfrac{144}{3} = 48;\quad \dfrac{48}{3} = 16;$

$\dfrac{16}{2} = 8;\quad \dfrac{8}{2} = 4;\quad \dfrac{4}{2} = 2$

$720 = \mathbf{2 \cdot 2 \cdot 2 \cdot 2 \cdot 3 \cdot 3 \cdot 5}$

(c) $\dfrac{1440}{5} = 288;\quad \dfrac{288}{3} = 96;\quad \dfrac{96}{3} = 32;$

$\dfrac{32}{2} = 16;\quad \dfrac{16}{2} = 8;\quad \dfrac{8}{2} = 4;\quad \dfrac{4}{2} = 2$

$1440 = \mathbf{2 \cdot 2 \cdot 2 \cdot 2 \cdot 2 \cdot 3 \cdot 3 \cdot 5}$

14. Factors of 15: 1, 3, 5, 15

Factors of 18: 1, 2, 3, 6, 9, 18

Common factors: **1, 3**

15. Factors of 36: 1, 2, 3, 4, 6, 9, 12, 18, 36

Factors of 45: 1, 3, 5, 9, 15, 45

Common factors: **1, 3, 9**

16. Factors of 9: 1, 3, 9

Factors of 10: 1, 2, 5, 10

The only common factor is **1.**

17.

	$643.28	$643.28	Check:	$643.28
−	N	− $257.29		− $385.99
	$257.29	**$385.99**		$257.29

18.

	1019.05	2364.41	Check:	1019.05
+	S	− 1019.05		+ 1345.36
	2364.41	**1345.36**		2364.41

19. $245 \div V = 9.8$

$$\begin{array}{r} 25 \\ 98\overline{)2450} \\ 196 \\ \hline 490 \\ 490 \\ \hline 0 \end{array} \qquad \text{Check:} \begin{array}{r} 9.8 \\ 25\overline{)245.0} \\ 225 \\ \hline 20\,0 \\ 20\,0 \\ \hline 0 \end{array}$$

20. $8 = ②\cdot②\cdot 2$

$36 = ②\cdot②\cdot 3 \cdot 3$

GCF $(8, 36) = 2 \cdot 2 = $ **4**

21. $50 = 2 \cdot ⑤\cdot⑤$

$75 = 3 \cdot ⑤\cdot⑤$

GCF $(50, 75) = 5 \cdot 5 = $ **25**

22. $\dfrac{311.12}{1.2}$

$$\begin{array}{r} 259.266 \\ 12\overline{)3111.200} \\ 24 \\ \hline 71 \\ 60 \\ \hline 111 \\ 108 \\ \hline 3\,2 \\ 2\,4 \\ \hline 80 \\ 72 \\ \hline 80 \\ 72 \\ \hline 8 \end{array}$$

The rounded number is **259.27.**

23. $\dfrac{621}{0.31}$

$$\begin{array}{r} 2003.225 \\ 31\overline{)62100.000} \\ 62 \\ \hline 0100 \\ 93 \\ \hline 7\,0 \\ 6\,2 \\ \hline 80 \\ 62 \\ \hline 180 \\ 155 \\ \hline 25 \end{array}$$

The rounded number is **2003.23.**

24.

$$1231.62⑤67 \;\downarrow$$

The rounded number is **1231.626.**

25. **321,617,212.231**

26. **straight angle (180° angle)**

27. 161.016 is **one hundred sixty-one and sixteen thousandths**

28. 613.162 is **six hundred thirteen and one hundred sixty-two thousandths**

29.

	621.81
+	31.62
	653.43

30. **0.901, 1.091, 1.119, 1.191, 1.911**

PROBLEM SET 14

1. $12 \times N = 5282$

$$\begin{array}{r} 440 \\ 12\overline{)5282} \\ 48 \\ \hline 48 \\ 48 \\ \hline 02 \end{array}$$

440 family meals can be sold.

2. $14 \times N = 780$

$$\begin{array}{r} 55 \\ 14\overline{)780} \\ 70 \\ \hline 80 \\ 70 \\ \hline 10 \end{array}$$

56 toadstools would be needed.

3. 0.00001620
    − 0.00000423
    0.00001197 m

4. 32,015,032 units
    − 14,000,642 units
    18,014,390 units

5. 3000.000787
    ×　　　14
    12000003148
    3000000787
    42,000.011018

    42,000.011018 second wind
    + 3,000.000787 first wind
    45,000.011805 ft

6. (a) The following numbers are divisible by 10, since the last digit of each is 0: **120** and **1620.**

    (b) The following numbers are divisible by 5, since the last digit of each is either 5 or 0: **120, 135,** and **1620.**

    (c) The following numbers are divisible by 2, since the last digit of each is even: **120, 122, 1332,** and **1620.**

    (d) The following numbers are divisible by 3, since the sum of the digits of each is divisible by 3: **120, 135, 1332,** and **1620.**

7. $(10 + 18 + 11 + 18 + 10 + 26 + 31 + 26)$ km
    = **150 km**

8. $(16 + 18 + 9 + 18 + 6 + 26 + 31 + 26)$ km
    = **150 km**

9. (a) $\dfrac{45}{60} = \dfrac{45 \div 15}{60 \div 15} = \dfrac{3}{4}$

    (b) $\dfrac{30}{75} = \dfrac{30 \div 15}{75 \div 15} = \dfrac{2}{5}$

    (c) $\dfrac{80}{220} = \dfrac{80 \div 20}{220 \div 20} = \dfrac{4}{11}$

    (d) $\dfrac{16}{80} = \dfrac{16 \div 16}{80 \div 16} = \dfrac{1}{5}$

10. (a) $\dfrac{1}{2} = \dfrac{1 \cdot 10}{2 \cdot 10} = \dfrac{10}{20}$

    (b) $7 = \dfrac{7 \cdot 20}{1 \cdot 20} = \dfrac{140}{20}$

11. $12.361 \div 1000 =$ **0.012361**

12. $11.3 \times 1000 =$ **11,300**

13. 　113
    × 0.009
    1.017

14. 　2.14
    × 11.6
    1284
    214
    214
    24.824

15. (a) $\dfrac{1800}{5} = 360$; $\dfrac{360}{5} = 72$; $\dfrac{72}{3} = 24$;

    $\dfrac{24}{3} = 8$; $\dfrac{8}{2} = 4$; $\dfrac{4}{2} = 2$

    $1800 =$ **2 · 2 · 2 · 3 · 3 · 5 · 5**

    (b) $\dfrac{900}{5} = 180$; $\dfrac{180}{5} = 36$; $\dfrac{36}{3} = 12$;

    $\dfrac{12}{3} = 4$; $\dfrac{4}{2} = 2$

    $900 =$ **2 · 2 · 3 · 3 · 5 · 5**

    (c) $\dfrac{450}{5} = 90$; $\dfrac{90}{5} = 18$; $\dfrac{18}{3} = 6$; $\dfrac{6}{3} = 2$

    $450 =$ **2 · 3 · 3 · 5 · 5**

16. 131.61　　Check:　119.74
    − 11.87　　　　 + 11.87
    119.74　　　　　131.61

17. 181.811　　Check:　175.482
    − 6.329　　　　 + 6.329
    175.482　　　　　181.811

18. 　　　M　$657.06　　Check:　$193.94
    + $463.12　− $463.12　　　 + $463.12
    $657.06　**$193.94**　　　　 $657.06

19. $653.28　　$653.28　　Check:　$653.28
    −　　S　− $468.29　　　 − $184.99
    $468.29　**$184.99**　　　　 $468.29

20. $\dfrac{24.455}{N} = 6.7$

    $$
    \begin{array}{r}
    3.65 \\
    67\overline{)244.55} \\
    201 \\
    \hline
    43\,5 \\
    40\,2 \\
    \hline
    3\,35 \\
    3\,35 \\
    \hline
    0
    \end{array}
    $$

    Check:
    $$
    \begin{array}{r}
    6.7 \\
    365\overline{)2445.5} \\
    2190 \\
    \hline
    255\,5 \\
    255\,5 \\
    \hline
    0
    \end{array}
    $$

21. $40 = ②·②·2·⑤$
$60 = ②·②·3·⑤$
GCF $(40, 60) = 2·2·5 = \textbf{20}$

22. $7 = ①·⑦$
$42 = ①·2·3·⑦$
GCF $(7, 42) = 1·7 = \textbf{7}$

23. $6111.12 ÷ 7.5$

```
        814.816
  75)61111.200
     600
     ‾‾‾
     111
      75
     ‾‾‾
     361
     300
     ‾‾‾
      61 2
      60 0
     ‾‾‾‾
       1 20
         75
       ‾‾‾‾
        450
        450
       ‾‾‾
          0
```

The rounded number is **814.82.**

24. $\dfrac{611.21}{0.2}$

```
    3056.05
  2)6112.10
    6
    ‾
    011
     10
    ‾‾
     12
     12
    ‾‾
      0 10
        10
      ‾‾‾
         0
```

25.

$1612.316②89$
The rounded number is **1612.3163.**

26. 1231.161 is **one thousand, two hundred thirty-one and one hundred sixty-one thousandths**

27. 3.121 is **three and one hundred twenty-one thousandths**

28. Factors of 48: 1, 2, 3, 4, 6, 8, 12, 16, 24, 48
Factors of 72: 1, 2, 3, 4, 6, 8, 9, 12, 18, 24, 36, 72
Common factors: **1, 2, 3, 4, 6, 8, 12, 24**

29.
```
  1093.0600
   113.1016
+  915.0900
‾‾‾‾‾‾‾‾‾‾‾
  2121.2516
```

30.
```
   647.1120
+ 9158.0109
‾‾‾‾‾‾‾‾‾‾‾
  9805.1229
```

PROBLEM SET 15

1.
```
  14,007,920
+    900,067
‾‾‾‾‾‾‾‾‾‾‾‾
  14,907,987 hungry locusts
```

2.
```
   9043      9043      108,516
 ×    8    ×   12       72,344
 ‾‾‾‾‾‾    ‾‾‾‾‾‾     +  9,043
 72,344     18086     ‾‾‾‾‾‾‾
             9043      189,903 votes
           ‾‾‾‾‾‾
           108,516
```

3. $20 × N = 9042$

```
     452
  20)9042
     80
     ‾‾
     104
     100
     ‾‾‾
      42
      40
      ‾‾
       2
```

They formed **452 bunches with 2 not in a bunch.**

4. (a) The following numbers are divisible by 3, since the sum of the digits of each is divisible by 3: **2133, 312,** and **630.**

(b) The following numbers are divisible by 2, since the last digit of each is even: **212, 312, 610,** and **630.**

5. $31 + 37 + 41 = \textbf{109}$

6. $(32 + 31 + 32 + 13 + 21 + 18 + 21)$ ft
$= \textbf{168 ft}$

7. $\dfrac{5}{9}$

```
    0.555
  9)5.000
    4 5
    ‾‾‾
     50
     45
     ‾‾
     50
     45
     ‾‾
      5
```

The repeating decimal is written $0.\overline{5}.$

Algebra ½, Third Edition

21

8. $\dfrac{1}{20}$

$$\begin{array}{r} 0.05 \\ 20\overline{)1.00} \\ \underline{1\ 00} \\ 0 \end{array}$$

9. $\dfrac{3}{14} = \dfrac{3 \cdot 3}{14 \cdot 3} = \dfrac{9}{42}$

10. (a) $\dfrac{70}{60} = \dfrac{70 \div 10}{60 \div 10} = \dfrac{7}{6}$

(b) $\dfrac{16}{24} = \dfrac{16 \div 8}{24 \div 8} = \dfrac{2}{3}$

(c) $\dfrac{24}{36} = \dfrac{24 \div 12}{36 \div 12} = \dfrac{2}{3}$

11. $169{,}211 \div 10{,}000 = \mathbf{16.9211}$

12. $123.61311 \times 100 = \mathbf{12{,}361.311}$

13.
$$\begin{array}{r} 89.21 \\ \times\ 62.1 \\ \hline 8921 \\ 17842\ \\ 53526\ \ \\ \hline \mathbf{5539.941} \end{array}$$

14.
$$\begin{array}{r} 2.16 \\ \times\ 32.8 \\ \hline 1728 \\ 432\ \\ 648\ \ \\ \hline \mathbf{70.848} \end{array}$$

15. (a) $\dfrac{3600}{5} = 720;\ \dfrac{720}{5} = 144;\ \dfrac{144}{3} = 48;$

$\dfrac{48}{3} = 16;\ \dfrac{16}{2} = 8;\ \dfrac{8}{2} = 4;\ \dfrac{4}{2} = 2$

$3600 = \mathbf{2 \cdot 2 \cdot 2 \cdot 2 \cdot 3 \cdot 3 \cdot 5 \cdot 5}$

(b) $\dfrac{450}{5} = 90;\ \dfrac{90}{5} = 18;\ \dfrac{18}{3} = 6;\ \dfrac{6}{3} = 2$

$450 = \mathbf{2 \cdot 3 \cdot 3 \cdot 5 \cdot 5}$

(c) $\dfrac{4500}{5} = 900;\ \dfrac{900}{5} = 180;\ \dfrac{180}{5} = 36;$

$\dfrac{36}{3} = 12;\ \dfrac{12}{3} = 4;\ \dfrac{4}{2} = 2$

$4500 = \mathbf{2 \cdot 2 \cdot 3 \cdot 3 \cdot 5 \cdot 5 \cdot 5}$

16.
$$\begin{array}{r} 1131.13 \\ -\ 131.98 \\ \hline \mathbf{999.15} \end{array} \qquad \begin{array}{l} \text{Check:} \end{array} \begin{array}{r} 999.15 \\ +\ 131.98 \\ \hline 1131.13 \end{array}$$

17.
$$\begin{array}{r} 192.680 \\ -\ 6.321 \\ \hline \mathbf{186.359} \end{array} \qquad \begin{array}{l} \text{Check:} \end{array} \begin{array}{r} 186.359 \\ +\ 6.321 \\ \hline 192.680 \end{array}$$

18.
$$\begin{array}{r} N \\ -\ 364.82 \\ \hline 59.69 \end{array} \qquad \begin{array}{r} 364.82 \\ +\ 59.69 \\ \hline \mathbf{424.51} \end{array} \qquad \begin{array}{l} \text{Check:} \end{array} \begin{array}{r} 424.51 \\ -\ 364.82 \\ \hline 59.69 \end{array}$$

19.
$$\begin{array}{r} 16.0375 \\ +\ A \\ \hline 17.0094 \end{array} \qquad \begin{array}{r} 17.0094 \\ -\ 16.0375 \\ \hline \mathbf{0.9719} \end{array} \qquad \begin{array}{l} \text{Check:} \end{array} \begin{array}{r} 16.0375 \\ +\ 0.9719 \\ \hline 17.0094 \end{array}$$

20. $\dfrac{2.76}{F} = 12$

$$\begin{array}{r} 0.23 \\ 12\overline{)2.76} \\ \underline{2\ 4} \\ 36 \\ \underline{36} \\ 0 \end{array} \qquad \text{Check: } \begin{array}{r} 12 \\ 23\overline{)276} \\ \underline{23} \\ 46 \\ \underline{46} \\ 0 \end{array}$$

21. $34 = 2 \cdot \boxed{17}$

$51 = 3 \cdot \boxed{17}$

GCF $(34, 51) = \mathbf{17}$

22. $26 = 2 \cdot \boxed{13}$

$117 = 3 \cdot 3 \cdot \boxed{13}$

GCF $(26, 117) = \mathbf{13}$

23. $7.81 \div 3.1$

$$\begin{array}{r} 2.519 \\ 31\overline{)78.100} \\ \underline{62\ \ } \\ 16\ 1 \\ \underline{15\ 5} \\ 60 \\ \underline{31} \\ 290 \\ \underline{279} \\ 11 \end{array}$$

The rounded number is **2.52**.

24. $\dfrac{2310}{13}$

$$
\begin{array}{r}
177.692 \\
13\overline{)2310.000} \\
\underline{13} \\
101 \\
\underline{91} \\
100 \\
\underline{91} \\
9\,0 \\
\underline{7\,8} \\
1\,20 \\
\underline{1\,17} \\
30 \\
\underline{26} \\
4
\end{array}
$$

The rounded number is **177.69.**

25.

$$\downarrow$$

4017.33633⑥3

The rounded number is **4017.336336.**

26.

$$\downarrow$$

946.0545454⑤4

The rounded number is **946.05454545.**

27. $0.85 = \dfrac{85}{100} = \dfrac{85 \div 5}{100 \div 5} = \dfrac{\mathbf{17}}{\mathbf{20}}$

28. $0.76 = \dfrac{76}{100} = \dfrac{76 \div 4}{100 \div 4} = \dfrac{\mathbf{19}}{\mathbf{25}}$

29.
$$
\begin{array}{r}
613.1 \\
7214.6 \\
11.2 \\
+3.1 \\
\hline
\mathbf{7842.0}
\end{array}
$$

30. **0.0119, 0.0191, 0.091, 0.9**

PROBLEM SET 16

1. $14 \times N = 3808$

$$
\begin{array}{r}
\mathbf{272\ spaces} \\
14\overline{)3808} \\
\underline{28} \\
100 \\
\underline{98} \\
28 \\
\underline{28} \\
0
\end{array}
$$

2.
$$
\begin{array}{r}
14.0742 \\
\times 7 \\
\hline
98.5194
\end{array}
\qquad
\begin{array}{r}
14.0742 \\
+\ 98.5194 \\
\hline
\mathbf{112.5936\ units}
\end{array}
$$

3.
$$
\begin{array}{r}
91{,}042 \\
-\ 12{,}015 \\
\hline
\mathbf{79{,}027\ genes}
\end{array}
$$

4.
$$
\begin{array}{r}
142 \\
\times\ 432 \\
\hline
284 \\
426 \\
568 \\
\hline
61{,}344
\end{array}
\qquad
\begin{array}{r}
61{,}344 \\
+5 \\
\hline
\mathbf{61{,}349\ apples}
\end{array}
$$

5. (a) The following numbers are divisible by 5, since the last digit of each is either 5 or 0: **650, 625, 15, 20,** and **30.**

(b) The following numbers are divisible by 10, since the last digit of each is 0: **650, 20,** and **30.**

6. **41, 43, 47, 53, 59, 61**

7. $(25 + 42 + 25 + 14 + 14 + 42)\ yd = \mathbf{162\ yd}$

8. $\dfrac{19}{25}$

$$
\begin{array}{r}
0.76 \\
25\overline{)19.00} \\
\underline{17\,5} \\
1\,50 \\
\underline{1\,50} \\
0
\end{array}
$$

9. $\dfrac{4}{7}$

$$
\begin{array}{r}
0.571 \\
7\overline{)4.000} \\
\underline{3\,5} \\
50 \\
\underline{49} \\
10 \\
\underline{7} \\
3
\end{array}
$$

The rounded number is **0.57.**

10. $\dfrac{36}{42} = \dfrac{36 \div 6}{42 \div 6} = \dfrac{\mathbf{6}}{\mathbf{7}}$

11. $\dfrac{72}{120} = \dfrac{72 \div 24}{120 \div 24} = \dfrac{\mathbf{3}}{\mathbf{5}}$

12. $\dfrac{9}{27} = \dfrac{9 \cdot 9}{27 \cdot 9} = \dfrac{\mathbf{81}}{\mathbf{243}}$

13. $(0.5)^2 = 0.5 \cdot 0.5 = \mathbf{0.25}$

14. $9^3 = 9 \cdot 9 \cdot 9 = 81 \cdot 9 = \mathbf{729}$

15.
$$
\begin{array}{r}
13.61 \\
\times\ \ 71.3 \\
\hline
4083 \\
1361\ \ \\
9527\ \ \ \ \\
\hline
\mathbf{970.393}
\end{array}
$$

16. $12{,}389.32 \div 100 = \mathbf{123.8932}$

17.
$$
\begin{array}{r}
181{,}131.6200 \\
-\ \ \ \ \ \ 1.9876 \\
\hline
\mathbf{181{,}129.6324}
\end{array}
\qquad
\begin{array}{r}
\text{Check:}\quad 181{,}129.6324 \\
+\ \ \ \ \ \ \ 1.9876 \\
\hline
181{,}131.6200
\end{array}
$$

18. $\dfrac{613.15}{3.1}$

$$
\begin{array}{r}
197.790 \\
31\overline{)6131.500} \\
\underline{31\ \ \ \ \ \ \ \ } \\
303\ \ \ \ \ \ \\
\underline{279\ \ \ \ \ \ } \\
241\ \ \ \ \\
\underline{217\ \ \ \ } \\
24\ 5\ \ \\
\underline{21\ 7\ \ } \\
2\ 80 \\
\underline{2\ 79} \\
10 \\
\underline{0} \\
10
\end{array}
$$

The rounded number is **197.79.**

19. Factors of 20 = 1, 2, 4, 5, 10, 20

Factors of 30 = 1, 2, 3, 5, 6, 10, 15, 30

Common factors = **1, 2, 5, 10**

20. $20 = ②\cdot 2 \cdot ⑤$

$30 = ②\cdot 3 \cdot ⑤$

GCF (20, 30) = $2 \cdot 5 = \mathbf{10}$

21. $\dfrac{288}{3} = 96;\ \dfrac{96}{3} = 32;\ \dfrac{32}{2} = 16;$

$\dfrac{16}{2} = 8;\ \dfrac{8}{2} = 4;\ \dfrac{4}{2} = 2$

$288 = 2 \cdot 2 \cdot 2 \cdot 2 \cdot 2 \cdot 3 \cdot 3 = \mathbf{2^5 \cdot 3^2}$

22. $\dfrac{1080}{5} = 216;\ \dfrac{216}{3} = 72;\ \dfrac{72}{3} = 24;\ \dfrac{24}{3} = 8;$

$\dfrac{8}{2} = 4;\ \dfrac{4}{2} = 2$

$1080 = 2 \cdot 2 \cdot 2 \cdot 3 \cdot 3 \cdot 3 \cdot 5$

$\qquad = \mathbf{2^3 \cdot 3^3 \cdot 5}$

23. $\dfrac{10{,}800}{5} = 2160;\ \dfrac{2160}{5} = 432;\ \dfrac{432}{3} = 144;$

$\dfrac{144}{3} = 48;\ \dfrac{48}{3} = 16;\ \dfrac{16}{2} = 8;\ \dfrac{8}{2} = 4;$

$\dfrac{4}{2} = 2$

$10{,}800 = 2 \cdot 2 \cdot 2 \cdot 2 \cdot 3 \cdot 3 \cdot 3 \cdot 5 \cdot 5$

$\qquad = \mathbf{2^4 \cdot 3^3 \cdot 5^2}$

24.
$$\downarrow$$
$87{,}621.32178⑨39$

The rounded number is **87,621.321789.**

25.
$$\downarrow$$
$437.0062162①6$

The rounded number is **437.00621622.**

26. 172.312 is **one hundred seventy-two and three hundred twelve thousandths**

27.
$$
\begin{array}{r}
1361.31 \\
21.14 \\
112.17 \\
+\ \ \ \ 1.18 \\
\hline
\mathbf{1495.80}
\end{array}
$$

28. $0.65 = \dfrac{65}{100} = \dfrac{65 \div 5}{100 \div 5} = \dfrac{\mathbf{13}}{\mathbf{20}}$

29.
$$
\begin{array}{r}
14{,}392.091 \\
+\ \ \ \ \ \ \ \ A \\
\hline
107{,}072.060
\end{array}
\qquad
\begin{array}{r}
107{,}072.060 \\
-\ \ 14{,}392.091 \\
\hline
\mathbf{92{,}679.969}
\end{array}
$$

Check:
$$
\begin{array}{r}
14{,}392.091 \\
+\ 92{,}679.969 \\
\hline
107{,}072.060
\end{array}
$$

30. $\dfrac{321}{B} = 0.04$

$$
\begin{array}{r}
8025 \\
4\overline{)32100} \\
\underline{32\ \ \ \ \ } \\
01\ \ \ \\
\underline{0\ \ \ } \\
10\ \\
\underline{8\ } \\
20 \\
\underline{20} \\
0
\end{array}
\qquad
\begin{array}{r}
0.04 \\
\text{Check: } 8025\overline{)321.00} \\
\underline{321.00} \\
0
\end{array}
$$

PROBLEM SET 17

1. $11 \times N = 10{,}802$

$$
\begin{array}{r}
\textbf{982 mi} \\
11\overline{)10{,}802} \\
\underline{9\,9} \\
90 \\
\underline{88} \\
22 \\
\underline{22} \\
0
\end{array}
$$

2.
$$
\begin{array}{r}
4842 \\
\times\ \ 330 \\
\hline
145260 \\
14526 \\
\hline
\textbf{1,597,860 ants}
\end{array}
$$

3.
$$
\begin{array}{r}
108{,}015 \\
-\ \ 4{,}842 \\
\hline
\textbf{103,173 ants}
\end{array}
$$

4. (a) The following numbers are divisible by 3, since the sum of the digits of each is divisible by 3: **135** and **1050**.

(b) The following numbers are divisible by 5, since the last digit of each is either 5 or 0: **135, 1050, 335, 4145,** and **1010**.

5. $2 + 3 + 5 + 7 + 11 + 13 + 17 + 19 + 23$
$+ 29 + 31 + 37 + 41 = \textbf{238}$

6. $(29 + 46 + 19 + 46 + 11 + 21)\,\text{m} = \textbf{172 m}$

7. $(6\,\text{cm} \times 6\,\text{cm}) + (3\,\text{cm} \times 14\,\text{cm})$
$= 36\,\text{cm}^2 + 42\,\text{cm}^2 = \textbf{78 cm}^2$

8. $(2\,\text{in.} \times 2\,\text{in.}) + (2\,\text{in.} \times 2\,\text{in.}) + (2\,\text{in.} \times 8\,\text{in.})$
$= 4\,\text{in.}^2 + 4\,\text{in.}^2 + 16\,\text{in.}^2 = 24\,\text{in.}^2$

So, **24 tiles,** each one-inch-square, are needed.

9. $\dfrac{3}{17}$

$$
\begin{array}{r}
0.176 \\
17\overline{)3.000} \\
\underline{1\,7} \\
1\,30 \\
\underline{1\,19} \\
110 \\
\underline{102} \\
8
\end{array}
$$

The rounded number is **0.18.**
(An ambitious student may find that this one does repeat after 16 digits: **0.$\overline{1764705882352941}$**)

10. $\dfrac{7}{11}$

$$
\begin{array}{r}
0.636 \\
11\overline{)7.000} \\
\underline{6\,6} \\
40 \\
\underline{33} \\
70 \\
\underline{66} \\
4
\end{array}
$$

The repeating decimal is **0.$\overline{63}$.**

11. $\dfrac{7}{500}$

$$
\begin{array}{r}
0.014 \\
500\overline{)7.000} \\
\underline{5\,00} \\
2\,000 \\
\underline{2\,000} \\
0
\end{array}
$$

The rounded number is **0.01.**

12. $\dfrac{540}{5} = 108;\ \dfrac{108}{3} = 36;\ \dfrac{36}{3} = 12;\ \dfrac{12}{3} = 4;$

$\dfrac{4}{2} = 2$

$540 = 2 \cdot 2 \cdot 3 \cdot 3 \cdot 3 \cdot 5 = \mathbf{2^2 \cdot 3^3 \cdot 5}$

13. $\dfrac{30}{5} = 6;\ \dfrac{6}{3} = 2$

$30 = \mathbf{2 \cdot 3 \cdot 5}$

14. $\dfrac{210}{5} = 42;\ \dfrac{42}{3} = 14;\ \dfrac{14}{2} = 7$

$210 = \mathbf{2 \cdot 3 \cdot 5 \cdot 7}$

15. $36 = ②\cdot②\cdot③\cdot③$
$216 = ②\cdot②\cdot 2 \cdot③\cdot③\cdot 3$
GCF $(36, 216) = 2 \cdot 2 \cdot 3 \cdot 3 = \textbf{36}$

16. $\dfrac{21}{49} = \dfrac{21 \div 7}{49 \div 7} = \dfrac{\textbf{3}}{\textbf{7}}$

17. $\dfrac{36}{48} = \dfrac{36 \div 12}{48 \div 12} = \dfrac{\textbf{3}}{\textbf{4}}$

18. $\dfrac{3}{7} = \dfrac{3 \cdot 30}{7 \cdot 30} = \dfrac{\textbf{90}}{\textbf{210}}$

19.
$$
\begin{array}{r}
62.13 \\
\times\ \ 7.8 \\
\hline
49704 \\
43491 \\
\hline
\textbf{484.614}
\end{array}
$$

20. $625.3 \times 100 = \mathbf{62{,}530}$

21.
$$\begin{array}{r} 185.3617 \\ -\ 17.3169 \\ \hline \mathbf{168.0448} \end{array} \qquad \text{Check:} \begin{array}{r} 168.0448 \\ +\ 17.3169 \\ \hline 185.3617 \end{array}$$

22.
$$\begin{array}{r} 179.3622 \\ -\quad Z \\ \hline 57.4643 \end{array} \qquad \begin{array}{r} 179.3622 \\ -\ 57.4643 \\ \hline \mathbf{121.8979} \end{array}$$

Check:
$$\begin{array}{r} 179.3622 \\ -\ 121.8979 \\ \hline 57.4643 \end{array}$$

23. $\dfrac{V}{0.6} = 44.4$

$$\begin{array}{r} 44.4 \\ \times\ 0.6 \\ \hline \mathbf{26.64} \end{array} \qquad \text{Check: } 6\overline{)266.4}$$

$$\begin{array}{r} 44.4 \\ 6\overline{)266.4} \\ 24 \\ \hline 26 \\ 24 \\ \hline 2\ 4 \\ 2\ 4 \\ \hline 0 \end{array}$$

24. $\dfrac{231.2}{0.048}$

$$\begin{array}{r} 4816.666 \\ 48\overline{)231200.000} \\ 192 \\ \hline 392 \\ 384 \\ \hline 80 \\ 48 \\ \hline 320 \\ 288 \\ \hline 32\ 0 \\ 28\ 8 \\ \hline 3\ 20 \\ 2\ 88 \\ \hline 320 \\ 288 \\ \hline 32 \end{array}$$

The rounded number is **4816.67**.

25.
$$\downarrow$$
3.141592⑥54
The rounded number is **3.1415927**.

26.
$$\downarrow$$
2.070707⓪7
The rounded number is **2.0707071**.

27. $0.134 = \dfrac{134}{1000} = \dfrac{134 \div 2}{1000 \div 2} = \dfrac{\mathbf{67}}{\mathbf{500}}$

28. $0.025 = \dfrac{25}{1000} = \dfrac{25 \div 25}{1000 \div 25} = \dfrac{\mathbf{1}}{\mathbf{40}}$

29. $(0.8)^3 = 0.8 \cdot 0.8 \cdot 0.8 = 0.64 \cdot 0.8 = \mathbf{0.512}$

30. $1^{24} = \mathbf{1}$ (since 1 raised to any power is 1)

Problem Set 18

1.
$$\begin{array}{r} 5482 \\ \times\quad 4 \\ \hline 21{,}928 \end{array} \qquad \begin{array}{r} 21{,}928 \\ \times\quad 2 \\ \hline 43{,}856 \end{array} \qquad \begin{array}{r} 43{,}856 \\ 21{,}928 \\ +\ 5{,}482 \\ \hline \mathbf{71{,}266 \text{ peaches}} \end{array}$$

2.
$$\begin{array}{r} 19{,}000{,}000.00047 \\ +\quad 17{,}005.02370 \\ \hline \mathbf{19{,}017{,}005.02417 \text{ cm}} \end{array}$$

3.
$$\begin{array}{r} 700{,}033 \\ \times\quad 63 \\ \hline 2100099 \\ 4200198 \\ \hline \mathbf{44{,}102{,}079 \text{ rocks}} \end{array}$$

4.
$$\begin{array}{r} 416{,}943 \\ \times\quad 63 \\ \hline 1250829 \\ 2501658 \\ \hline \mathbf{26{,}267{,}409 \text{ rocks}} \end{array}$$

5. **29, 31, 37, 41, 43, 47**

6. $(28 + 40 + 20 + 40 + 11 + 19) \text{ cm}$
$= \mathbf{158 \text{ cm}}$

7. $(5 \text{ cm} \times 30 \text{ cm}) + (14 \text{ cm} \times 30 \text{ cm})$
$= 150 \text{ cm}^2 + 420 \text{ cm}^2 = \mathbf{570 \text{ cm}^2}$

8. $(7 \text{ cm} \times 26 \text{ cm}) + (18 \text{ cm} \times 11 \text{ cm})$
$= 182 \text{ cm}^2 + 198 \text{ cm}^2 = \mathbf{380 \text{ cm}^2}$

9. $\dfrac{480}{1000} = \dfrac{480 \div 10}{1000 \div 10} = \dfrac{48 \div 4}{100 \div 4} = \dfrac{\mathbf{12}}{\mathbf{25}}$

10. $\dfrac{66}{104} = \dfrac{66 \div 2}{104 \div 2} = \dfrac{\mathbf{33}}{\mathbf{52}}$

11. $\dfrac{210}{294} = \dfrac{210 \div 2}{294 \div 2} = \dfrac{105 \div 3}{147 \div 3} = \dfrac{35 \div 7}{49 \div 7} = \dfrac{\mathbf{5}}{\mathbf{7}}$

12. $\dfrac{738}{882} = \dfrac{738 \div 2}{882 \div 2} = \dfrac{369 \div 3}{441 \div 3} = \dfrac{123 \div 3}{147 \div 3}$

$= \dfrac{\mathbf{41}}{\mathbf{49}}$

13. $\dfrac{16}{17}$

$$\begin{array}{r} 0.9411 \\ 17\overline{)16.0000} \\ \underline{15\,3} \\ 70 \\ \underline{68} \\ 20 \\ \underline{17} \\ 30 \\ \underline{17} \\ 13 \end{array}$$

The rounded number is **0.941.**

14. $\dfrac{7}{250}$

$$\begin{array}{r} \mathbf{0.028} \\ 250\overline{)7.000} \\ \underline{5\,00} \\ 2\,000 \\ \underline{2\,000} \\ 0 \end{array}$$

15. $\dfrac{11}{13}$

$$\begin{array}{r} 0.8461 \\ 13\overline{)11.0000} \\ \underline{10\,4} \\ 60 \\ \underline{52} \\ 80 \\ \underline{78} \\ 20 \\ \underline{13} \\ 7 \end{array}$$

The rounded number is **0.846.**

16. $\dfrac{3}{10} = \dfrac{3 \cdot 9}{10 \cdot 9} = \dfrac{\mathbf{27}}{\mathbf{90}}$

17. $\dfrac{14}{15} = \dfrac{14 \cdot 6}{15 \cdot 6} = \dfrac{\mathbf{84}}{\mathbf{90}}$

18.
$$\begin{array}{r} 62.13 \\ \times\ 11.31 \\ \hline 6213 \\ 18639 \\ 6213 \\ 6213 \\ \hline \mathbf{702.6903} \end{array}$$

19. $621.1378 \div 10,000 = \mathbf{0.06211378}$

20. $4^3 = 4 \cdot 4 \cdot 4 = 16 \cdot 4 = \mathbf{64}$

21. $13^2 = 13 \cdot 13 = \mathbf{169}$

22.
$$\begin{array}{r} 183.1782 \\ -\ 1.8999 \\ \hline \mathbf{181.2783} \end{array} \qquad \begin{array}{r} \text{Check:} \\ \\ \end{array} \qquad \begin{array}{r} 181.2783 \\ +\ 1.8999 \\ \hline 183.1782 \end{array}$$

23. $\dfrac{2}{3} \cdot \dfrac{15}{1} = \dfrac{30}{3} = \mathbf{10}$

24. $\dfrac{5}{9} \cdot \dfrac{10}{1} = \dfrac{\mathbf{50}}{\mathbf{9}}$

25. $\dfrac{3}{5} \cdot \dfrac{100}{1} = \dfrac{300}{5} = \mathbf{60}$

26. $\dfrac{4}{7} \cdot \dfrac{12}{1} = \dfrac{\mathbf{48}}{\mathbf{7}}$

27.
$$\begin{array}{r} 0.11111 \\ -\ \ K \\ \hline 0.01121 \end{array} \qquad \begin{array}{r} 0.11111 \\ -\ 0.01121 \\ \hline \mathbf{0.09990} \end{array} \qquad \begin{array}{r} \text{Check:} \\ \\ \end{array} \qquad \begin{array}{r} 0.11111 \\ -\ 0.09990 \\ \hline 0.01121 \end{array}$$

28. $P \times 78 = 43.68$

$$\begin{array}{r} \mathbf{0.56} \\ 78\overline{)43.68} \\ \underline{39\,0} \\ 4\,68 \\ \underline{4\,68} \\ 0 \end{array} \qquad \begin{array}{r} \text{Check:} \\ \\ \\ \\ \\ \end{array} \qquad \begin{array}{r} 0.56 \\ \times\ 78 \\ \hline 448 \\ 392 \\ \hline 43.68 \end{array}$$

29. $0.095 = \dfrac{95}{1000} = \dfrac{95 \div 5}{1000 \div 5} = \dfrac{\mathbf{19}}{\mathbf{200}}$

30. $121 = 11 \cdot 11$

$132 = 2 \cdot 2 \cdot 3 \cdot 11$

GCF $(121, 132) = \mathbf{11}$

PROBLEM SET 19

1. Charles:
$$\begin{array}{r} 14,016.2163 \text{ kg} \\ -\ 13,041.0006 \text{ kg} \\ \hline \mathbf{975.2157 \text{ kg}} \end{array}$$

Mary:
$$\begin{array}{r} 14,991.2300 \text{ kg} \\ -\ 14,016.2163 \text{ kg} \\ \hline \mathbf{975.0137 \text{ kg}} \end{array}$$

Mary's guess was closer.

2.
$$\begin{array}{r} 51,416 \\ \times\ \ \ \ 4 \\ \hline 205,664 \end{array} \qquad \begin{array}{r} 205,664 \\ +\ 51,416 \\ \hline \mathbf{257,080} \text{ mice} \end{array}$$

3. $\begin{array}{r} 19,007 \\ -\ \ 5,715 \\ \hline \textbf{13,292 Romans} \end{array}$

4. $315 \times N = 14,321$

$$\begin{array}{r} 45 \\ 315\overline{)14,321} \\ \underline{12\ 60} \\ 1\ 721 \\ \underline{1\ 575} \\ 146 \end{array}$$

46 rooms are needed.

5.   **37**

6. $\dfrac{5}{8} \cdot \dfrac{4}{7} = \dfrac{5}{\cancel{8}} \cdot \dfrac{\overset{1}{\cancel{4}}}{7} = \dfrac{\mathbf{5}}{\mathbf{14}}$
     $\phantom{\dfrac{5}{8}}_{2}$

7. $\dfrac{5}{9} \cdot \dfrac{15}{20} = \dfrac{5}{\underset{3}{\cancel{9}}} \cdot \dfrac{\overset{5}{\cancel{15}}}{\underset{4}{\cancel{20}}} = \dfrac{\mathbf{5}}{\mathbf{12}}$

8. $\dfrac{3}{8} \div \dfrac{2}{3} = \dfrac{3}{8} \cdot \dfrac{3}{2} = \dfrac{\mathbf{9}}{\mathbf{16}}$

9. $\dfrac{9}{4} \div \dfrac{14}{3} = \dfrac{9}{4} \cdot \dfrac{3}{14} = \dfrac{\mathbf{27}}{\mathbf{56}}$

10. $(20 + 31 + 16 + 9 + 13 + 31)\ m = \mathbf{120\ m}$

11. $(50\ yd \times 10\ yd) + (20\ yd \times 10\ yd)$
     $= 500\ yd^2 + 200\ yd^2 = \mathbf{700\ yd^2}$

12. $(22\ yd \times 15\ yd) + (10\ yd \times 8\ yd)$
     $\quad + (22\ yd \times 10\ yd)$
     $= 330\ yd^2 + 80\ yd^2 + 220\ yd^2 = \mathbf{630\ yd^2}$

13. $\dfrac{360}{540} = \dfrac{360 \div 10}{540 \div 10} = \dfrac{36 \div 6}{54 \div 6} = \dfrac{6 \div 3}{9 \div 3} = \dfrac{\mathbf{2}}{\mathbf{3}}$

14. $\dfrac{196}{360} = \dfrac{196 \div 4}{360 \div 4} = \dfrac{\mathbf{49}}{\mathbf{90}}$

15. $\dfrac{256}{720} = \dfrac{256 \div 4}{720 \div 4} = \dfrac{64 \div 4}{180 \div 4} = \dfrac{\mathbf{16}}{\mathbf{45}}$

16. $\dfrac{17}{23}$

$$\begin{array}{r} 0.739 \\ 23\overline{)17.000} \\ \underline{16\ 1} \\ 90 \\ \underline{69} \\ 210 \\ \underline{207} \\ 3 \end{array}$$

The rounded number is **0.74.**

17. $\dfrac{5}{17}$

$$\begin{array}{r} 0.294 \\ 17\overline{)5.000} \\ \underline{3\ 4} \\ 1\ 60 \\ \underline{1\ 53} \\ 70 \\ \underline{68} \\ 2 \end{array}$$

The rounded number is **0.29.**

18. $\dfrac{4200}{7} = 600;\ \dfrac{600}{5} = 120;\ \dfrac{120}{5} = 24;$

$\dfrac{24}{3} = 8;\ \dfrac{8}{2} = 4;\ \dfrac{4}{2} = 2$

$4200 = 2 \cdot 2 \cdot 2 \cdot 3 \cdot 5 \cdot 5 \cdot 7$
$\quad\quad = \mathbf{2^3 \cdot 3 \cdot 5^2 \cdot 7}$

19. $\dfrac{10,080}{7} = 1440;\ \dfrac{1440}{5} = 288;\ \dfrac{288}{3} = 96;$

$\dfrac{96}{3} = 32;\ \dfrac{32}{2} = 16;\ \dfrac{16}{2} = 8;\ \dfrac{8}{2} = 4;$

$\dfrac{4}{2} = 2$

$10,080 = 2 \cdot 2 \cdot 2 \cdot 2 \cdot 2 \cdot 3 \cdot 3 \cdot 5 \cdot 7$
$\quad\quad\ = \mathbf{2^5 \cdot 3^2 \cdot 5 \cdot 7}$

20. $\begin{array}{r} 1.28 \\ \times\ 62.1 \\ \hline 128 \\ 256\ \ \\ 768\ \ \ \ \\ \hline \textbf{79.488} \end{array}$

21. $(1.2)^2 = 1.2 \times 1.2 = \mathbf{1.44}$

22. $\left(\dfrac{3}{8}\right)^2 = \dfrac{3}{8} \cdot \dfrac{3}{8} = \dfrac{\mathbf{9}}{\mathbf{64}}$

23. $\dfrac{7.32}{0.12}$

$$12\overline{)732} \quad \begin{array}{r} \mathbf{61} \\ \hline \end{array}$$

$$\begin{array}{r} \mathbf{61} \\ 12\overline{)732} \\ \underline{72} \\ 12 \\ \underline{12} \\ 0 \end{array}$$

24. $\dfrac{61.31}{3.1}$

$$\begin{array}{r} 19.777 \\ 31\overline{)613.100} \\ \underline{31} \\ 303 \\ \underline{279} \\ 24\,1 \\ \underline{21\,7} \\ 2\,40 \\ \underline{2\,17} \\ 230 \\ \underline{217} \\ 13 \end{array}$$

The rounded number is **19.78.**

25.

$$\downarrow$$

28.305757⑤7

The rounded number is **28.3057576.**

26. $0.362 = \dfrac{362}{1000} = \dfrac{362 \div 2}{1000 \div 2} = \dfrac{\mathbf{181}}{\mathbf{500}}$

27. $\dfrac{3}{7} \cdot \dfrac{140}{1} = \dfrac{3}{\cancel{7}} \cdot \dfrac{\overset{20}{\cancel{140}}}{1} = \mathbf{60}$

28. $\dfrac{2}{9} \cdot \dfrac{81}{1} = \dfrac{2}{\cancel{9}} \cdot \dfrac{\overset{9}{\cancel{81}}}{1} = \mathbf{18}$

29.
$$\begin{array}{r} 18.320 \\ + \quad A \\ \hline 1921.761 \end{array} \qquad \begin{array}{r} 1921.761 \\ - \quad 18.320 \\ \hline \mathbf{1903.441} \end{array} \qquad \text{Check:} \quad \begin{array}{r} 18.320 \\ + 1903.441 \\ \hline 1921.761 \end{array}$$

30. $\dfrac{4.5}{M} = 3$

$$\begin{array}{r} \mathbf{1.5} \\ 3\overline{)4.5} \\ \underline{3} \\ 1\,5 \\ \underline{1\,5} \\ 0 \end{array} \qquad \text{Check:} \quad \begin{array}{r} 3 \\ 15\overline{)45} \\ \underline{45} \\ 0 \end{array}$$

PROBLEM SET 20

1.
$$\begin{array}{r} 7{,}042 \\ + \ 93{,}975 \\ \hline 101{,}017 \end{array} \qquad \begin{array}{r} 137{,}842 \\ - \ 101{,}017 \\ \hline \mathbf{36{,}825} \textbf{ volksmarchers} \end{array}$$

2. $1903 \times N = 47{,}300$

$$\begin{array}{r} 24 \\ 1903\overline{)47{,}300} \\ \underline{38\,06} \\ 9\,240 \\ \underline{7\,612} \\ 1\,628 \end{array}$$

25 boxes would be required.

3.
$$\begin{array}{r} 19{,}002 \\ \times \qquad 5 \\ \hline 95{,}010 \end{array} \qquad \begin{array}{r} 95{,}010 \\ + \ 19{,}002 \\ \hline \mathbf{114{,}012} \textbf{ ragweed plants} \end{array}$$

4.
$$\begin{array}{r} 636.57 \\ \times \qquad 3 \\ \hline 1909.71 \end{array} \qquad \begin{array}{r} 636.57 \\ \times \qquad 2 \\ \hline 1273.14 \end{array} \qquad \begin{array}{r} 636.57 \\ 1909.71 \\ + \ 1273.14 \\ \hline \mathbf{3819.42} \textbf{ mi} \end{array}$$

5. Count by fours: **4, 8, 12, 16, 20, 24, 28, 32, 36, 40**

6. $\dfrac{42}{15} \cdot \dfrac{3}{7} = \dfrac{\overset{6}{\cancel{42}}}{\underset{5}{\cancel{15}}} \cdot \dfrac{\overset{1}{\cancel{3}}}{\underset{1}{\cancel{7}}} = \dfrac{\mathbf{6}}{\mathbf{5}}$

7. $\dfrac{5}{18} \cdot \dfrac{90}{120} = \dfrac{5}{\underset{1}{\cancel{18}}} \cdot \dfrac{\overset{\overset{1}{\cancel{5}}}{\cancel{90}}}{\underset{24}{\cancel{120}}} = \dfrac{\mathbf{5}}{\mathbf{24}}$

8. $\dfrac{5}{8} \div \dfrac{7}{4} = \dfrac{5}{\underset{2}{\cancel{8}}} \cdot \dfrac{\overset{1}{\cancel{4}}}{7} = \dfrac{\mathbf{5}}{\mathbf{14}}$

9. $\dfrac{7}{18} \div \dfrac{14}{9} = \dfrac{\overset{1}{\cancel{7}}}{\underset{2}{\cancel{18}}} \cdot \dfrac{\overset{1}{\cancel{9}}}{\underset{2}{\cancel{14}}} = \dfrac{\mathbf{1}}{\mathbf{4}}$

10. $(20 + 50 + 20 + 50) \text{ ft} = \mathbf{140\ ft}$

11. $(20\text{ m} \times 20\text{ m}) + (12\text{ m} \times 7\text{ m})$
$= 400\text{ m}^2 + 84\text{ m}^2 = \mathbf{484\ m^2}$

12. $(10\text{ m} \times 10\text{ m}) + (20\text{ m} \times 31\text{ m})$
$= 100\text{ m}^2 + 620\text{ m}^2 = \mathbf{720\ m^2}$

13. $\dfrac{540}{720} = \dfrac{540 \div 90}{720 \div 90} = \dfrac{6 \div 2}{8 \div 2} = \dfrac{\mathbf{3}}{\mathbf{4}}$

14. $\dfrac{144}{360} = \dfrac{144 \div 12}{360 \div 12} = \dfrac{12 \div 6}{30 \div 6} = \dfrac{2}{5}$

15. $\left(\dfrac{7}{13}\right)^2 = \dfrac{7}{13} \cdot \dfrac{7}{13} = \dfrac{49}{169}$

16. $\dfrac{21}{23}$

$$
\begin{array}{r}
0.913 \\
23\overline{)21.000} \\
\underline{20\,7} \\
30 \\
\underline{23} \\
70 \\
\underline{69} \\
1
\end{array}
$$

The rounded number is **0.91.**

17. $\dfrac{6}{13}$

$$
\begin{array}{r}
0.461 \\
13\overline{)6.000} \\
\underline{5\,2} \\
80 \\
\underline{78} \\
20 \\
\underline{13} \\
7
\end{array}
$$

The rounded number is **0.46.**

18. Choice **D** is correct because **9924** is divisible by 2, since its last digit is even, and by 3, since the sum of its digits, 24, is divisible by 3.

19. **0.0049, 0.0096, 0.04, 0.1**

20. $\dfrac{2}{7} \cdot C = \dfrac{4}{21}$

$\dfrac{4}{21} \div \dfrac{2}{7} = \dfrac{\cancel{4}}{\cancel{21}} \cdot \dfrac{\cancel{7}}{\cancel{2}} = \dfrac{2}{3}$

21. $87(M) = 28.71$

$$
\begin{array}{r}
0.33 \\
87\overline{)28.71} \\
\underline{26\,1} \\
2\,61 \\
\underline{2\,61} \\
0
\end{array}
$$

22. $75 = 3 \cdot 5 \cdot 5$

$180 = 2 \cdot 2 \cdot 3 \cdot 3 \cdot 5$

LCM $(75, 180) = 2 \cdot 2 \cdot 3 \cdot 3 \cdot 5 \cdot 5 = $ **900**

23. $22 = 2 \cdot 11$

$121 = 11 \cdot 11$

$200 = 2 \cdot 2 \cdot 2 \cdot 5 \cdot 5$

LCM $(22, 121, 200)$

$= 2 \cdot 2 \cdot 2 \cdot 5 \cdot 5 \cdot 11 \cdot 11$

$= $ **24,200**

24. $0.024 = \dfrac{24}{1000} = \dfrac{24 \div 4}{1000 \div 4} = \dfrac{6 \div 2}{250 \div 2}$

$= \dfrac{3}{125}$

25. $0.00064 = \dfrac{64}{100,000} = \dfrac{64 \div 8}{100,000 \div 8}$

$= \dfrac{8 \div 4}{12,500 \div 4} = \dfrac{2}{3125}$

26.

$19\,\textcircled{4}\,,591,014.62$

The rounded number is **195,000,000.**

27. 111,546,435 is **one hundred eleven million, five hundred forty-six thousand, four hundred thirty-five**

28. $8361.2361 \div 100 = $ **83.612361**

29. $\dfrac{3}{4} \cdot \dfrac{42}{1} = \dfrac{3}{\cancel{4}} \cdot \dfrac{\cancel{42}^{\,21}}{1} = \dfrac{63}{2}$

30. $\dfrac{4}{7} \cdot \dfrac{4}{1} = \dfrac{16}{7}$

PROBLEM SET 21

1. $\begin{array}{r} 10,000,475,015.0923 \\ -\ \ 8,473,000,042.0750 \\ \hline \mathbf{1,527,474,973.0173} \end{array}$

2. $\begin{array}{r} 4012.06000 \text{ m} \\ 0.00418 \text{ m} \\ 732.05000 \text{ m} \\ +\ \ \ \ 9.01600 \text{ m} \\ \hline \mathbf{4753.13018 \text{ m}} \end{array}$

3. $243 \times N = 416{,}202$

$$
\begin{array}{r}
1712 \\
243{\overline{\smash{\big)}\,416{,}202}} \\
\underline{243\phantom{{,}202}} \\
173\,2 \\
\underline{170\,1} \\
3\,10 \\
\underline{2\,43} \\
672 \\
\underline{486} \\
186
\end{array}
$$

It would take **1713 times.**

4. (a) The following numbers are divisible by 3, since the sum of the digits of each is divisible by 3: **11,682, 2193,** and **4200.**

   (b) The following number is divisible by 5, since the last digit is 0: **4200.**

5. **31, 37, 41, 43, 47**

6. $\text{Avg} = \dfrac{1242 + 87 + 521 + 169{,}810}{4}$

   $= \dfrac{171{,}660}{4} = \mathbf{42{,}915}$

7. $\$48.20 + \$40.60 + \$63.75$
   $+ \$70.15 + N = 5(\$316.05)$
   $\$222.70 + N = \$1580.25$

$$
\begin{array}{r}
\$1580.25 \\
- \;\$\;\,222.70 \\
\hline
\mathbf{\$1357.55}
\end{array}
$$

8. $\dfrac{36}{28} \cdot \dfrac{7}{6} = \dfrac{\overset{}{\cancel{36}}}{\underset{4}{\cancel{28}}} \cdot \dfrac{\overset{1}{\cancel{7}}}{\underset{1}{\cancel{6}}} = \dfrac{\overset{3}{\cancel{6}}}{\underset{2}{\cancel{4}}} = \dfrac{3}{2}$

9. $\dfrac{21}{32} \cdot \dfrac{4}{7} = \dfrac{\overset{3}{\cancel{21}}}{\underset{8}{\cancel{32}}} \cdot \dfrac{\overset{1}{\cancel{4}}}{\underset{1}{\cancel{7}}} = \dfrac{3}{8}$

10. $\dfrac{5}{7} \div \dfrac{15}{21} = \dfrac{\overset{1}{\cancel{5}}}{\underset{1}{\cancel{7}}} \cdot \dfrac{\overset{3}{\cancel{21}}}{\underset{3}{\cancel{15}}} = \dfrac{\overset{1}{\cancel{3}}}{\underset{1}{\cancel{3}}} = 1$

11. $\dfrac{9}{25} \div \dfrac{27}{15} = \dfrac{\overset{1}{\cancel{9}}}{\underset{5}{\cancel{25}}} \cdot \dfrac{\overset{3}{\cancel{15}}}{\underset{3}{\cancel{27}}} = \dfrac{\overset{1}{\cancel{3}}}{\underset{5}{\cancel{15}}} = \dfrac{1}{5}$

12. $(18 + 20 + 31 + 25 + 12 + 20)\,\text{m} = \mathbf{126\ m}$

13. $(20\,\text{cm} \times 10\,\text{cm}) + (20\,\text{cm} \times 9\,\text{cm})$
   $= 200\,\text{cm}^2 + 180\,\text{cm}^2 = \mathbf{380\ cm^2}$

14. $\dfrac{270}{360} = \dfrac{270 \div 10}{360 \div 10} = \dfrac{27 \div 9}{36 \div 9} = \dfrac{3}{4}$

15. $\dfrac{35}{42} = \dfrac{35 \div 7}{42 \div 7} = \dfrac{5}{6}$

16. Count by threes: **3, 6, 9, 12, 15, 18, 21, 24, 27, 30, 33, 36, 39, 42, 45, 48**

17. $\dfrac{11}{17}$

$$
\begin{array}{r}
0.647 \\
17{\overline{\smash{\big)}\,11.000}} \\
\underline{10\,2} \\
80 \\
\underline{68} \\
120 \\
\underline{119} \\
1
\end{array}
$$

The rounded number is **0.65.**

18. $\dfrac{5}{13}$

$$
\begin{array}{r}
0.384 \\
13{\overline{\smash{\big)}\,5.000}} \\
\underline{3\,9} \\
1\,10 \\
\underline{1\,04} \\
60 \\
\underline{52} \\
8
\end{array}
$$

The rounded number is **0.38.**

19. $\dfrac{2520}{5} = 504; \quad \dfrac{504}{3} = 168; \quad \dfrac{168}{3} = 56;$

   $\dfrac{56}{2} = 28; \quad \dfrac{28}{2} = 14; \quad \dfrac{14}{2} = 7$

   $2520 = 2 \cdot 2 \cdot 2 \cdot 3 \cdot 3 \cdot 5 \cdot 7$
   $ = 2^3 \cdot 3^2 \cdot 5 \cdot 7$

20. $\dfrac{8400}{7} = 1200; \quad \dfrac{1200}{5} = 240; \quad \dfrac{240}{5} = 48;$

   $\dfrac{48}{3} = 16; \quad \dfrac{16}{2} = 8; \quad \dfrac{8}{2} = 4; \quad \dfrac{4}{2} = 2$

   $8400 = 2 \cdot 2 \cdot 2 \cdot 2 \cdot 3 \cdot 5 \cdot 5 \cdot 7$
   $ = 2^4 \cdot 3 \cdot 5^2 \cdot 7$

21.
$$66.12$$
$$\times \quad 1.7$$
$$\overline{\quad 46284}$$
$$\underline{\quad 6612}$$
$$\mathbf{112.404}$$

22. $\left(\dfrac{3}{4}\right)^3 = \dfrac{3}{4} \cdot \dfrac{3}{4} \cdot \dfrac{3}{4} = \dfrac{9}{16} \cdot \dfrac{3}{4} = \dfrac{\mathbf{27}}{\mathbf{64}}$

23. $0.95 = \dfrac{95}{100} = \dfrac{95 \div 5}{100 \div 5} = \dfrac{\mathbf{19}}{\mathbf{20}}$

24.
	12.8760		12.8760	Check:	12.8760
−	N	−	3.0931		− 9.7829
	3.0931		**9.7829**		3.0931

25. $x \div \dfrac{1}{4} = \dfrac{5}{6}$

$\dfrac{5}{6} \cdot \dfrac{1}{4} = \dfrac{\mathbf{5}}{\mathbf{24}}$

26. $\dfrac{6.25}{0.87}$

$$
\begin{array}{r}
7.183 \\
87\overline{)625.000} \\
\underline{609} \\
16\,0 \\
\underline{8\,7} \\
7\,30 \\
\underline{6\,96} \\
340 \\
\underline{261} \\
79
\end{array}
$$

The rounded number is **7.18.**

27. $\dfrac{71.3}{3.7}$

$$
\begin{array}{r}
19.270 \\
37\overline{)713.000} \\
\underline{37} \\
343 \\
\underline{333} \\
10\,0 \\
\underline{7\,4} \\
2\,60 \\
\underline{2\,59} \\
10
\end{array}
$$

The rounded number is **19.27.**

28. $45 = 3 \cdot 3 \cdot 5$

$50 = 2 \cdot 5 \cdot 5$

$60 = 2 \cdot 2 \cdot 3 \cdot 5$

LCM $(45, 50, 60) = 2 \cdot 2 \cdot 3 \cdot 3 \cdot 5 \cdot 5$
$= \mathbf{900}$

29. $45 = 3 \cdot 3 \cdot 5$

$50 = 2 \cdot 5 \cdot 5$

$60 = 2 \cdot 2 \cdot 3 \cdot 5$

GCF $(45, 50, 60) = \mathbf{5}$

30. $\dfrac{3}{8} \cdot \dfrac{64}{1} = \dfrac{3}{\cancel{8}} \cdot \dfrac{\overset{8}{\cancel{64}}}{1} = \mathbf{24}$

PROBLEM SET 22

1.
$$90{,}004{,}000.000062$$
$$\underline{-\;86{,}427{,}000.001400}$$
$$\mathbf{3{,}576{,}999.998662}$$

2.
$$742{,}000$$
$$96{,}016$$
$$\underline{+\;1{,}001{,}892}$$
$$\mathbf{1{,}839{,}908 \text{ grains of sand}}$$

3. $45.2 + 51.8 + N = 3(49.8)$

$\qquad\qquad 97 + N = 149.4$

$149.4 - 97 = \mathbf{52.4 \text{ s}}$

4.
146	146
$\times \quad 5$	$+ \ 730$
730	**876 Komodo dragons**

5. $\text{Avg} = \dfrac{1862 + 1430 + 276}{3} = \dfrac{3568}{3}$

$\qquad = \mathbf{1189.\overline{3}}$

6. (a) **83, 89**

(b) **81, 84, 87**

7. $\dfrac{63}{54} \cdot \dfrac{21}{35} \cdot \dfrac{15}{22} = \dfrac{\overset{7}{\cancel{63}}}{\underset{\underset{2}{6}}{\cancel{54}}} \cdot \dfrac{\overset{3}{\cancel{21}}}{\underset{1}{\cancel{35}}} \cdot \dfrac{\overset{\overset{1}{\cancel{3}}}{\cancel{15}}}{22} = \dfrac{\mathbf{21}}{\mathbf{44}}$

8. $\dfrac{26}{90} \cdot \dfrac{24}{25} \div \dfrac{4}{15} = \dfrac{26}{\underset{\underset{1}{6}}{\cancel{90}}} \cdot \dfrac{24}{25} \cdot \dfrac{\overset{\overset{1}{\cancel{6}}}{\cancel{15}}}{\underset{1}{\cancel{4}}} = \dfrac{\mathbf{26}}{\mathbf{25}}$

9. $\dfrac{36}{24} \cdot \dfrac{8}{6} \div \dfrac{3}{6} = \dfrac{\overset{12}{\cancel{36}}}{\underset{3}{\cancel{24}}} \cdot \dfrac{\overset{1}{\cancel{8}}}{\underset{1}{\cancel{6}}} \cdot \dfrac{\overset{1}{\cancel{6}}}{\underset{1}{\cancel{3}}} = \dfrac{\overset{4}{\cancel{12}}}{\underset{1}{\cancel{3}}} = \mathbf{4}$

10. $(25 + 51 + 25 + 51)$ in. $= \mathbf{152 \text{ in.}}$

11. $[(23 \times 9) + (15 \times 10)]$ ft^2 = **357 ft^2**

12. $100 = 2 \cdot 2 \cdot 5 \cdot 5$

$14 = 2 \cdot 7$

$20 = 2 \cdot 2 \cdot 5$

LCM $(100, 14, 20) = 2 \cdot 2 \cdot 5 \cdot 5 \cdot 7 =$ **700**

13. $10 = 2 \cdot 5$

$15 = 3 \cdot 5$

$20 = 2 \cdot 2 \cdot 5$

GCF $(10, 15, 20) =$ **5**

14. $\dfrac{13}{17}$

$$\begin{array}{r} 0.764 \\ 17\overline{)13.000} \\ \underline{11\,9} \\ 1\,10 \\ \underline{1\,02} \\ 80 \\ \underline{68} \\ 12 \end{array}$$

The rounded number is **0.76.**

15. $\dfrac{3}{11}$

$$\begin{array}{r} 0.272 \\ 11\overline{)3.000} \\ \underline{2\,2} \\ 80 \\ \underline{77} \\ 30 \\ \underline{22} \\ 8 \end{array}$$

The rounded number is **0.27.**

16. $\dfrac{7}{20}$

$$\begin{array}{r} \mathbf{0.35} \\ 20\overline{)7.00} \\ \underline{6\,0} \\ 1\,00 \\ \underline{1\,00} \\ 0 \end{array}$$

17. $\dfrac{540}{1440} = \dfrac{540 \div 10}{1440 \div 10} = \dfrac{54 \div 9}{144 \div 9}$

$= \dfrac{6 \div 2}{16 \div 2} = \dfrac{\mathbf{3}}{\mathbf{8}}$

18. $\dfrac{128}{360} = \dfrac{128 \div 8}{360 \div 8} = \dfrac{\mathbf{16}}{\mathbf{45}}$

19. $\dfrac{720}{840} = \dfrac{720 \div 10}{840 \div 10} = \dfrac{72 \div 12}{84 \div 12} = \dfrac{\mathbf{6}}{\mathbf{7}}$

20. $\dfrac{240}{256} = \dfrac{240 \div 4}{256 \div 4} = \dfrac{60 \div 4}{64 \div 4} = \dfrac{\mathbf{15}}{\mathbf{16}}$

21. $(0.4)^3 = 0.4 \times 0.4 \times 0.4 = 0.16 \times 0.4$

$= \mathbf{0.064}$

22.

$$\begin{array}{r} 581 \\ \times\ 0.163 \\ \hline 1743 \\ 3486 \\ \underline{581} \\ \mathbf{94.703} \end{array}$$

23.

$$\begin{array}{r} 6854.320 \\ -\quad 1.871 \\ \hline \mathbf{6852.449} \end{array} \qquad \text{Check:} \begin{array}{r} 6852.449 \\ +\quad 1.871 \\ \hline 6854.320 \end{array}$$

24. $4.56 + F = 12.3$

$$\begin{array}{r} 12.30 \\ -\ 4.56 \\ \hline \mathbf{7.74} \end{array} \qquad \text{Check:} \begin{array}{r} 4.56 \\ +\ 7.74 \\ \hline 12.30 \end{array}$$

25. $\dfrac{1}{4}(D) = 16$

$16 \div \dfrac{1}{4} = \dfrac{16}{1} \cdot \dfrac{4}{1} = \mathbf{64}$

26. $\dfrac{11.7}{3.1}$

$$\begin{array}{r} 3.774 \\ 31\overline{)117.000} \\ \underline{93} \\ 24\,0 \\ \underline{21\,7} \\ 2\,30 \\ \underline{2\,17} \\ 130 \\ \underline{124} \\ 6 \end{array}$$

The rounded number is **3.77.**

27.
$$\frac{1132.1}{0.7}$$

$$
\begin{array}{r}
1617.285 \\
7\overline{)11{,}321.000} \\
\underline{7} \\
4\,3 \\
\underline{4\,2} \\
12 \\
\underline{7} \\
51 \\
\underline{49} \\
2\,0 \\
\underline{1\,4} \\
60 \\
\underline{56} \\
40 \\
\underline{35} \\
5
\end{array}
$$

The rounded number is **1617.29.**

28.

$$0.3\,\overset{\downarrow}{\textcircled{0}}\,1029996$$

The rounded number is **0.30.**

29. $\dfrac{3}{11} \cdot \dfrac{132}{1} = \dfrac{3}{\cancel{11}} \cdot \dfrac{\overset{12}{\cancel{132}}}{1} = \mathbf{36}$

30. $0.015 = \dfrac{15}{1000} = \dfrac{15 \div 5}{1000 \div 5} = \dfrac{\mathbf{3}}{\mathbf{200}}$

PROBLEM SET 23

1.
$$
\begin{array}{r}
19{,}007.4230 \text{ cm} \\
+\ \ 8{,}042.0765 \text{ cm} \\
\hline
\mathbf{27{,}049.4995 \text{ cm}}
\end{array}
$$

2.
$$
\begin{array}{r}
2{,}046{,}021 \\
\times\ \ \ \ \ \ \ 3 \\
\hline
6{,}138{,}063
\end{array}
\qquad
\begin{array}{r}
2{,}046{,}021 \\
+\ 6{,}138{,}063 \\
\hline
\mathbf{8{,}184{,}084 \text{ microbes}}
\end{array}
$$

3. $243 \times N = 19{,}197$

$$
\begin{array}{r}
\mathbf{79 \text{ work periods}} \\
243\overline{)19{,}197} \\
\underline{17\,01} \\
2\,187 \\
\underline{2\,187} \\
0
\end{array}
$$

4.
$$
\begin{array}{r}
90{,}663 \\
+\ 20{,}163 \\
\hline
\mathbf{110{,}826}
\end{array}
$$

5. (a) **61, 67** (b) **63**

6. $216 + 159.8 + 301.25 + N = 4(638.4)$

$677.05 + N = 2553.6$

$2553.6 - 677.05 = \mathbf{1876.55}$

7. $120 \;\cancel{\text{in.}} \times \dfrac{1 \text{ ft}}{12 \;\cancel{\text{in.}}} = \dfrac{120}{12} \text{ ft} = \mathbf{10 \text{ ft}}$

8. $999 \;\cancel{\text{ft}} \times \dfrac{1 \text{ yd}}{3 \;\cancel{\text{ft}}} = \dfrac{999}{3} \text{ yd} = \mathbf{333 \text{ yd}}$

9. $336 \;\cancel{\text{oz}} \cdot \dfrac{1 \text{ lb}}{16 \;\cancel{\text{oz}}} = \dfrac{336}{16} \text{ lb} = \mathbf{21 \text{ lb}}$

10. $\dfrac{42}{12} \cdot \dfrac{36}{14} \div \dfrac{18}{2} = \dfrac{42}{12} \cdot \dfrac{36}{14} \cdot \dfrac{2}{18}$

$$= \dfrac{\overset{3}{\cancel{42}}}{\underset{1}{\cancel{12}}} \cdot \dfrac{\overset{3}{\cancel{36}}}{\underset{1}{\cancel{14}}} \cdot \dfrac{\overset{1}{\cancel{2}}}{\underset{9}{\cancel{18}}} = \dfrac{3}{1} \cdot \dfrac{3}{1} \cdot \dfrac{1}{9} = \mathbf{1}$$

11. $\dfrac{12}{16} \cdot \dfrac{4}{3} \div \dfrac{5}{6} = \dfrac{\overset{1}{\cancel{12}}}{\underset{1}{\cancel{16}}} \cdot \dfrac{\overset{1}{\cancel{4}}}{3} \cdot \dfrac{6}{5} = \dfrac{\mathbf{6}}{\mathbf{5}}$

12. $\dfrac{16}{24} \cdot \dfrac{12}{4} \cdot \dfrac{1}{2} = \dfrac{\overset{4}{\cancel{16}}}{\underset{2}{\cancel{24}}} \cdot \dfrac{\overset{1}{\cancel{12}}}{\underset{1}{\cancel{4}}} \cdot \dfrac{1}{2} = \dfrac{\overset{2}{\cancel{4}}}{\underset{1}{\cancel{2}}} \cdot \dfrac{1}{\underset{1}{\cancel{2}}} = \mathbf{1}$

13. $(30 + 10 + 19 + 6 + 16 + 4 + 65 + 20) \text{ yd}$
$= \mathbf{170 \text{ yd}}$

14. $[(21 \times 23) + (39 \times 10)] \text{ m}^2 = \mathbf{873 \text{ m}^2}$

15. $\dfrac{2880}{5} = 576; \quad \dfrac{576}{3} = 192; \quad \dfrac{192}{3} = 64;$

$\dfrac{64}{2} = 32; \quad \dfrac{32}{2} = 16; \quad \dfrac{16}{2} = 8; \quad \dfrac{8}{2} = 4;$

$\dfrac{4}{2} = 2$

$2880 = 2 \cdot 2 \cdot 2 \cdot 2 \cdot 2 \cdot 2 \cdot 3 \cdot 3 \cdot 5$
$ = 2^6 \cdot 3^2 \cdot 5$

16. $\dfrac{6750}{5} = 1350; \quad \dfrac{1350}{5} = 270; \quad \dfrac{270}{5} = 54;$

$\dfrac{54}{3} = 18; \quad \dfrac{18}{3} = 6; \quad \dfrac{6}{3} = 2$

$6750 = 2 \cdot 3 \cdot 3 \cdot 3 \cdot 5 \cdot 5 \cdot 5$
$ = \mathbf{2 \cdot 3^3 \cdot 5^3}$

17. $\dfrac{13}{19}$

$$
\begin{array}{r}
0.684 \\
19\overline{)13.000} \\
11\,4 \\
\hline
1\,60 \\
1\,52 \\
\hline
80 \\
76 \\
\hline
4
\end{array}
$$

The rounded number is **0.68.**

18. $\dfrac{17}{20}$

$$
\begin{array}{r}
0.85 \\
20\overline{)17.00} \\
16\,0 \\
\hline
1\,00 \\
1\,00 \\
\hline
0
\end{array}
$$

19. $\dfrac{180}{256} = \dfrac{180 \div 4}{256 \div 4} = \dfrac{45}{64}$

20. $\dfrac{240}{600} = \dfrac{240 \div 10}{600 \div 10} = \dfrac{24 \div 12}{60 \div 12} = \dfrac{2}{5}$

21. $\dfrac{540}{2160} = \dfrac{540 \div 10}{2160 \div 10} = \dfrac{54 \div 9}{216 \div 9}$

$= \dfrac{6 \div 6}{24 \div 6} = \dfrac{1}{4}$

22.
$$
\begin{array}{r}
71.2 \\
\times\ 0.173 \\
\hline
2136 \\
4984 \\
712 \\
\hline
12.3176
\end{array}
$$

23.
$$
\begin{array}{r}
61.7 \\
\times\ 11.2 \\
\hline
1234 \\
617 \\
617 \\
\hline
691.04
\end{array}
$$

24. $18 = 2 \cdot 3 \cdot 3$

$81 = 3 \cdot 3 \cdot 3 \cdot 3$

$270 = 2 \cdot 3 \cdot 3 \cdot 3 \cdot 5$

LCM $(18, 81, 270) = 2 \cdot 3 \cdot 3 \cdot 3 \cdot 3 \cdot 5$

$= \mathbf{810}$

25. The first 3 common multiples are equivalent to the first 3 multiples of the LCM.

$10 = 2 \cdot 5$

$15 = 3 \cdot 5$

$20 = 2 \cdot 2 \cdot 5$

LCM $(10, 15, 20) = 2 \cdot 2 \cdot 3 \cdot 5 = 60$

The first 3 common multiples are **60, 120, 180.**

26. $\dfrac{713.7}{2.5}$

$$
\begin{array}{r}
285.48 \\
25\overline{)7137.00} \\
50 \\
\hline
213 \\
200 \\
\hline
137 \\
125 \\
\hline
12\,0 \\
10\,0 \\
\hline
2\,00 \\
2\,00 \\
\hline
0
\end{array}
$$

27. $\dfrac{7181.3}{0.3}$

$$
\begin{array}{r}
23{,}937.666 \\
3\overline{)71{,}813.000} \\
6 \\
\hline
11 \\
9 \\
\hline
2\,8 \\
2\,7 \\
\hline
11 \\
9 \\
\hline
23 \\
21 \\
\hline
2\,0 \\
1\,8 \\
\hline
20 \\
18 \\
\hline
20 \\
18 \\
\hline
2
\end{array}
$$

The repeating decimal is $\mathbf{23{,}937.\overline{6}.}$

28. $\dfrac{3}{4} \cdot \dfrac{400}{1} = \dfrac{1200}{4} = 300$

29.
$$
\begin{array}{r}
71.1610 \\
-\quad Q \\
\hline
0.8121
\end{array}
\qquad
\begin{array}{r}
71.1610 \\
-\ 0.8121 \\
\hline
\mathbf{70.3489}
\end{array}
$$
Check:
$$
\begin{array}{r}
71.1610 \\
-\ 70.3489 \\
\hline
0.8121
\end{array}
$$

30. $\dfrac{2}{3} \cdot B = \dfrac{3}{8}$

$\dfrac{3}{8} \div \dfrac{2}{3} = \dfrac{3}{8} \cdot \dfrac{3}{2} = \dfrac{9}{16}$

PROBLEM SET 24

1.
$$\begin{array}{r} 14{,}007 \\ -\ \ 7{,}942 \\ \hline \mathbf{6{,}065}\ \textbf{items} \end{array}$$

2.
$$\begin{array}{r} 147.000923\ \text{in.} \\ -\ 146.314200\ \text{in.} \\ \hline \mathbf{0.686723}\ \textbf{in.} \end{array}$$

3. $47 \times N = 1982$

$$\begin{array}{r} 42 \\ 47\overline{)1982} \\ \underline{188} \\ 102 \\ \underline{94} \\ 8 \end{array}$$

43 slots were needed.

4. (a) The following numbers are divisible by 2, since the last digit of each is even: **90,528** and **4020.**

  (b) The following number is divisible by 10, since the last digit is 0: **4020.**

5. (a) **23, 29**

  (b) **24, 28**

6. $\text{Avg} = \dfrac{98 + 142 + 76 + 81 + 6}{5}$

  $= \dfrac{403}{5} = \mathbf{80.6}$

7. $859\ \cancel{\text{cm}} \times \dfrac{1\ \text{m}}{100\ \cancel{\text{cm}}} = \mathbf{8.59\ m}$

8. $6400\ \cancel{\text{km}} \times \dfrac{1000\ \text{m}}{1\ \cancel{\text{km}}} = \mathbf{6{,}400{,}000\ m}$

9. $204\ \cancel{\text{in.}} \times \dfrac{1\ \text{ft}}{12\ \cancel{\text{in.}}} = \mathbf{17\ ft}$

10. $17\ \cancel{\text{tons}} \cdot \dfrac{2000\ \text{lb}}{1\ \cancel{\text{ton}}} = \mathbf{34{,}000\ lb}$

11. $50{,}000\ \cancel{\text{cg}} \cdot \dfrac{1\ \text{g}}{100\ \cancel{\text{cg}}} = \mathbf{500\ g}$

12. $\dfrac{16}{24} \cdot \dfrac{12}{4} \cdot \dfrac{1}{3} = \dfrac{\cancel{16}}{\cancel{24}} \cdot \dfrac{\cancel{12}}{\cancel{4}} \cdot \dfrac{1}{\cancel{3}} = \mathbf{\dfrac{2}{3}}$

13. $\dfrac{18}{24} \cdot \dfrac{8}{9} \cdot \dfrac{3}{2} = \dfrac{\cancel{18}}{\cancel{24}} \cdot \dfrac{\cancel{8}}{\cancel{9}} \cdot \dfrac{\cancel{3}}{\cancel{2}} = \mathbf{1}$

14. $\dfrac{14}{21} \cdot \dfrac{7}{2} \div \dfrac{14}{6} = \dfrac{\cancel{14}}{\cancel{21}} \cdot \dfrac{\cancel{7}}{\cancel{2}} \cdot \dfrac{6}{14} = \dfrac{\cancel{7}}{\cancel{3}} \cdot \dfrac{\cancel{6}}{\cancel{14}} = \mathbf{1}$

15. $(10 + 11 + 31 + 11 + 10 + 31)\ \text{ft} = \mathbf{104\ ft}$

16. $[(10 \cdot 40) + (10 \cdot 12)]\ \text{in.}^2 = \mathbf{520\ in.^2}$

17. **15, 30, 45**

18. Factors of 12: 1, 2, 3, 4, 6, 12

  Factors of 18: 1, 2, 3, 6, 9, 18

  Common Factors: **1, 2, 3, 6**

19. $\dfrac{3}{14}$

$$\begin{array}{r} 0.214 \\ 14\overline{)3.000} \\ \underline{2\ 8} \\ 20 \\ \underline{14} \\ 60 \\ \underline{56} \\ 4 \end{array}$$

The rounded number is **0.21.**

20. $\dfrac{19}{25}$

$$\begin{array}{r} \mathbf{0.76} \\ 25\overline{)19.00} \\ \underline{17\ 5} \\ 1\ 50 \\ \underline{1\ 50} \\ 0 \end{array}$$

21. $\dfrac{81}{135}$

$$\begin{array}{r} \mathbf{0.6} \\ 135\overline{)81.0} \\ \underline{81\ 0} \\ 0 \end{array}$$

22. $\dfrac{24}{72} = \dfrac{24 \div 24}{72 \div 24} = \mathbf{\dfrac{1}{3}}$

23. $\dfrac{18}{24} = \dfrac{18 \div 6}{24 \div 6} = \dfrac{3}{4}$

24. $180 = 2 \cdot 2 \cdot 3 \cdot 3 \cdot 5$

$300 = 2 \cdot 2 \cdot 3 \cdot 5 \cdot 5$

GCF $(180, 300) = 2 \cdot 2 \cdot 3 \cdot 5 = \mathbf{60}$

25. $0.035 = \dfrac{35}{1000} = \dfrac{35 \div 5}{1000 \div 5} = \dfrac{7}{200}$

26. $\dfrac{X}{0.7} = 10.23$

$\begin{array}{r} 10.23 \\ \times \quad 0.7 \\ \hline \mathbf{7.161} \end{array}$

27. (a) $\dfrac{3}{14} = \dfrac{3 \cdot 3}{14 \cdot 3} = \dfrac{9}{42}$

(b) $\dfrac{4}{7} = \dfrac{4 \cdot 6}{7 \cdot 6} = \dfrac{24}{42}$

(c) $\dfrac{33}{126} = \dfrac{33 \div 3}{126 \div 3} = \dfrac{11}{42}$

28.

$218.09797\textcircled{9}7$

The rounded number is **218.097980.**

29. $20 = 2 \cdot 2 \cdot 5$

$25 = 5 \cdot 5$

$36 = 2 \cdot 2 \cdot 3 \cdot 3$

LCM $(20, 25, 36) = 2 \cdot 2 \cdot 3 \cdot 3 \cdot 5 \cdot 5$

$= \mathbf{900}$

30. $\dfrac{3}{10} \cdot \dfrac{35}{1} = \dfrac{105}{10} = \dfrac{21}{2}$

PROBLEM SET 25

1.
$\begin{array}{r} 10{,}042 \\ \times \quad 4 \\ \hline 40{,}168 \end{array}$
$\begin{array}{r} 10{,}042 \\ 40{,}168 \\ + \quad 4{,}075 \\ \hline \mathbf{54{,}285 \text{ students}} \end{array}$

2.
$\begin{array}{r} 17{,}842 \\ - \quad 9{,}085 \\ \hline \mathbf{8{,}757 \text{ flowers}} \end{array}$

3.
$\begin{array}{r} 482 \\ \times \quad 5 \\ \hline 2410 \end{array}$
$\begin{array}{r} 482 \\ 2410 \\ + \quad 392 \\ \hline \mathbf{3284 \text{ monkeys}} \end{array}$

4. The following numbers are divisible by 3, since the sum of the digits of each is divisible by 3: **315,234** and **21,387.**

5. (a) **41, 43, 47**

(b) **42, 45, 48**

6. $6.2 + 5.1 + 9.8 + 8.8 + 10 + N = 6(48.6)$

$39.9 + N = 291.6$

$291.6 - 39.9 = \mathbf{251.7}$

7. $46.31 \ \cancel{m} \times \dfrac{100 \text{ cm}}{1 \ \cancel{m}} = \mathbf{4631 \text{ cm}}$

8. $48 \ \cancel{mi} \times \dfrac{5280 \text{ ft}}{1 \ \cancel{mi}} = \mathbf{253{,}440 \text{ ft}}$

9. $416 \ \cancel{g} \times \dfrac{1 \text{ kg}}{1000 \ \cancel{g}} = \mathbf{0.416 \text{ kg}}$

10. $2.54 \ \cancel{cm} \cdot \dfrac{1 \text{ m}}{100 \ \cancel{cm}} = \dfrac{2.54}{100} \text{ m} = \mathbf{0.0254 \text{ m}}$

11. $\dfrac{42}{24} \cdot \dfrac{6}{7} \cdot \dfrac{2}{3} = \dfrac{\overset{6}{\cancel{42}}}{\underset{4}{\cancel{24}}} \cdot \dfrac{\overset{1}{\cancel{6}}}{\cancel{7}} \cdot \dfrac{2}{3} = \dfrac{\overset{2}{\cancel{6}}}{\underset{2}{\cancel{4}}} \cdot \dfrac{\overset{1}{\cancel{2}}}{\cancel{3}} \cdot \dfrac{\overset{1}{\cancel{2}}}{\underset{1}{\cancel{2}}}$

$= \mathbf{1}$

12. $\dfrac{72}{48} \cdot \dfrac{8}{9} \cdot \dfrac{3}{4} = \dfrac{\overset{8}{\cancel{72}}}{\underset{6}{\cancel{48}}} \cdot \dfrac{\overset{1}{\cancel{8}}}{\cancel{9}} \cdot \dfrac{3}{4} = \dfrac{\overset{2}{\cancel{8}}}{\underset{2}{\cancel{6}}} \cdot \dfrac{\overset{1}{\cancel{3}}}{\cancel{4}} \cdot \dfrac{\overset{1}{\cancel{2}}}{\underset{1}{\cancel{2}}}$

$= \mathbf{1}$

13. $\dfrac{2}{3} \cdot \dfrac{27}{64} \div \dfrac{3}{16} = \dfrac{2}{3} \cdot \dfrac{\overset{9}{\cancel{27}}}{\underset{4}{\cancel{64}}} \cdot \dfrac{\overset{1}{\cancel{16}}}{\cancel{3}} = \dfrac{\overset{1}{\cancel{2}}}{\cancel{3}} \cdot \dfrac{\overset{3}{\cancel{9}}}{\underset{2}{\cancel{4}}} = \dfrac{3}{2}$

14. $(11 + 10 + 23 + 20 + 23 + 10 + 11 + 20) \text{ ft}$
$= \mathbf{128 \text{ ft}}$

15.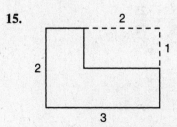

$A_{\text{Figure}} = A_{\text{Rectangle}} - A_{\text{Missing}}$

$= (3 \text{ m})(2 \text{ m}) - (1 \text{ m})(2 \text{ m})$

$= 6 \text{ m}^2 - 2 \text{ m}^2 = \mathbf{4 \text{ m}^2}$

16. $\dfrac{288}{3} = 96; \quad \dfrac{96}{3} = 32; \quad \dfrac{32}{2} = 16; \quad \dfrac{16}{2} = 8;$

$\dfrac{8}{2} = 4; \quad \dfrac{4}{2} = 2$

$288 = 2 \cdot 2 \cdot 2 \cdot 2 \cdot 3 \cdot 3 = \mathbf{2^5 \cdot 3^2}$

17. $\dfrac{630}{7} = 90; \quad \dfrac{90}{5} = 18; \quad \dfrac{18}{3} = 6; \quad \dfrac{6}{3} = 2$

$630 = 2 \cdot 3 \cdot 3 \cdot 5 \cdot 7 = \mathbf{2 \cdot 3^2 \cdot 5 \cdot 7}$

18. $\dfrac{3}{19}$

$$
\begin{array}{r}
0.157 \\
19\overline{)3.000} \\
\underline{1\,9} \\
1\,10 \\
\underline{95} \\
150 \\
\underline{133} \\
17
\end{array}
$$

The rounded number is **0.16.**

19. $\dfrac{8}{23}$

$$
\begin{array}{r}
0.347 \\
23\overline{)8.000} \\
\underline{6\,9} \\
1\,10 \\
\underline{92} \\
180 \\
\underline{161} \\
19
\end{array}
$$

The rounded number is **0.35.**

20. $\dfrac{39}{72} = \dfrac{39 \div 3}{72 \div 3} = \dfrac{\mathbf{13}}{\mathbf{24}}$

21. $\dfrac{210}{420} = \dfrac{210 \div 10}{420 \div 10} = \dfrac{21 \div 21}{42 \div 21} = \dfrac{\mathbf{1}}{\mathbf{2}}$

22. $\dfrac{125}{360} = \dfrac{125 \div 5}{360 \div 5} = \dfrac{\mathbf{25}}{\mathbf{72}}$

23. $42 = 2 \cdot 3 \cdot 7$

$63 = 3 \cdot 3 \cdot 7$

GCF $(42, 63) = 3 \cdot 7 = \mathbf{21}$

24. $0.18 = \dfrac{18}{100} = \dfrac{18 \div 2}{100 \div 2} = \dfrac{\mathbf{9}}{\mathbf{50}}$

25. $\dfrac{7}{8} \cdot \dfrac{32}{1} = \dfrac{7}{\cancel{8}} \cdot \dfrac{\overset{4}{\cancel{32}}}{1} = \mathbf{28}$

26.
$$
\begin{array}{rr}
M & \quad 9.93 \\
-\,1.38 & \quad +\,1.38 \\
\hline
9.93 & \quad \mathbf{11.31}
\end{array}
\qquad
\text{Check:}
\begin{array}{r}
11.31 \\
-\,1.38 \\
\hline
9.93
\end{array}
$$

27. $\dfrac{7}{20} \div Z = \dfrac{6}{5}$

$\dfrac{7}{20} \div \dfrac{6}{5} = \dfrac{7}{\underset{4}{\cancel{20}}} \cdot \dfrac{\overset{1}{\cancel{5}}}{6} = \dfrac{\mathbf{7}}{\mathbf{24}}$

28. $\dfrac{625.8}{1.7}$

$$
\begin{array}{r}
368.117 \\
17\overline{)6258.000} \\
\underline{51} \\
115 \\
\underline{102} \\
138 \\
\underline{136} \\
2\,0 \\
\underline{1\,7} \\
30 \\
\underline{17} \\
130 \\
\underline{119} \\
11
\end{array}
$$

The rounded number is **368.12.**

29. $\dfrac{1361.4}{0.4}$

$$
\begin{array}{r}
3403.5 \\
4\overline{)13{,}614.0} \\
\underline{12} \\
1\,6 \\
\underline{1\,6} \\
014 \\
\underline{12} \\
2\,0 \\
\underline{2\,0} \\
0
\end{array}
$$

30. $24 = 2 \cdot 2 \cdot 2 \cdot 3$

$100 = 2 \cdot 2 \cdot 5 \cdot 5$

LCM $(24, 100) = 2 \cdot 2 \cdot 2 \cdot 3 \cdot 5 \cdot 5 = \mathbf{600}$

PROBLEM SET 26

1. $3250 + 3890 + 3640 + N = 4(3700)$

$\qquad\quad 10{,}780 + N = 14{,}800$

$14{,}800 - 10{,}780 = \textbf{4020 lb}$

2. Mean $= \dfrac{214.063 + 435.09}{2} = \dfrac{649.153}{2}$

$\qquad\quad = \textbf{324.5765 m}$

3. Mean $= \dfrac{2 + 3 + 5 + 7 + 11 + 13 + 17 + 19}{8}$

$\qquad\quad = \dfrac{77}{8} = \textbf{9.625}$

4. (a) The following numbers are divisible by 3:
   41,625, 9081, 20,733, and 10,662.

   (b) Mean $= \dfrac{41{,}625 + 9081 + 20{,}733 + 10{,}662}{4}$

   $\qquad\qquad = \dfrac{82{,}101}{4} = \textbf{20,525.25}$

5. $325 \times N = 495{,}625$

1525 beanbags

$$
\begin{array}{r}
325\overline{)495{,}625} \\
\underline{325} \\
170\ 6 \\
\underline{162\ 5} \\
8\ 12 \\
\underline{6\ 50} \\
1\ 625 \\
\underline{1\ 625} \\
0
\end{array}
$$

6. $\begin{array}{r} 82{,}006.012 \\ -\ 60{,}152.035 \\ \hline \textbf{21,853.977 m} \end{array}$

7. $136.15\ \cancel{\text{cm}} \times \dfrac{1\ \text{m}}{100\ \cancel{\text{cm}}} = \textbf{1.3615 m}$

8. $12\ \cancel{\text{mi}} \times \dfrac{5280\ \text{ft}}{1\ \cancel{\text{mi}}} = \textbf{63,360 ft}$

9. $1899\ \cancel{\text{m}} \times \dfrac{1\ \text{km}}{1000\ \cancel{\text{m}}} = \textbf{1.899 km}$

10. $\dfrac{21}{36} \cdot \dfrac{12}{14} \cdot \dfrac{2}{3} = \dfrac{\overset{}{\cancel{21}}}{\underset{3}{\cancel{36}}} \cdot \dfrac{\overset{}{\cancel{12}}}{\underset{2}{\cancel{14}}} \cdot \dfrac{2}{3} = \dfrac{\overset{}{\cancel{3}}}{\underset{1}{\cancel{3}}} \cdot \dfrac{1}{\cancel{2}} \cdot \dfrac{\cancel{2}}{3}$

$\qquad = \dfrac{1}{3}$

11. $\dfrac{36}{42} \cdot \dfrac{21}{9} \cdot \dfrac{1}{2} = \dfrac{\overset{4}{\cancel{36}}}{\underset{2}{\cancel{42}}} \cdot \dfrac{\overset{1}{\cancel{21}}}{\underset{1}{\cancel{9}}} \cdot \dfrac{1}{2} = \dfrac{\overset{2}{\cancel{4}}}{\underset{1}{\cancel{2}}} \cdot \dfrac{1}{\underset{1}{\cancel{2}}} = 1$

12. $\dfrac{28}{36} \cdot \dfrac{8}{21} \div \dfrac{16}{27} = \dfrac{\overset{7}{\cancel{28}}}{\underset{9}{\cancel{36}}} \cdot \dfrac{\overset{1}{\cancel{8}}}{\underset{7}{\cancel{21}}} \cdot \dfrac{\overset{9}{\cancel{27}}}{\underset{2}{\cancel{16}}} = \dfrac{\overset{1}{\cancel{7}}}{\underset{1}{\cancel{9}}} \cdot \dfrac{1}{\cancel{7}} \cdot \dfrac{\overset{1}{\cancel{9}}}{2}$

$= \dfrac{1}{2}$

13. $[(25 \cdot 11) + (30 \cdot 10)]\ \text{m}^2 = \textbf{575 m}^2$

14.

$A_{\text{Figure}} = A_{\text{Rectangle}} - A_{\text{Missing}}$

$\qquad\quad = (25\ \text{ft})(41\ \text{ft}) - (15\ \text{ft})(26\text{ft})$

$\qquad\quad = 1025\ \text{ft}^2 - 390\ \text{ft}^2$

$\qquad\quad = \textbf{635 ft}^2$

15. $\left(\dfrac{8}{3}\right)^3 = \dfrac{8}{3} \cdot \dfrac{8}{3} \cdot \dfrac{8}{3} = \dfrac{64}{9} \cdot \dfrac{8}{3} = \dfrac{\textbf{512}}{\textbf{27}}$

16. $12 = 2 \cdot 2 \cdot 3$

$\quad 26 = 2 \cdot 13$

$\quad 39 = 3 \cdot 13$

$\quad \text{LCM}\ (12, 26, 39) = 2 \cdot 2 \cdot 3 \cdot 13 = \textbf{156}$

17. $\dfrac{7}{13}$

$$
\begin{array}{r}
0.538 \\
13\overline{)7.000} \\
\underline{6\ 5} \\
50 \\
\underline{39} \\
110 \\
\underline{104} \\
6
\end{array}
$$

The rounded number is **0.54.**

18. $\dfrac{11}{17}$

$$
\begin{array}{r}
0.647 \\
17\overline{)11.000} \\
\underline{10\ 2} \\
80 \\
\underline{68} \\
120 \\
\underline{119} \\
1
\end{array}
$$

The rounded number is **0.65**.

19. $\dfrac{24}{28} = \dfrac{24 \div 4}{28 \div 4} = \dfrac{6}{7}$

20. $\dfrac{240}{300} = \dfrac{240 \div 60}{300 \div 60} = \dfrac{4}{5}$

21. $\dfrac{144}{256} = \dfrac{144 \div 4}{256 \div 4} = \dfrac{36 \div 4}{64 \div 4} = \dfrac{9}{16}$

22. $0.004 = \dfrac{4}{1000} = \dfrac{1}{250}$

23. $\dfrac{4}{11} \cdot \dfrac{132}{1} = \dfrac{4}{\cancel{11}} \cdot \dfrac{\overset{12}{\cancel{132}}}{1} = \mathbf{48}$

24.
$$
\begin{array}{r}
1361.780 \\
-\ \ \ 31.921 \\
\hline
\mathbf{1329.859}
\end{array}
\qquad
\text{Check:}
\begin{array}{r}
1329.859 \\
+\ \ \ 31.921 \\
\hline
1361.780
\end{array}
$$

25.
$$
\begin{array}{r}
789.891 \\
-\ \ 77.892 \\
\hline
\mathbf{711.999}
\end{array}
\qquad
\text{Check:}
\begin{array}{r}
711.999 \\
+\ \ 77.892 \\
\hline
789.891
\end{array}
$$

26. $\dfrac{7621.1}{2.1}$

$$
\begin{array}{r}
3629.095 \\
21\overline{)76,211.000} \\
\underline{63} \\
13\ 2 \\
\underline{12\ 6} \\
61 \\
\underline{42} \\
191 \\
\underline{189} \\
2\ 00 \\
\underline{1\ 89} \\
110 \\
\underline{105} \\
5
\end{array}
$$

The rounded number is **3629.10**.

27. $\dfrac{3.623}{0.02}$

$$
\begin{array}{r}
\mathbf{181.15} \\
2\overline{)362.30} \\
\underline{2} \\
16 \\
\underline{16} \\
02 \\
\underline{2} \\
0\ 3 \\
\underline{2} \\
10 \\
\underline{10} \\
0
\end{array}
$$

28. (a) Range $= 7 - 1 = \mathbf{6}$

(b) Mode $= \mathbf{6}$

(c) Median of 1, 2, 4, 6, 6, 6, 7 $= \mathbf{6}$

(d) Mean $= \dfrac{1 + 2 + 4 + 6 + 6 + 6 + 7}{7}$

$\qquad\quad = \dfrac{32}{7} \approx \mathbf{4.57}$

29. $\dfrac{B}{0.556} = 32$

$$
\begin{array}{r}
0.556 \\
\times\ \ \ \ \ 32 \\
\hline
1112 \\
1668\ \ \\
\hline
\mathbf{17.792}
\end{array}
$$

30. $48 = 2 \cdot 2 \cdot 2 \cdot 2 \cdot 3$

$60 = 2 \cdot 2 \cdot 3 \cdot 5$

$78 = 2 \cdot 3 \cdot 13$

GCF (48, 60, 78) $= 2 \cdot 3 = \mathbf{6}$

PROBLEM SET 27

1.
$$
\begin{array}{r}
14{,}782 \\
+\ \ 2{,}085 \\
\hline
16{,}867
\end{array}
\qquad
\begin{array}{r}
18{,}962 \\
-\ 16{,}867 \\
\hline
\mathbf{2{,}095}\ \textbf{units}
\end{array}
$$

2.
$$
\begin{array}{r}
0.65820 \\
-\ 0.07408 \\
\hline
0.58412
\end{array}
$$

Tom's guess was larger by 0.58412.

3.
$$
\begin{array}{r}
147{,}017 \\
+\ \ \ \ \ 984 \\
\hline
148{,}001
\end{array}
\qquad
\begin{array}{r}
148{,}001 \\
\times\ \ \ \ \ \ \ 2 \\
\hline
\mathbf{296{,}002}
\end{array}
$$

4. Average $= \dfrac{23 + 29 + 31 + 37 + 41}{5}$

$\qquad\qquad = \dfrac{161}{5} = \mathbf{32.2}$

5.
$$\begin{array}{r} 743 \\ +\ 486 \\ \hline 1229 \end{array} \qquad \begin{array}{r} 1229 \\ \times\quad 2 \\ \hline 2458 \end{array}$$

Mean $= \dfrac{743 + 486 + 2458}{3} = \dfrac{3687}{3} = \mathbf{1229}$

6. The following numbers are divisible by 5:
41,320 and **9405.**

7. $45\ \cancel{lb} \times \dfrac{16\ oz}{1\ \cancel{lb}} = \mathbf{720\ oz}$

8. $81\ \cancel{ft} \times \dfrac{1\ yd}{3\ \cancel{ft}} = \mathbf{27\ yd}$

9. $4899\ \cancel{m} \times \dfrac{100\ cm}{1\ \cancel{m}} = \mathbf{489{,}900\ cm}$

10. $\dfrac{20}{24} \cdot \dfrac{6}{15} \cdot \dfrac{2}{3} = \dfrac{\cancel{20}^{4}}{\cancel{24}_{4}} \cdot \dfrac{\cancel{6}^{1}}{\cancel{15}_{3}} \cdot \dfrac{2}{3} = \dfrac{\cancel{4}}{\cancel{4}} \cdot \dfrac{1}{3} \cdot \dfrac{2}{3}$

$\qquad = \mathbf{\dfrac{2}{9}}$

11. $\dfrac{16}{20} \cdot \dfrac{5}{8} \div \dfrac{1}{4} = \dfrac{\cancel{16}^{2}}{\cancel{20}_{4}} \cdot \dfrac{\cancel{5}^{1}}{\cancel{8}} \cdot \dfrac{4}{1} = \dfrac{2}{\cancel{4}} \cdot \dfrac{\cancel{4}^{1}}{1} = \mathbf{2}$

12. $\dfrac{18}{20} \cdot \dfrac{5}{3} \div \dfrac{6}{9} = \dfrac{\cancel{18}^{3}}{\cancel{20}_{4}} \cdot \dfrac{\cancel{5}^{1}}{\cancel{3}} \cdot \dfrac{\cancel{9}^{3}}{\cancel{6}} = \dfrac{3}{4} \cdot \dfrac{3}{1} = \mathbf{\dfrac{9}{4}}$

13. $(45 + 45 + 10 + 15 + 10 + 15)\ ft = \mathbf{140\ ft}$

14. $\dfrac{30\ ft \cdot 27\ ft}{2} = \dfrac{810}{2}\ ft^2 = \mathbf{405\ ft^2}$

15. $\dfrac{32\ ft \cdot 25\ ft}{2} = \dfrac{800}{2}\ ft^2 = \mathbf{400\ ft^2}$

16. $\dfrac{324}{3} = 108;\ \dfrac{108}{3} = 36;\ \dfrac{36}{3} = 12;\ \dfrac{12}{3} = 4;$

$\dfrac{4}{2} = 2$

$324 = 2 \cdot 2 \cdot 3 \cdot 3 \cdot 3 \cdot 3 = \mathbf{2^2 \cdot 3^4}$

17. $\dfrac{2916}{3} = 972;\ \dfrac{972}{3} = 324;\ \dfrac{324}{3} = 108;$

$\dfrac{108}{3} = 36;\ \dfrac{36}{3} = 12;\ \dfrac{12}{3} = 4;\ \dfrac{4}{2} = 2$

$2916 = 2 \cdot 2 \cdot 3 \cdot 3 \cdot 3 \cdot 3 \cdot 3 \cdot 3$

$\qquad = \mathbf{2^2 \cdot 3^6}$

18. $\dfrac{7}{17}$

$$\begin{array}{r} 0.411 \\ 17\overline{)7.000} \\ \underline{6\ 8} \\ 20 \\ \underline{17} \\ 30 \\ \underline{17} \\ 13 \end{array}$$

The rounded number is **0.41.**

19. $\dfrac{7}{13}$

$$\begin{array}{r} 0.538 \\ 13\overline{)7.000} \\ \underline{6\ 5} \\ 50 \\ \underline{39} \\ 110 \\ \underline{104} \\ 6 \end{array}$$

The rounded number is **0.54.**

20. $\dfrac{7}{200}$

$$\begin{array}{r} 0.035 \\ 200\overline{)7.000} \\ \underline{6\ 00} \\ 1\ 000 \\ \underline{1\ 000} \\ 0 \end{array}$$

The rounded number is **0.04.**

21. $\dfrac{72}{120} = \dfrac{72 \div 12}{120 \div 12} = \dfrac{6 \div 2}{10 \div 2} = \mathbf{\dfrac{3}{5}}$

22. $\dfrac{280}{360} = \dfrac{280 \div 10}{360 \div 10} = \dfrac{28 \div 4}{36 \div 4} = \mathbf{\dfrac{7}{9}}$

23. $0.125 = \dfrac{125}{1000} = \dfrac{125 \div 25}{1000 \div 25} = \dfrac{5}{40} = \mathbf{\dfrac{1}{8}}$

24. (a) Range = $23 - 2 =$ **21**

(b) Mode = **9**

(c) Median of 2, 3, 3, 9, 9, 9, 10, 23 = **9**

(d) Mean

$$= \frac{2 + 3 + 3 + 9 + 9 + 9 + 10 + 23}{8}$$

$$= \frac{68}{8} = \textbf{8.5}$$

25. $\frac{4}{9} \cdot \frac{270}{1} = \frac{4}{\cancel{9}} \cdot \frac{\overset{30}{\cancel{270}}}{1} = \textbf{120}$

26. **112.00956, 112.091, 112.9, 1129.0**

27. $\dfrac{611.32}{0.04}$

$$\begin{array}{r} 15,283 \\ 4\overline{)61,132} \\ \underline{4} \\ 21 \\ \underline{20} \\ 1\,1 \\ \underline{8} \\ 33 \\ \underline{32} \\ 12 \\ \underline{12} \\ 0 \end{array}$$

28. $\dfrac{0.0062}{0.07}$

$$\begin{array}{r} 0.088 \\ 7\overline{)0.620} \\ \underline{56} \\ 60 \\ \underline{56} \\ 4 \end{array}$$

The rounded number is **0.09.**

29. $10 = 2 \cdot 5$

$14 = 2 \cdot 7$

$25 = 5 \cdot 5$

LCM (10, 14, 25) = $2 \cdot 5 \cdot 5 \cdot 7 =$ **350**

30. $\left(\dfrac{3}{2}\right)F = \dfrac{67}{4}$

$$F = \frac{67}{4} \div \frac{3}{2} = \frac{67}{4} \cdot \frac{\overset{1}{\cancel{2}}}{\underset{2}{\cancel{4}}} \cdot \frac{1}{3} = \textbf{\dfrac{67}{6}}$$

PROBLEM SET 28

1. $\begin{array}{r} 0.04170 \\ -\ 0.00417 \\ \hline \textbf{0.03753} \end{array}$

2. $\begin{array}{r} 5283 \\ \times\quad 5 \\ \hline 26,415 \end{array}$ $\quad \begin{array}{r} 26,415 \\ \times\quad 2 \\ \hline 52,830 \end{array}$ $\quad \begin{array}{r} 5,283 \text{ 1st flock} \\ 26,415 \text{ 2nd flock} \\ +\ 52,830 \text{ 3rd flock} \\ \hline \textbf{84,528 birds} \end{array}$

3. $\begin{array}{r} 4,820,718 \\ \times\qquad 7 \\ \hline 33,745,026 \end{array}$

33,745,026 is **thirty-three million, seven hundred forty-five thousand, twenty-six**

4. Average $= \dfrac{11 + 13 + 17 + 19 + 23}{5}$

$$= \frac{83}{5} = \textbf{16.6}$$

5. $\dfrac{17}{9} = \mathbf{1\dfrac{8}{9}}$

$$\begin{array}{r} 1 \\ 9\overline{)17} \\ \underline{9} \\ 8 \end{array}$$

6. $\dfrac{24}{7} = \mathbf{3\dfrac{3}{7}}$

$$\begin{array}{r} 3 \\ 7\overline{)24} \\ \underline{21} \\ 3 \end{array}$$

7. The following diagram shows that $4\dfrac{1}{4} = \dfrac{17}{4}$.

Four $\dfrac{1}{4}$'s $\quad+\quad$ Four $\dfrac{1}{4}$'s $\quad+\quad$ Four $\dfrac{1}{4}$'s

$+\quad$ Four $\dfrac{1}{4}$'s $\quad+\quad$ One $\dfrac{1}{4}$

$=$ Seventeen $\dfrac{1}{4}$'s or $\dfrac{17}{4}$

8. $6\dfrac{2}{9} = \dfrac{9 \times 6 + 2}{9} = \dfrac{54 + 2}{9} = \mathbf{\dfrac{56}{9}}$

9. $8\dfrac{4}{7} = \dfrac{7 \times 8 + 4}{7} = \dfrac{56 + 4}{7} = \mathbf{\dfrac{60}{7}}$

Algebra $\frac{1}{2}$, Third Edition

10. $5\frac{2}{8} = \frac{8 \times 5 + 2}{8} = \frac{40 + 2}{8} = \frac{42}{8} = \mathbf{\frac{21}{4}}$

11. $192.72 \, \cancel{cm} \times \dfrac{1 \, m}{100 \, \cancel{cm}} = \mathbf{1.9272 \, m}$

12. $17 \, \cancel{mi} \times \dfrac{5280 \, ft}{1 \, \cancel{mi}} = \mathbf{89{,}760 \, ft}$

13. $\dfrac{28}{36} \cdot \dfrac{24}{21} \cdot \dfrac{3}{16} = \dfrac{\overset{7}{\cancel{28}}}{\underset{9}{\cancel{36}}} \cdot \dfrac{\overset{8}{\cancel{24}}}{\underset{7}{\cancel{21}}} \cdot \dfrac{3}{16} = \dfrac{\overset{1}{\cancel{8}}}{9} \cdot \dfrac{\cancel{3}}{\underset{2}{\cancel{16}}} = \mathbf{\dfrac{1}{6}}$

14. $\dfrac{18}{24} \cdot \dfrac{36}{28} \cdot \dfrac{14}{27} = \dfrac{\overset{3}{\cancel{18}}}{\underset{4}{\cancel{24}}} \cdot \dfrac{\overset{9}{\cancel{36}}}{\underset{7}{\cancel{28}}} \cdot \dfrac{\overset{}{14}}{27} = \dfrac{\overset{}{3}}{4} \cdot \dfrac{9}{\underset{1}{7}} \cdot \dfrac{\overset{2}{14}}{\underset{9}{27}}$

$= \dfrac{1}{\underset{2}{\cancel{4}}} \cdot \dfrac{\overset{1}{\cancel{9}}}{1} \cdot \dfrac{\overset{1}{\cancel{2}}}{\underset{1}{\cancel{9}}} = \mathbf{\dfrac{1}{2}}$

15. $\dfrac{16}{18} \cdot \dfrac{16}{12} \div \dfrac{8}{9} = \dfrac{16}{\underset{2}{\cancel{18}}} \cdot \dfrac{16}{\underset{3}{\cancel{12}}} \cdot \dfrac{\overset{1}{\cancel{9}}}{8} = \dfrac{\overset{4}{\cancel{16}}}{2} \cdot \dfrac{\overset{2}{\cancel{16}} \; \overset{1}{\cancel{9}}}{3 \cdot \underset{1}{\cancel{8}}} = \mathbf{\dfrac{4}{3}}$

16. $A_{\text{Figure}} = A_{\text{Rectangle}} + A_{\text{Triangle}}$

$= \left[(31)(26) + \dfrac{(17)(26)}{2} \right] ft^2$

$= [806 + 221] \, ft^2$

$= \mathbf{1027 \, ft^2}$

17. $P = (2 + 3 + 2 + 3) \, m = 10 \, m$

$10 \, \cancel{m} \times \dfrac{100 \, cm}{1 \, \cancel{m}} = \mathbf{1000 \, cm}$

18. $\dfrac{1440}{5} = 288; \quad \dfrac{288}{3} = 96; \quad \dfrac{96}{3} = 32; \quad \dfrac{32}{2} = 16;$

$\dfrac{16}{2} = 8; \quad \dfrac{8}{2} = 4; \quad \dfrac{4}{2} = 2$

$1440 = 2 \cdot 2 \cdot 2 \cdot 2 \cdot 2 \cdot 3 \cdot 3 \cdot 5$

$\qquad = \mathbf{2^5 \cdot 3^2 \cdot 5}$

19. $5\dfrac{16}{23}$

$\begin{array}{r} 0.695 \\ 23\overline{)16.000} \\ \underline{13\,8} \\ 2\,20 \\ \underline{2\,07} \\ 130 \\ \underline{115} \\ 15 \end{array}$

The rounded number is **5.70.**

20. $9\dfrac{7}{19}$

$\begin{array}{r} 0.368 \\ 19\overline{)7.000} \\ \underline{5\,7} \\ 1\,30 \\ \underline{1\,14} \\ 160 \\ \underline{152} \\ 8 \end{array}$

The rounded number is **9.37.**

21. $\dfrac{36}{42} = \dfrac{36 \div 6}{42 \div 6} = \mathbf{\dfrac{6}{7}}$

22. $\dfrac{280}{320} = \dfrac{280 \div 10}{320 \div 10} = \dfrac{28 \div 4}{32 \div 4} = \mathbf{\dfrac{7}{8}}$

23. $\dfrac{125}{175} = \dfrac{125 \div 25}{175 \div 25} = \mathbf{\dfrac{5}{7}}$

24. **18, 21, 24, 27**

25. $9 = 3 \cdot 3$

$18 = 2 \cdot 3 \cdot 3$

$60 = 2 \cdot 2 \cdot 3 \cdot 5$

LCM $(9, 18, 60) = 2 \cdot 2 \cdot 3 \cdot 3 \cdot 5 = \mathbf{180}$

26. $\begin{array}{r} 172.325 \\ -\ 61.890 \\ \hline \mathbf{110.435} \end{array}$ Check: $\begin{array}{r} 110.435 \\ +\ 61.890 \\ \hline 172.325 \end{array}$

27. $\dfrac{7811.3}{0.03}$

$\begin{array}{r} 260{,}376.666 \\ 3\overline{)781{,}130.000} \\ \underline{6} \\ 18 \\ \underline{18} \\ 01\,1 \\ \underline{9} \\ 23 \\ \underline{21} \\ 20 \\ \underline{18} \\ 2\,0 \\ \underline{1\,8} \\ 20 \\ \underline{18} \\ 20 \\ \underline{18} \\ 2 \end{array}$

The rounded number is **260,376.67.**

28.

$$\begin{array}{ccc} 21.1101 & 21.1101 & \text{Check:} \quad 21.1101 \\ -\underline{\quad H\quad} & -\underline{\;\;0.1001} & -\underline{\;\;21.0100} \\ 0.1001 & \mathbf{21.0100} & 0.1001 \end{array}$$

29. (a) Range = $100 - 72 = \mathbf{28}$

(b) Mode = **85**

(c) Median = $\dfrac{(85 + 86)}{2} = \mathbf{85.5}$

(d) Mean = $\dfrac{72 + 82 + 85 + 85}{8}$
$$+ \dfrac{86 + 92 + 97 + 100}{8}$$
$$= \dfrac{699}{8} = \mathbf{87.375}$$

30. $531.25 = 531\dfrac{25}{100} = \mathbf{531\dfrac{1}{4}}$

PROBLEM SET 29

1.

Daily Low Temperatures

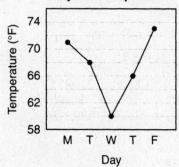

Day

2.
$$\begin{array}{r} 19{,}000.000075 \text{ ft} \\ +\;21{,}000.001003 \text{ ft} \\ \hline \mathbf{40{,}000.001078 \text{ ft}} \end{array}$$

3.
$$\begin{array}{r} 14{,}982 \\ \times\quad 10 \\ \hline 149{,}820 \end{array} \qquad \begin{array}{r} 149{,}820 \\ +\;14{,}982 \\ \hline \mathbf{164{,}802} \text{ visitors} \end{array}$$

4. Mean = $\dfrac{31 + 37 + 41 + 43}{4} = \dfrac{152}{4} = \mathbf{38}$

5. $840 \times N = 19{,}420$

$$\begin{array}{r} 23 \\ 840\overline{)19{,}420} \\ \underline{16\;80} \\ 2\;620 \\ \underline{2\;520} \\ 100 \end{array}$$

24 trucks were required.

6.
$$\begin{array}{r} 840 \\ \times\quad 18 \\ \hline 6720 \\ 840 \\ \hline 15{,}120 \end{array} \qquad \begin{array}{r} 19{,}420 \\ -\;15{,}120 \\ \hline \mathbf{4300} \text{ toasters} \end{array}$$

7. $\dfrac{5}{6} \cdot 48 = \dfrac{240}{6} = \mathbf{40}$

8. $\dfrac{6}{7} \cdot 35 = \dfrac{210}{7} = \mathbf{30}$

9. $\dfrac{15}{7} = \mathbf{2\dfrac{1}{7}}$

$$\begin{array}{r} 2 \\ 7\overline{)15} \\ \underline{14} \\ 1 \end{array}$$

10. $\dfrac{21}{5} = \mathbf{4\dfrac{1}{5}}$

$$\begin{array}{r} 4 \\ 5\overline{)21} \\ \underline{20} \\ 1 \end{array}$$

11. The following diagram shows that $2\dfrac{1}{4} = \dfrac{9}{4}$.

Four $\dfrac{1}{4}$'s + Four $\dfrac{1}{4}$'s + One $\dfrac{1}{4}$

$= $ Nine $\dfrac{1}{4}$'s or $\dfrac{9}{4}$

12. $7\dfrac{3}{8} = \dfrac{8 \times 7 + 3}{8} = \dfrac{56 + 3}{8} = \mathbf{\dfrac{59}{8}}$

13. $6\dfrac{2}{3} = \dfrac{3 \times 6 + 2}{3} = \dfrac{18 + 2}{3} = \mathbf{\dfrac{20}{3}}$

14. $5\dfrac{7}{11} = \dfrac{11 \times 5 + 7}{11} = \dfrac{55 + 7}{11} = \mathbf{\dfrac{62}{11}}$

15. $199.62 \; \cancel{dg} \times \dfrac{1 \text{ g}}{10 \; \cancel{dg}} = \mathbf{19.962 \text{ g}}$

16. $18 \; \cancel{mi} \times \dfrac{5280 \text{ ft}}{1 \; \cancel{mi}} = \mathbf{95{,}040 \text{ ft}}$

17. $\dfrac{42}{48} \cdot \dfrac{32}{14} \cdot \dfrac{3}{4} = \dfrac{\cancel{42}^{6}}{\cancel{48}_{6}} \cdot \dfrac{\cancel{32}^{4}}{\cancel{14}_{2}} \cdot \dfrac{3}{4} = \dfrac{\cancel{6}^{1}}{\cancel{6}_{1}} \cdot \dfrac{\cancel{4}^{1}}{\cancel{2}} \cdot \dfrac{3}{\cancel{4}_{1}}$

$= \mathbf{\dfrac{3}{2}}$

18. $\dfrac{18}{20} \cdot \dfrac{16}{8} \div \dfrac{27}{2} = \dfrac{\overset{6}{\cancel{18}}}{\underset{5}{\cancel{20}}} \cdot \dfrac{\overset{4}{\cancel{16}}}{8} \cdot \dfrac{2}{\underset{9}{\cancel{27}}}$

$= \dfrac{\overset{2}{\cancel{6}}}{5} \cdot \dfrac{\overset{1}{\cancel{4}}}{\underset{2}{\cancel{8}}} \cdot \dfrac{2}{\underset{3}{\cancel{9}}} = \dfrac{\overset{1}{\cancel{2}}}{5} \cdot \dfrac{1}{\cancel{2}} \cdot \dfrac{2}{3} = \dfrac{\mathbf{2}}{\mathbf{15}}$

19. $4000 \,\cancel{cm} \times \dfrac{1\text{ m}}{100 \,\cancel{cm}} = 40\text{ m}$

$6500 \,\cancel{cm} \times \dfrac{1\text{ m}}{100 \,\cancel{cm}} = 65\text{ m}$

$A_{\text{Figure}} = A_{\text{Rectangle}} + A_{\text{Triangle}}$

$= (40\text{ m})(20\text{ m}) + \dfrac{(25\text{ m})(20\text{ m})}{2}$

$= 800\text{ m}^2 + 250\text{ m}^2$

$= \mathbf{1050\ m^2}$

20. $\dfrac{2}{5}R = \dfrac{2}{5}$

$R = \dfrac{2}{5} \div \dfrac{2}{5} = \dfrac{2}{5} \cdot \dfrac{5}{2} = \dfrac{10}{10} = \mathbf{1}$

21. $8\dfrac{16}{25} = \mathbf{8.64}$

$$
\begin{array}{r}
0.64 \\
25\overline{)16.00} \\
\underline{15\ 0} \\
1\ 00 \\
\underline{1\ 00} \\
0
\end{array}
$$

22. $1\dfrac{8}{13}$

$$
\begin{array}{r}
0.615 \\
13\overline{)8.000} \\
\underline{7\ 8} \\
20 \\
\underline{13} \\
70 \\
\underline{65} \\
5
\end{array}
$$

The rounded number is **1.62.**

23. $\dfrac{183}{270} = \dfrac{183 \div 3}{270 \div 3} = \dfrac{\mathbf{61}}{\mathbf{90}}$

24. $\dfrac{360}{420} = \dfrac{360 \div 60}{420 \div 60} = \dfrac{\mathbf{6}}{\mathbf{7}}$

25.
$$
\begin{array}{r}
31.25 \\
\times\ 0.0012 \\
\hline
6250 \\
3125 \\
\hline
\mathbf{0.037500}
\end{array}
$$

26.
$$
\begin{array}{r}
17.01 \\
\times\ 0.12 \\
\hline
3402 \\
1701 \\
\hline
\mathbf{2.0412}
\end{array}
$$

27.
$$
\begin{array}{r}
61.892 \\
-\ 9.299 \\
\hline
\mathbf{52.593}
\end{array}
\qquad
\begin{array}{r}
\text{Check:}\quad 52.593 \\
+\ 9.299 \\
\hline
61.892
\end{array}
$$

28. $5.85 = 5\dfrac{85}{100} = 5\dfrac{\mathbf{17}}{\mathbf{20}}$

29. (a) Range $= 9 - 4 = \mathbf{5}$

(b) Mode $= \mathbf{7, 9}$

(c) Median $= \mathbf{7}$

(d) Average $= \dfrac{4 + 5 + 7 + 7 + 8 + 9 + 9}{7}$

$= \dfrac{49}{7} = \mathbf{7}$

30. $12 = 2 \cdot 2 \cdot 3$

$30 = 2 \cdot 3 \cdot 5$

$50 = 2 \cdot 5 \cdot 5$

LCM $(12, 30, 50) = 2 \cdot 2 \cdot 3 \cdot 5 \cdot 5 = \mathbf{300}$

PROBLEM SET 30

1. $350 - 150 = \mathbf{200\ cars}$

2.

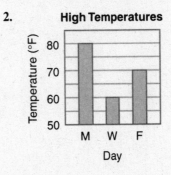

High Temperatures

3. $8 + 7 + N = 3(12)$

$15 + N = 36$

$36 - 15 =$ **21 elephants**

4. $53 \times N = 5062$

$$
\begin{array}{r}
95 \\
53\overline{)5062} \\
\underline{477} \\
292 \\
\underline{265} \\
27
\end{array}
$$

Hannibal needed **96 boxes.**

5.
$$
\begin{array}{r}
5142 \\
\times \quad 2 \\
\hline
10{,}284
\end{array}
\qquad
\begin{array}{r}
7821 \\
\times \quad 3 \\
\hline
23{,}463
\end{array}
\qquad
\begin{array}{r}
5142 \\
+ 7821 \\
\hline
12{,}963
\end{array}
$$

$$
\begin{array}{r}
10{,}284 \\
+ 23{,}463 \\
\hline
33{,}747
\end{array}
\qquad
\begin{array}{r}
33{,}747 \\
+ 12{,}963 \\
\hline
\mathbf{46{,}710 \text{ marbles}}
\end{array}
$$

6. (a) Range $= 62 - 46 =$ **16**

(b) Mode = **58**

(c) Median $= \dfrac{(54 + 58)}{2} =$ **56**

(d) Average

$= \dfrac{46 + 53 + 54 + 58 + 58 + 62}{6}$

$= 55.1\overline{6} \approx$ **55.17**

7. $\dfrac{3}{8} \cdot 42 = \dfrac{126}{8} = \dfrac{\mathbf{63}}{\mathbf{4}}$

8. $\dfrac{6}{7} \cdot 49 = \dfrac{294}{7} =$ **42**

9. $\dfrac{17}{8} = 2\dfrac{\mathbf{1}}{\mathbf{8}}$

$$
\begin{array}{r}
2 \\
8\overline{)17} \\
\underline{16} \\
1
\end{array}
$$

10. $\dfrac{22}{5} = 4\dfrac{\mathbf{2}}{\mathbf{5}}$

$$
\begin{array}{r}
4 \\
5\overline{)22} \\
\underline{20} \\
2
\end{array}
$$

11. The following diagram shows that $3\dfrac{1}{5} = \dfrac{16}{5}$.

Five $\dfrac{1}{5}$'s $\quad + \quad$ Five $\dfrac{1}{5}$'s

$+$

$+ \quad$ Five $\dfrac{1}{5}$'s $\quad + \quad$ One $\dfrac{1}{5}$

$=$ Sixteen $\dfrac{1}{5}$'s or $\dfrac{16}{5}$

12. $7\dfrac{2}{3} = \dfrac{3 \times 7 + 2}{3} = \dfrac{\mathbf{23}}{\mathbf{3}}$

13. $7\dfrac{6}{7} = \dfrac{7 \times 7 + 6}{7} = \dfrac{\mathbf{55}}{\mathbf{7}}$

14. $\dfrac{4}{5} - \dfrac{3}{4} = \dfrac{16}{20} - \dfrac{15}{20} = \dfrac{\mathbf{1}}{\mathbf{20}}$

15. $\dfrac{2}{3} + \dfrac{5}{8} - \dfrac{3}{4} = \dfrac{16}{24} + \dfrac{15}{24} - \dfrac{18}{24} = \dfrac{\mathbf{13}}{\mathbf{24}}$

16. $\dfrac{3}{8} + \dfrac{5}{6} - \dfrac{1}{4} = \dfrac{9}{24} + \dfrac{20}{24} - \dfrac{6}{24} = \dfrac{\mathbf{23}}{\mathbf{24}}$

17. $\dfrac{2}{3} + \dfrac{5}{8} + \dfrac{3}{4} = \dfrac{16}{24} + \dfrac{15}{24} + \dfrac{18}{24} = \dfrac{\mathbf{49}}{\mathbf{24}}$

18. $721 \cancel{\text{yd}} \times \dfrac{3 \text{ ft}}{1 \cancel{\text{yd}}} =$ **2163 ft**

19. $19{,}262 \cancel{\text{cm}} \times \dfrac{1 \text{ m}}{100 \cancel{\text{cm}}} =$ **192.62 m**

20. $\dfrac{28}{32} \cdot \dfrac{24}{21} \cdot \dfrac{3}{4} = \dfrac{\overset{4}{\cancel{28}}}{\underset{4}{\cancel{32}}} \cdot \dfrac{\overset{3}{\cancel{24}}}{\underset{3}{\cancel{21}}} \cdot \dfrac{3}{4} = \dfrac{\overset{1}{\cancel{4}}}{\underset{1}{\cancel{4}}} \cdot \dfrac{\overset{1}{\cancel{3}}}{\underset{1}{\cancel{3}}} \cdot \dfrac{3}{4}$

$= \dfrac{\mathbf{3}}{\mathbf{4}}$

21. $\dfrac{16}{18} \cdot \dfrac{20}{24} \div \dfrac{10}{9} = \dfrac{\overset{8}{\cancel{16}}}{\underset{9}{\cancel{18}}} \cdot \dfrac{\overset{5}{\cancel{20}}}{\underset{6}{\cancel{24}}} \cdot \dfrac{9}{10}$

$= \dfrac{\overset{4}{\cancel{8}}}{\underset{1}{\cancel{9}}} \cdot \dfrac{\overset{1}{\cancel{8}}}{\underset{3}{\cancel{6}}} \cdot \dfrac{\overset{1}{\cancel{9}}}{\underset{2}{\cancel{10}}} = \dfrac{\overset{2}{\cancel{4}}}{1} \cdot \dfrac{1}{3} \cdot \dfrac{1}{\underset{1}{\cancel{2}}} = \dfrac{\mathbf{2}}{\mathbf{3}}$

22. $2\,\cancel{m} \times \dfrac{100\text{ cm}}{1\,\cancel{m}} = 200\text{ cm}$

$1\,\cancel{m} \times \dfrac{100\text{ cm}}{1\,\cancel{m}} = 100\text{ cm}$

$A_{\text{Figure}} = A_{\text{Rectangle}} + 2A_{\text{Triangle}}$

$\qquad = \left[300(100) + 2\left(\dfrac{(200)(100)}{2}\right)\right]\text{ cm}^2$

$\qquad = (30{,}000 + 20{,}000)\text{ cm}^2$

$\qquad = \textbf{50,000 cm}^2$

23. $200\,\cancel{cm} \times \dfrac{1\text{ m}}{100\,\cancel{cm}} = 2\text{ m}$

$100\,\cancel{cm} \times \dfrac{1\text{ m}}{100\,\cancel{cm}} = 1\text{ m}$

$80\,\cancel{cm} \times \dfrac{1\text{ m}}{100\,\cancel{cm}} = 0.8\text{ m}$

$P = (2 + 1 + 0.8 + 1 + 2 + 1 + 0.8 + 1)\text{ m}$

$\quad = \textbf{9.6 m}$

24. $\dfrac{180}{200} = \dfrac{180 \div 10}{200 \div 10} = \dfrac{18 \div 2}{20 \div 2} = \dfrac{\textbf{9}}{\textbf{10}}$

25. $\dfrac{256}{720} = \dfrac{256 \div 4}{720 \div 4} = \dfrac{64 \div 4}{180 \div 4} = \dfrac{\textbf{16}}{\textbf{45}}$

26. $5\dfrac{16}{21}$

$$
\begin{array}{r}
0.761 \\
21\overline{)16.000} \\
\underline{14\,7} \\
1\,30 \\
\underline{1\,26} \\
40 \\
\underline{21} \\
19
\end{array}
$$

The rounded number is **5.76.**

27. $2\dfrac{9}{17}$

$$
\begin{array}{r}
0.529 \\
17\overline{)9.000} \\
\underline{8\,5} \\
50 \\
\underline{34} \\
160 \\
\underline{153} \\
7
\end{array}
$$

The rounded number is **2.53.**

28. $\dfrac{2450}{7} = 350;\ \dfrac{350}{7} = 50;\ \dfrac{50}{5} = 10;\ \dfrac{10}{5} = 2$

$2450 = 2 \cdot 5 \cdot 5 \cdot 7 \cdot 7 = \textbf{2} \cdot \textbf{5}^2 \cdot \textbf{7}^2$

29. $301.008 = 301\dfrac{8}{1000} = \textbf{301}\dfrac{\textbf{1}}{\textbf{125}}$

30. $\dfrac{4}{5} + G = \dfrac{8}{5}$

$\dfrac{8}{5} - \dfrac{4}{5} = \dfrac{\textbf{4}}{\textbf{5}}$

PROBLEM SET 31

1.
$$
\begin{array}{r}
\$2.50 \\
\times\quad 20 \\
\hline
\$50.00
\end{array}
\qquad
\begin{array}{r}
\$100.00 \\
-\ \$50.00 \\
\hline
\mathbf{\$50.00}
\end{array}
$$

2. $30{,}000 - 20{,}000 = \textbf{10,000 cars}$

3. $175 \times N = 2975$

$$
\begin{array}{r}
\textbf{17 hr} \\
175\overline{)2975} \\
\underline{175} \\
1225 \\
\underline{1225} \\
0
\end{array}
$$

4. $\$11.25 + \$8.75 + \$9.35$
$\qquad + \$8.90 + N = 5(\$9.50)$

$\qquad \$38.25 + N = \$47.50$

$\$47.50 - \$38.25 = \textbf{\$9.25}$

5. (a) $\dfrac{\textbf{20 kursh}}{\textbf{1 riyal}}, \dfrac{\textbf{1 riyal}}{\textbf{20 kursh}}$

(b) $16{,}000\,\cancel{\text{kursh}} \times \dfrac{1\text{ riyal}}{20\,\cancel{\text{kursh}}} = \textbf{800 riyals}$

6. $\dfrac{5}{6} \times 960 = \dfrac{4800}{6} = \textbf{800}$

7. $\dfrac{11}{5} \times 55 = \dfrac{605}{5} = \textbf{121}$

8. $\dfrac{41}{3} = 13\dfrac{2}{3}$

$$
\begin{array}{r}
13 \\
3\overline{)41} \\
\underline{3} \\
11 \\
\underline{9} \\
2
\end{array}
$$

9. $\frac{52}{16} = \frac{13}{4} = 3\frac{1}{4}$

$$4\overline{)13}^{3}$$
$$\underline{12}$$
$$1$$

10. The following diagram shows that $4\frac{1}{3} = \frac{13}{3}$.

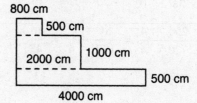

Three $\frac{1}{3}$'s + Three $\frac{1}{3}$'s + Three $\frac{1}{3}$'s

+ Three $\frac{1}{3}$'s + One $\frac{1}{3}$ = Thirteen $\frac{1}{3}$'s or $\frac{13}{3}$

11. $14 = 2 \cdot 7$
$94 = 2 \cdot 47$
$300 = 2 \cdot 2 \cdot 3 \cdot 5 \cdot 5$
LCM $(14, 94, 300) = 2 \cdot 2 \cdot 3 \cdot 5 \cdot 5 \cdot 7 \cdot 47$
$ = \mathbf{98,700}$

12. $27 = 3 \cdot 3 \cdot 3$
$66 = 2 \cdot 3 \cdot 11$
$90 = 2 \cdot 3 \cdot 3 \cdot 5$
LCM $(27, 66, 90) = 2 \cdot 3 \cdot 3 \cdot 3 \cdot 5 \cdot 11$
$ = \mathbf{2970}$

13. $15\frac{2}{3} = \frac{3 \times 15 + 2}{3} = \frac{\mathbf{47}}{\mathbf{3}}$

14. $14,780,000 \, \cancel{mg} \times \dfrac{1 \, g}{1000 \, \cancel{mg}} = \mathbf{14,780 \, g}$

15. $\dfrac{15}{18} \cdot \dfrac{6}{30} \div \dfrac{3}{2} = \dfrac{\overset{5}{\cancel{15}}}{\underset{6}{\cancel{18}}} \cdot \dfrac{\overset{1}{\cancel{6}}}{\underset{5}{\cancel{30}}} \cdot \dfrac{2}{3} = \dfrac{\overset{1}{\cancel{5}}}{\underset{3}{\cancel{6}}} \cdot \dfrac{1}{\cancel{5}} \cdot \dfrac{\cancel{2}}{3}$

$= \dfrac{\mathbf{1}}{\mathbf{9}}$

16. $\dfrac{160}{180} \cdot \dfrac{10}{12} \div \dfrac{20}{18} = \dfrac{\overset{8}{\cancel{160}}}{\underset{10}{\cancel{180}}} \cdot \dfrac{10}{12} \cdot \dfrac{\overset{1}{\cancel{18}}}{\underset{1}{\cancel{20}}} = \dfrac{\overset{2}{\cancel{8}}}{\underset{1}{\cancel{10}}} \cdot \dfrac{\overset{1}{\cancel{10}}}{\underset{3}{\cancel{12}}}$

$= \dfrac{\mathbf{2}}{\mathbf{3}}$

17. $\dfrac{8}{9} + \dfrac{5}{6} = \dfrac{16}{18} + \dfrac{15}{18} = \dfrac{\mathbf{31}}{\mathbf{18}}$

18. $\dfrac{5}{8} + \dfrac{11}{16} + \dfrac{1}{2} = \dfrac{10}{16} + \dfrac{11}{16} + \dfrac{8}{16} = \dfrac{\mathbf{29}}{\mathbf{16}}$

19. $\dfrac{7}{27} + \dfrac{1}{3} - \dfrac{1}{9} = \dfrac{7}{27} + \dfrac{9}{27} - \dfrac{3}{27} = \dfrac{\mathbf{13}}{\mathbf{27}}$

20. $\dfrac{1}{4} + \dfrac{2}{3} - \dfrac{6}{7} = \dfrac{21}{84} + \dfrac{56}{84} - \dfrac{72}{84} = \dfrac{\mathbf{5}}{\mathbf{84}}$

21. $7 + 5 \times 9 = 7 + 45 = \mathbf{52}$

22. $5 \cdot 8 - 2 \cdot 7 + 5 = 40 - 14 + 5 = 26 + 5$
$= \mathbf{31}$

23. $4 \cdot 8 - 8 \cdot 2 = 32 - 16 = \mathbf{16}$

24. $6 \cdot 7 + 6 - 2 \cdot 9 = 42 + 6 - 18$
$= 48 - 18 = \mathbf{30}$

25. $20 \, \cancel{m} \times \dfrac{100 \, cm}{1 \, \cancel{m}} = 2000 \, cm$

$42 \, \cancel{m} \times \dfrac{100 \, cm}{1 \, \cancel{m}} = 4200 \, cm$

$A_{\text{Triangle}} = \dfrac{bh}{2} = \dfrac{(4200)(2000)}{2} \, cm^2$

$\phantom{A_{\text{Triangle}}} = \mathbf{4,200,000 \, cm^2}$

26. $20 \, m = 2000 \, cm; \quad 8 \, m = 800 \, cm;$
$10 \, m = 1000 \, cm; \quad 40 \, m = 4000 \, cm;$
$5 \, m = 500 \, cm$

800 cm
```
      500 cm
 ┌───┐
 │   └──────┐
 │          │ 1000 cm
 │ 2000 cm  │
 │          └──────────┐
 │                     │ 500 cm
 └─────────────────────┘
      4000 cm
```

$A = [4000(500) + 1000(2000)$
$ + 800(500)] \, cm^2$
$ = \mathbf{4,400,000 \, cm^2}$

27. $x - \dfrac{5}{17} = \dfrac{10}{17}$

$\dfrac{10}{17} + \dfrac{5}{17} = \dfrac{\mathbf{15}}{\mathbf{17}}$

28. $\dfrac{42}{17}$

$$17\overline{)42.000}^{2.470}$$
$$\underline{34}$$
$$8\,0$$
$$\underline{6\,8}$$
$$1\,20$$
$$\underline{1\,19}$$
$$10$$

The rounded number is **2.47**.

29. $\dfrac{40}{25}$

$$\begin{array}{r} 1.6 \\ 25\overline{)40.0} \\ \underline{25} \\ 15\,0 \\ \underline{15\,0} \\ 0 \end{array}$$

30. $12.12 = 12\dfrac{12}{100} = \mathbf{12\dfrac{3}{25}}$

PROBLEM SET 32

1. $\$28.50 + \$41.25 + \$50 + N = 4(\$39.96)$
$\qquad\qquad \$119.75 + N = \$159.84$

$\$159.84 - \$119.75 = \mathbf{\$40.09}$

2.

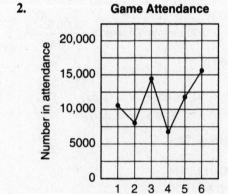

Game Attendance

3. (a) Range = $15{,}700 - 7000 = \mathbf{8700}$

(b) None of the figures are repeated, so there is **no mode.**

(c) 7000; 8000; 10,400; 12,000; 14,600; 15,700

$$\text{Median} = \frac{10{,}400 + 12{,}000}{2} = \mathbf{11{,}200}$$

(d) $\text{Mean} = \dfrac{7000 + 8000 + 10{,}400 + 12{,}000}{6}$

$\qquad\qquad + \dfrac{14{,}600 + 15{,}700}{6}$

$\qquad = \dfrac{67{,}700}{6} \approx \mathbf{11{,}283.33}$

4. $\text{Avg} = \dfrac{1481 + 1300 + 300}{3} = \dfrac{3081}{3}$

$\qquad = \mathbf{1027 \ tokens}$

5.
$$\begin{array}{r} 4057 \\ \times \quad 17 \\ \hline 28399 \\ 4057 \quad \\ \hline \mathbf{68{,}969} \ \textbf{fans} \end{array}$$

6. $\dfrac{4}{13} \times 39 = \dfrac{156}{13} = \mathbf{12}$

7. $\dfrac{11}{5} \times 500 = \dfrac{5500}{5} = \mathbf{1100}$

8. $\dfrac{93}{13} = \mathbf{7\dfrac{2}{13}}$

$$\begin{array}{r} 7 \\ 13\overline{)93} \\ \underline{91} \\ 2 \end{array}$$

9. $\dfrac{41}{7} = \mathbf{5\dfrac{6}{7}}$

$$\begin{array}{r} 5 \\ 7\overline{)41} \\ \underline{35} \\ 6 \end{array}$$

10. $p + gp = (1) + (22)(1) = 1 + 22 = \mathbf{23}$

11. $xyz + yz = (3)(4)(5) + (4)(5) = 60 + 20 = \mathbf{80}$

12. $200 = 2 \cdot 2 \cdot 2 \cdot 5 \cdot 5$
$120 = 2 \cdot 2 \cdot 2 \cdot 3 \cdot 5$
$180 = 2 \cdot 2 \cdot 3 \cdot 3 \cdot 5$

$\text{LCM}\,(200, 120, 180) = 2 \cdot 2 \cdot 2 \cdot 3 \cdot 3$
$\qquad\qquad\qquad\qquad\qquad \cdot 5 \cdot 5$
$\qquad\qquad\qquad\qquad = \mathbf{1800}$

13. $\dfrac{14}{31} + x = \dfrac{21}{31}$

$\dfrac{21}{31} - \dfrac{14}{31} = \mathbf{\dfrac{7}{31}}$

14. $\dfrac{2}{3} + \dfrac{5}{8} + \dfrac{1}{2} = \dfrac{16}{24} + \dfrac{15}{24} + \dfrac{12}{24} = \mathbf{\dfrac{43}{24}}$

15. $\dfrac{6}{7} - \dfrac{4}{5} = \dfrac{30}{35} - \dfrac{28}{35} = \mathbf{\dfrac{2}{35}}$

16. $\dfrac{3}{5} + \dfrac{4}{7} - \dfrac{1}{3} = \dfrac{63}{105} + \dfrac{60}{105} - \dfrac{35}{105} = \mathbf{\dfrac{88}{105}}$

17. $\dfrac{4}{5} \cdot \dfrac{20}{40} \div \dfrac{5}{10} = \dfrac{4}{\cancel{8}} \cdot \dfrac{\cancel{20}}{\cancel{40}} \cdot \dfrac{\cancel{10}}{5} = \dfrac{4}{\cancel{10}} \cdot \dfrac{\cancel{10}}{5} = \mathbf{\dfrac{4}{5}}$

18. $\dfrac{2}{8} \cdot \dfrac{16}{24} \div \dfrac{15}{8} = \dfrac{2}{\cancel{8}} \cdot \dfrac{\cancel{16}}{\cancel{24}} \cdot \dfrac{8}{15} = \dfrac{2}{\cancel{12}} \cdot \dfrac{8}{15}$

$\qquad = \mathbf{\dfrac{4}{45}}$

19.
$$\begin{array}{r} 0.0023 \\ \times\ \ 1.047 \\ \hline 161 \\ 92\ \ \\ 230\ \ \ \\ \hline 0.0024081 \end{array}$$

20. $15\dfrac{2}{25} = 15\dfrac{8}{100} = \textbf{15.08}$

21. $16.16 = 16\dfrac{16}{100} = \textbf{16}\dfrac{\textbf{4}}{\textbf{25}}$

22. (a) $\dfrac{\textbf{20 shillings}}{\textbf{1 pound}}, \dfrac{\textbf{1 pound}}{\textbf{20 shillings}}$

(b) $1000 \text{ ~~pounds~~} \times \dfrac{20 \text{ shillings}}{1 \text{ ~~pound~~}}$
$= \textbf{20,000 shillings}$

23. $A_{\text{Figure}} = A_{\text{Triangle 1}} + A_{\text{Triangle 2}}$
$= \left[\dfrac{(22)(19)}{2} + \dfrac{(18)(24)}{2}\right] \text{ft}^2$
$= [209 + 216] \text{ ft}^2$
$= \textbf{425 ft}^2$

24. $[20 + 20 + 2(15 + 19 + 11)] \text{ cm} = \textbf{130 cm}$

25. $\dfrac{260}{305} = \dfrac{260 \div 5}{305 \div 5} = \dfrac{\textbf{52}}{\textbf{61}}$

26. $\dfrac{52}{63}$

$$\begin{array}{r} 0.825 \\ 63\overline{)52.000} \\ 50\,4\ \ \ \ \\ \hline 1\,60\ \ \\ 1\,26\ \ \\ \hline 340 \\ 315 \\ \hline 25 \end{array}$$

The rounded number is **0.83**.

27. $\dfrac{14}{19}$

$$\begin{array}{r} 0.736 \\ 19\overline{)14.000} \\ 13\,3\ \ \ \ \\ \hline 70\ \ \\ 57\ \ \\ \hline 130 \\ 114 \\ \hline 16 \end{array}$$

The rounded number is **0.74**.

28. $8 \cdot 10 - 2 - 10 \cdot 6 = 80 - 2 - 60$
$= 78 - 60 = \textbf{18}$

29. $\dfrac{1}{8} + \dfrac{\overset{1}{\cancel{3}}}{4} \cdot \dfrac{1}{\underset{2}{\cancel{6}}} = \dfrac{1}{8} + \dfrac{1}{8} = \dfrac{2}{8} = \dfrac{\textbf{1}}{\textbf{4}}$

30. $3 \cdot 8 - 6 \cdot 4 + 24 \cdot 1 - 12 \cdot 2$
$= 24 - 24 + 24 - 24 = \textbf{0}$

PROBLEM SET 33

1.
$$\begin{array}{r} 175 \\ \times\ \ 37 \\ \hline 1225 \\ 525\ \ \\ \hline 6475 \text{ tickets} \end{array}$$

2. $280 \text{ tons} - 200 \text{ tons} = 80 \text{ ~~tons~~} \times \dfrac{2000 \text{ lb}}{1 \text{ ~~ton~~}}$
$= \textbf{160,000 lb}$

3. $85 + 73 + 92 + 66 + N = 5(82)$
$316 + N = 410$
$410 - 316 = \textbf{94}$

4.
$$\begin{array}{r} 22,000,013.0984 \\ -\ 19,484,000.0750 \\ \hline 2,516,013.0234 \end{array}$$

5. $\$5(7) + \$3.40(9) + \$1.30(20)$
$= \$35 + \$30.60 + \$26 = \textbf{\$91.60}$

6. $\dfrac{5}{16} \times 128 = \dfrac{640}{16} = \textbf{40}$

7. $\dfrac{14}{3} \times 30 = \dfrac{420}{3} = \textbf{140}$

8. $\dfrac{93}{12} = \dfrac{31}{4} = \textbf{7}\dfrac{\textbf{3}}{\textbf{4}}$

$$\begin{array}{r} 7 \\ 4\overline{)31} \\ 28 \\ \hline 3 \end{array}$$

9. $\dfrac{40}{7} = \textbf{5}\dfrac{\textbf{5}}{\textbf{7}}$

$$\begin{array}{r} 5 \\ 7\overline{)40} \\ 35 \\ \hline 5 \end{array}$$

10. $e + g + fg = (0.08) + (1.7) + (0.5)(1.7)$
$= 0.08 + 1.7 + 0.85$
$= 1.78 + 0.85$
$= \mathbf{2.63}$

11. $\dfrac{x}{y} + xy = \dfrac{(6)}{(2)} + (6)(2) = 3 + 12 = \mathbf{15}$

12. $\dfrac{3}{7} + \dfrac{2}{5} - \dfrac{3}{10} = \dfrac{30}{70} + \dfrac{28}{70} - \dfrac{21}{70} = \mathbf{\dfrac{37}{70}}$

13. $\dfrac{5}{8} + \dfrac{3}{5} - \dfrac{1}{4} = \dfrac{25}{40} + \dfrac{24}{40} - \dfrac{10}{40} = \mathbf{\dfrac{39}{40}}$

14. $\dfrac{7}{11} - \dfrac{1}{3} = \dfrac{21}{33} - \dfrac{11}{33} = \mathbf{\dfrac{10}{33}}$

15. $32 - 2 \cdot 5 + 3 \cdot 6 = 32 - 10 + 18 = \mathbf{40}$

16. $7 + 5 \cdot 3 - 2 \cdot 4 + 5 \cdot 3$
$= 7 + 15 - 8 + 15 = \mathbf{29}$

17. $3 + 3 \cdot 5 - 4 \cdot 2 = 3 + 15 - 8 = \mathbf{10}$

18. $\dfrac{4}{5} \cdot \dfrac{25}{20} \div \dfrac{5}{10} = \dfrac{\cancel{4}}{\cancel{8}} \cdot \dfrac{\cancel{25}}{\cancel{20}} \cdot \dfrac{\cancel{10}}{5} = \dfrac{\cancel{8}}{\cancel{8}} \cdot \dfrac{\cancel{10}}{\cancel{8}} = \mathbf{2}$

19. $\dfrac{3}{4} - \dfrac{7}{12} = \dfrac{9}{12} - \dfrac{7}{12} = \dfrac{2}{12} = \mathbf{\dfrac{1}{6}}$

20.
$$
\begin{array}{r}
0.016 \\
\times\ 0.0023 \\
\hline
48 \\
32 \\
\hline
\mathbf{0.0000368}
\end{array}
$$

21. $50.05 = 50\dfrac{5}{100} = \mathbf{50\dfrac{1}{20}}$

22. (a) $\dfrac{\mathbf{16\ ounces}}{\mathbf{1\ pint}}, \dfrac{\mathbf{1\ pint}}{\mathbf{16\ ounces}}$

(b) $640\ \cancel{oz} \times \dfrac{1\ pt}{16\ \cancel{oz}} = \mathbf{40\ pt}$

23. $90\ \cancel{yd} \times \dfrac{3\ \cancel{ft}}{1\ \cancel{yd}} \times \dfrac{12\ in.}{1\ \cancel{ft}} = \mathbf{90(3)(12)\ in.}$
$= \mathbf{3240\ in.}$

24. $7.5\ \cancel{ft^2} \times \dfrac{12\ in.}{1\ \cancel{ft}} \times \dfrac{12\ in.}{1\ \cancel{ft}} = \mathbf{7.5(12)(12)\ in.^2}$
$= \mathbf{1080\ in.^2}$

25. $450\ \cancel{ft^2} \times \dfrac{1\ yd}{3\ \cancel{ft}} \times \dfrac{1\ yd}{3\ \cancel{ft}} = \dfrac{\mathbf{450}}{\mathbf{(3)(3)}}\ \mathbf{yd^2} = \mathbf{50\ yd^2}$

26. $A = A_{\text{Rectangle}} + A_{\text{Triangle}}$
$= \left[30(10) + \dfrac{30(10)}{2} \right] ft^2$
$= \mathbf{450\ ft^2}$

27. $\dfrac{4160}{5} = 832; \dfrac{832}{2} = 416; \dfrac{416}{2} = 208;$

$\dfrac{208}{2} = 104; \dfrac{104}{2} = 52; \dfrac{52}{2} = 26; \dfrac{26}{2} = 13$

$4160 = 2 \cdot 2 \cdot 2 \cdot 2 \cdot 2 \cdot 2 \cdot 5 \cdot 13$
$= \mathbf{2^6 \cdot 5 \cdot 13}$

28. $\dfrac{13}{5} = 2\dfrac{3}{5} = 2\dfrac{6}{10} = \mathbf{2.6}$

29. $\dfrac{17}{4} = 4\dfrac{1}{4} = 4\dfrac{25}{100} = \mathbf{4.25}$

30. $\dfrac{7}{18}$

$$
\begin{array}{r}
0.388 \\
18\overline{)7.000} \\
5\,4 \\
\hline
1\,60 \\
1\,44 \\
\hline
160 \\
144 \\
\hline
16
\end{array}
$$

The rounded number is **0.39.**

PROBLEM SET 34

1. $420 \times N = 130,420$

$$
\begin{array}{r}
310 \\
420\overline{)130,420} \\
126\,0 \\
\hline
4\,42 \\
4\,20 \\
\hline
220
\end{array}
$$

Since there are 220 chips remaining, **311 boxes** are needed.

2. $\$15.95(3) + \$17.75(7) + \$11.95(4)$
$= \$47.85 + \$124.25 + \$47.80 = \mathbf{\$219.90}$

3. $\text{Avg} = \dfrac{50 + 150 + 100 + 350 + 100}{5}$
$= \dfrac{750}{5} = \mathbf{150\ cars}$

4. (a) Range = 18,200 − 12,000 = **6200 lb**

(b) The four weights are all different, so there is **no mode.**

(c) Median = $\dfrac{14{,}000 + 16{,}280}{2}$ = **15,140 lb**

(d) Mean

$= \dfrac{14{,}000 + 12{,}000 + 18{,}200 + 16{,}280}{4}$

$= \dfrac{60{,}480}{4}$ = **15,120 lb**

5. $\dfrac{64 \text{ ft}}{8 \text{ s}} = \dfrac{8 \text{ ft}}{1 \text{ s}}$ = **8 ft per s**

$\dfrac{8 \text{ s}}{64 \text{ ft}} = \dfrac{1 \text{ s}}{8 \text{ ft}} = \dfrac{1}{8}$ **s per ft**

6. $\dfrac{6 \text{ locusts}}{30 \text{ min}} = \dfrac{1 \text{ locust}}{5 \text{ min}}$

$= \dfrac{1}{5}$ **locust per min**

$\dfrac{30 \text{ min}}{6 \text{ locusts}} = \dfrac{5 \text{ min}}{1 \text{ locust}}$

$=$ **5 min per locust**

7. $0.375 + P = 1.2$

$\begin{array}{r} 1.200 \\ -\ 0.375 \\ \hline \mathbf{0.825} \end{array}$

8. $\dfrac{5}{17} \times 136 = \dfrac{680}{17}$ = **40**

9. $\dfrac{37}{3} = 12\dfrac{1}{3}$

$\begin{array}{r} 12 \\ 3\overline{)37} \\ \underline{3} \\ 07 \\ \underline{6} \\ 1 \end{array}$

10. $\dfrac{421}{5} = 84\dfrac{1}{5}$

$\begin{array}{r} 84 \\ 5\overline{)421} \\ \underline{40} \\ 21 \\ \underline{20} \\ 1 \end{array}$

11. $60 = 2 \cdot 2 \cdot 3 \cdot 5$

$84 = 2 \cdot 2 \cdot 3 \cdot 7$

$120 = 2 \cdot 2 \cdot 2 \cdot 3 \cdot 5$

LCM (60, 84, 120) $= 2 \cdot 2 \cdot 2 \cdot 3 \cdot 5 \cdot 7$

$=$ **840**

12. $\dfrac{3}{4} + \dfrac{5}{8} + \dfrac{2}{3} - \dfrac{1}{6} = \dfrac{18}{24} + \dfrac{15}{24} + \dfrac{16}{24} - \dfrac{4}{24}$

$= \dfrac{45}{24} = \dfrac{\mathbf{15}}{\mathbf{8}}$

13. $\dfrac{5}{8} + \dfrac{1}{16} + \dfrac{1}{2} - \dfrac{1}{4} = \dfrac{10}{16} + \dfrac{1}{16} + \dfrac{8}{16} - \dfrac{4}{16}$

$= \dfrac{\mathbf{15}}{\mathbf{16}}$

14. $3 + 2 \cdot 6 - 4 \cdot 3 = 3 + 12 - 12 =$ **3**

15. $5 \cdot 2 - 3 \cdot 2 + 4 \cdot 3 = 10 - 6 + 12 =$ **16**

16. $\dfrac{16}{25} \cdot \dfrac{15}{8} \div \dfrac{3}{2} = \dfrac{\overset{2}{\cancel{16}}}{\underset{5}{\cancel{25}}} \cdot \dfrac{\overset{3}{\cancel{15}}}{\underset{1}{\cancel{8}}} \cdot \dfrac{2}{3} = \dfrac{2}{5} \cdot \dfrac{\overset{1}{\cancel{3}}}{1} \cdot \dfrac{2}{\underset{1}{\cancel{3}}}$

$= \dfrac{\mathbf{4}}{\mathbf{5}}$

17. $\dfrac{30.03}{0.0021}$

$\begin{array}{r} \mathbf{14{,}300} \\ 21\overline{)300{,}300} \\ \underline{21} \\ 90 \\ \underline{84} \\ 6\ 3 \\ \underline{6\ 3} \\ 0 \end{array}$

18. $\dfrac{4}{3} - \dfrac{7}{10} = \dfrac{40}{30} - \dfrac{21}{30} = \dfrac{\mathbf{19}}{\mathbf{30}}$

19. $8\dfrac{1}{2} + 2\dfrac{5}{6} = 8\dfrac{3}{6} + 2\dfrac{5}{6} = 10\dfrac{8}{6} = 11\dfrac{2}{6} = \mathbf{11\dfrac{1}{3}}$

20. $57\dfrac{5}{13} = 57\dfrac{15}{39}$

$+\ \ 13\dfrac{2}{3} = 13\dfrac{26}{39}$

$= 70\dfrac{41}{39} = \mathbf{71\dfrac{2}{39}}$

21. $15.15 = 15\dfrac{15}{100} = \mathbf{15\dfrac{3}{20}}$

22. (a) $\dfrac{40 \text{ gallons}}{1 \text{ barrel}}, \dfrac{1 \text{ barrel}}{40 \text{ gallons}}$

(b) $2500 \text{ barrels} \times \dfrac{40 \text{ gal}}{1 \text{ barrel}} = \textbf{100,000 gal}$

23. $132 \text{ m}^2 \times \dfrac{100 \text{ cm}}{1 \text{ m}} \times \dfrac{100 \text{ cm}}{1 \text{ m}}$

$= 132(100)(100) \text{ cm}^2 = \textbf{1,320,000 cm}^2$

24. $xy + 2m = (2)(4) + (2)(3) = 8 + 6 = \textbf{14}$

25. $xym + xy = (3)(6)(4) + (3)(6)$

$= 72 + 18$

$= \textbf{90}$

26. $mx + 4m = (5)\left(\dfrac{2}{3}\right) + 4(5)$

$= \dfrac{10}{3} + 20$

$= \dfrac{10}{3} + \dfrac{60}{3}$

$= \dfrac{70}{3}$

$= \textbf{23}\dfrac{1}{3}$

27.

$A_{\text{Figure}} = A_{\text{Rectangle}} + A_{\text{Triangle}}$

$= \left[23(10) + \dfrac{23(10)}{2} \right] \text{ft}^2$

$= \textbf{345 ft}^2$

28. $[2(44) + 2(12 + 10)] \text{ m} = \textbf{132 m}$

29. $4\dfrac{3}{5} = 4\dfrac{6}{10} = \textbf{4.6}$

30.
```
      0.375
  8 ) 3.000
      2 4
      ---
        60
        56
        ---
         40
         40
         ---
          0
```

Since $\dfrac{3}{8} = 0.375,\ 301\dfrac{3}{8} = \textbf{301.375}$

Problem Set 35

1.
```
   0.0450
 - 0.0417
  -------
   0.0033
```

The second measurement was larger by 0.0033 m.

2. $2(\$50,000) - 2(\$30,000) = \textbf{\$40,000}$

3. $450 \times N = 9000$
```
      20 parades
 450 ) 9000
        900
        ---
          0
```

4. Avg

$= \dfrac{417 + 832 + 619 + 148 + 212 + 184}{6}$

$= \dfrac{2412}{6} = \textbf{402 lb}$

5. $\dfrac{105}{112} - x = \dfrac{98}{112}$

$\dfrac{105}{112} - \dfrac{98}{112} = \dfrac{7}{112} = \dfrac{1}{16}$

6. $\dfrac{5}{12} \times 48 = \dfrac{240}{12} = \textbf{20}$

7. $\dfrac{214}{5} = \textbf{42}\dfrac{4}{5}$
```
       42
  5 ) 214
      20
      ---
       14
       10
       ---
        4
```

8. $\dfrac{47}{2} = \textbf{23}\dfrac{1}{2}$
```
      23
  2 ) 47
      4
      --
      07
       6
       --
       1
```

9. $50 = 2 \cdot 5 \cdot 5$

$60 = 2 \cdot 2 \cdot 3 \cdot 5$

$72 = 2 \cdot 2 \cdot 2 \cdot 3 \cdot 3$

$\text{LCM }(50, 60, 72) = 2 \cdot 2 \cdot 2 \cdot 3 \cdot 3 \cdot 5 \cdot 5$

$= \textbf{1800}$

10. $\dfrac{3}{4} + \dfrac{5}{8} + \dfrac{3}{16} = \dfrac{12}{16} + \dfrac{10}{16} + \dfrac{3}{16} = \mathbf{\dfrac{25}{16}}$

11. $\dfrac{13}{15} - \dfrac{1}{5} = \dfrac{13}{15} - \dfrac{3}{15} = \dfrac{10}{15} = \mathbf{\dfrac{2}{3}}$

12. $3 \cdot 12 - 4 \cdot 2 + 3 \cdot 5 = 36 - 8 + 15 = \mathbf{43}$

13. $2 \cdot 5 \cdot 2 - 3 \cdot 5 + 2 - 5$
$= 20 - 15 + 2 - 5 = \mathbf{2}$

14. $\dfrac{4}{6} \cdot \dfrac{9}{14} \div \dfrac{2}{5} = \dfrac{\overset{2}{\cancel{4}}}{\cancel{6}} \cdot \dfrac{\overset{3}{\cancel{9}}}{\cancel{14}} \cdot \dfrac{5}{2} = \dfrac{\cancel{2}}{\cancel{2}} \cdot \dfrac{3}{7} \cdot \dfrac{5}{2}$

$= \dfrac{15}{14} = \mathbf{1\dfrac{1}{14}}$

15. $52\dfrac{3}{8} = 52\dfrac{9}{24}$

$+ \ 19\dfrac{2}{3} = 19\dfrac{16}{24}$
$\overline{\phantom{+ \ 19\dfrac{2}{3}}}$
$= 71\dfrac{25}{24} = \mathbf{72\dfrac{1}{24}}$

16. $5\dfrac{1}{2} = \overset{4}{\cancel{5}}\overset{15}{\cancel{\dfrac{5}{10}}}$

$- \ 2\dfrac{4}{5} = 2\dfrac{8}{10}$
$\overline{\phantom{- \ 2\dfrac{4}{5}}}$
$= \mathbf{2\dfrac{7}{10}}$

17. $9\dfrac{2}{14} = 9\overset{8}{\cancel{\dfrac{8}{7}}}$

$- \ 3\dfrac{15}{21} = 3\dfrac{5}{7}$
$\overline{\phantom{- \ 3\dfrac{15}{21}}}$
$= \mathbf{5\dfrac{3}{7}}$

18. $600\dfrac{2}{9} = \overset{599}{\cancel{600}}\overset{77}{\cancel{\dfrac{14}{63}}}$

$- \ 311\dfrac{3}{7} = 311\dfrac{27}{63}$
$\overline{\phantom{- \ 311\dfrac{3}{7}}}$
$= \mathbf{288\dfrac{50}{63}}$

19. $5\dfrac{2}{3} = \dfrac{3 \times 5 + 2}{3} = \mathbf{\dfrac{17}{3}}$

20. $xy - y = (5)(4) - (4) = 20 - 4 = \mathbf{16}$

21. $m - xy = (10) - (2)(3) = 10 - 6 = \mathbf{4}$

22. $xym - m = \left(\dfrac{1}{2}\right)(4)(2) - (2) = 4 - 2 = \mathbf{2}$

23. $400 \ \cancel{\text{in.}} \times \dfrac{1 \ \cancel{\text{ft}}}{12 \ \cancel{\text{in.}}} \times \dfrac{1 \ \text{yd}}{3 \ \cancel{\text{ft}}} = \mathbf{\dfrac{100}{9}} \ \textbf{yd} = \mathbf{11\dfrac{1}{9}} \ \textbf{yd}$

24. $3.6 \ \cancel{\text{km}} \times \dfrac{1000 \ \cancel{\text{m}}}{1 \ \cancel{\text{km}}} \times \dfrac{100 \ \text{cm}}{1 \ \cancel{\text{m}}} = \mathbf{360{,}000 \ cm}$

25. $4 \ \cancel{\text{mi}^2} \times \dfrac{5280 \ \text{ft}}{1 \ \cancel{\text{mi}}} \times \dfrac{5280 \ \text{ft}}{1 \ \cancel{\text{mi}}} = 4(5280)(5280) \ \text{ft}^2$

$= \mathbf{111{,}513{,}600 \ ft^2}$

26. $\dfrac{24 \ \text{apricots}}{2 \ \text{dollars}} = \dfrac{12 \ \text{apricots}}{1 \ \text{dollar}}$

$= \mathbf{12 \ apricots \ per \ dollar}$

$\dfrac{2 \ \text{dollars}}{24 \ \text{apricots}} = \dfrac{1 \ \text{dollar}}{12 \ \text{apricots}}$

$= \mathbf{\dfrac{1}{12}} \ \textbf{dollar per apricot}$

27. $0.4 \ \cancel{\text{m}} \times \dfrac{100 \ \text{cm}}{1 \ \cancel{\text{m}}} = \mathbf{40 \ cm}$

$A = A_{\text{Rectangle}} + A_{\text{Triangle}}$

$= \left[40(15) + \dfrac{40(7)}{2} \right] \text{cm}^2$

$= \mathbf{740 \ cm^2}$

28. $0.1 \ \cancel{\text{m}} \times \dfrac{100 \ \text{cm}}{1 \ \cancel{\text{m}}} = \mathbf{10 \ cm}$

$P = (50 + 10 + 15 + 15 + 10 + 50) \ \text{cm}$

$= \mathbf{150 \ cm}$

29.
$$
\begin{array}{r}
0.875 \\
8\overline{)7.000} \\
\underline{6\ 4} \\
60 \\
\underline{56} \\
40 \\
\underline{40} \\
0
\end{array}
$$

Since $\dfrac{7}{8} = 0.875$, $12\dfrac{7}{8} = \mathbf{12.875}$.

30.

$\downarrow$

$5.5555\text{⑤}5$

The rounded number is **5.55556**.

Problem Set 36

1. $215 + 305 + 265 + 196 + 221$
$\qquad + 236 + N = 7(236)$
$\qquad\quad 1438 + N = 1652$
$1652 - 1438 = \mathbf{214\ lb}$

2.

Crop Value Per Year

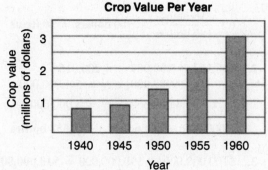

3. $7(\$5.40) + 200(\$0.30) + 40(\$22.50)$
$= \$37.80 + \$60 + \$900 = \mathbf{\$997.80}$

4. $460 \times N = 10{,}120$

$$
\begin{array}{r}
\mathbf{22\ shifts} \\
460\overline{)10{,}120} \\
9\ 20 \\
\hline
920 \\
920 \\
\hline
0
\end{array}
$$

5. (a) $\dfrac{\mathbf{40\ skins}}{\mathbf{8\ liras}}, \dfrac{\mathbf{8\ liras}}{\mathbf{40\ skins}}$

(b) $\dfrac{40\ skins}{8\ \cancel{liras}} \times 200\ \cancel{liras} = \mathbf{1000\ skins}$

(c) $\dfrac{8\ liras}{40\ \cancel{skins}} \times 200\ \cancel{skins} = \mathbf{40\ liras}$

6. (a) $\dfrac{\mathbf{3\ books}}{\mathbf{7\ CDs}}, \dfrac{\mathbf{7\ CDs}}{\mathbf{3\ books}}$

(b) $\dfrac{7\ CDs}{3\ \cancel{books}} \times 9\ \cancel{books} = \mathbf{21\ CDs}$

7.
$$
\begin{array}{r}
713.8910 \\
-\ 712.9993 \\
\hline
\mathbf{0.8917}
\end{array}
$$

8. $\dfrac{4}{5} \times 200 = \dfrac{800}{5} = \mathbf{160}$

9. $\dfrac{21}{4} = \mathbf{5\dfrac{1}{4}}$

$$
\begin{array}{r}
5 \\
4\overline{)21} \\
20 \\
\hline
1
\end{array}
$$

10. $\dfrac{86}{11} = \mathbf{7\dfrac{9}{11}}$

$$
\begin{array}{r}
7 \\
11\overline{)86} \\
77 \\
\hline
9
\end{array}
$$

11. $40 = 2 \cdot 2 \cdot 2 \cdot 5$
$50 = 2 \cdot 5 \cdot 5$
$70 = 2 \cdot 5 \cdot 7$

$\text{LCM}(40, 50, 70) = 2 \cdot 2 \cdot 2 \cdot 5 \cdot 5 \cdot 7$
$\qquad\qquad\qquad\quad = \mathbf{1400}$

12. $\dfrac{3}{5} - \dfrac{1}{15} = \dfrac{9}{15} - \dfrac{1}{15} = \mathbf{\dfrac{8}{15}}$

13. $\dfrac{1}{10} + \dfrac{3}{5} - \dfrac{1}{20} = \dfrac{2}{20} + \dfrac{12}{20} - \dfrac{1}{20} = \mathbf{\dfrac{13}{20}}$

14. $3 + 5 - 2 \cdot 4 + 3 \cdot 5 = 8 - 8 + 15 = \mathbf{15}$

15. $4 + 3(2) + 5 \cdot 4 = 4 + 6 + 20 = \mathbf{30}$

16. $2\dfrac{1}{4} + 3\dfrac{1}{8} = 2\dfrac{2}{8} + 3\dfrac{1}{8} = \mathbf{5\dfrac{3}{8}}$

17. $429\dfrac{1}{5} + 8162\dfrac{4}{15} = 429\dfrac{3}{15} + 8162\dfrac{4}{15}$

$= \mathbf{8591\dfrac{7}{15}}$

18. $534\dfrac{3}{8} + 371\dfrac{1}{40} = 534\dfrac{15}{40} + 371\dfrac{1}{40}$

$= 905\dfrac{16}{40} = \mathbf{905\dfrac{2}{5}}$

19. $7\dfrac{3}{7} = 7\overset{\overset{40}{\cancel{12}}}{\dfrac{\cancel{12}}{28}}$

$-\ 4\dfrac{3}{4} = 4\dfrac{21}{28}$
$\qquad\qquad\quad = \mathbf{2\dfrac{19}{28}}$

20. $79\dfrac{2}{7} = 79\overset{\overset{27}{\cancel{6}}}{\dfrac{\cancel{8}}{21}}$

$-\ 55\dfrac{1}{3} = 55\dfrac{7}{21}$
$\qquad\qquad\quad = \mathbf{23\dfrac{20}{21}}$

21. $\dfrac{716.2}{0.008}$

$$\begin{array}{r} 89{,}525 \\ 8\overline{)716{,}200} \\ \underline{64} \\ 76 \\ \underline{72} \\ 4\,2 \\ \underline{4\,0} \\ 20 \\ \underline{16} \\ 40 \\ \underline{40} \\ 0 \end{array}$$

22. $\dfrac{3}{8} \cdot \dfrac{24}{9} \div \dfrac{3}{7} = \dfrac{\cancel{3}^{1}}{\cancel{8}^{1}} \cdot \dfrac{\cancel{24}^{3}}{9} \cdot \dfrac{7}{\cancel{3}^{1}} = \dfrac{\cancel{3}^{1}}{\cancel{9}^{3}} \cdot \dfrac{7}{1} = \dfrac{7}{3}$

23. $40.04 = 40\dfrac{4}{100} = \mathbf{40\dfrac{1}{25}}$

24. $zy - z = (3)(4) - (3) = 12 - 3 = \mathbf{9}$

25. $xyz + yz\left(\dfrac{1}{3}\right)(9)(2) + (9)(2) = 6 + 18 = \mathbf{24}$

26. $540\,\cancel{\text{in.}} \times \dfrac{1\,\cancel{\text{ft}}}{12\,\cancel{\text{in.}}} \times \dfrac{1\,\text{yd}}{3\,\cancel{\text{ft}}} = \dfrac{540}{(12)(3)}\,\text{yd} = \mathbf{15\ yd}$

27. $187{,}625.8\,\cancel{\text{cg}} \times \dfrac{1\,\cancel{g}}{100\,\cancel{\text{cg}}} \times \dfrac{1\,\text{kg}}{1000\,\cancel{g}}$

$= \dfrac{187{,}625.8}{(100)(1000)}\,\text{kg} = \mathbf{1.876258\ kg}$

28.

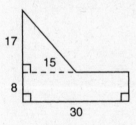

$A = A_{\text{Rectangle}} + A_{\text{Triangle}}$

$= \left[30(8) + \dfrac{15(17)}{2}\right]\text{in.}^2$

$= \mathbf{367.5\ in.^2}$

29. $P = (32 + 25 + 32 + 25)\,\text{cm}$

$= \mathbf{114\ cm}$

30. (a) **23, 29, 31**

(b) **21, 28, 35**

1. (a) $\dfrac{180\text{ dollars}}{10\text{ items}},\ \dfrac{10\text{ items}}{180\text{ dollars}}$

(b) $\dfrac{\$180}{10\,\cancel{\text{items}}} \times 25\,\cancel{\text{items}} = \mathbf{\$450}$

(c) $\dfrac{10\text{ items}}{180\,\cancel{\text{dollars}}} \times 900\,\cancel{\text{dollars}} = \mathbf{50\ items}$

2. $998{,}163 + 899{,}989 + 1{,}200{,}316$
$+\ 987{,}900 + N = 5(1{,}000{,}000)$

$4{,}086{,}368 + N = 5{,}000{,}000$

$5{,}000{,}000 - 4{,}086{,}368 = \mathbf{913{,}632\ bottles}$

3. $\$100{,}000{,}000 - \$90{,}000{,}000 = \mathbf{\$10{,}000{,}000}$

4. $415 \times N = 29{,}050$

$$\begin{array}{r} 70\text{ units} \\ 415\overline{)29{,}050} \\ \underline{29\ 05} \\ 0 \end{array}$$

5. $\dfrac{5}{16} \times 32 = \dfrac{160}{16} = \mathbf{10}$

6. $x - 17 = 20$

$37 - 17 = 20$

The solution is **37.**

7. $x + 4 = 10$

$6 + 4 = 10$

The solution is **6.**

8. $x - 0.0009 = 513.0011$

$$\begin{array}{r} 513.0011 \\ +\ \ \ 0.0009 \\ \hline 513.0020 \end{array}$$

9. $\dfrac{2}{7} + x = \dfrac{5}{7}$

$\dfrac{2}{7} + \dfrac{3}{7} = \dfrac{5}{7}$

The solution is $\dfrac{3}{7}$.

10. $\dfrac{31}{5} = 6\dfrac{1}{5}$

$$\begin{array}{r} 6 \\ 5\overline{)31} \\ \underline{30} \\ 1 \end{array}$$

11. $\dfrac{93}{7} = 13\dfrac{2}{7}$

$$\begin{array}{r} 13 \\ 7\overline{)93} \\ \underline{7} \\ 23 \\ \underline{21} \\ 2 \end{array}$$

12. $\dfrac{3}{5} - \dfrac{2}{10} = \dfrac{3}{5} - \dfrac{1}{5} = \dfrac{2}{5}$

13. $\dfrac{3}{8} + \dfrac{1}{2} - \dfrac{1}{4} = \dfrac{3}{8} + \dfrac{4}{8} - \dfrac{2}{8} = \dfrac{5}{8}$

14. $4 + 3 \cdot 2 - 5 = 4 + 6 - 5 = 5$

15. $3\dfrac{1}{8} + 2\dfrac{1}{4} + 5\dfrac{1}{2} = 3\dfrac{1}{8} + 2\dfrac{2}{8} + 5\dfrac{4}{8} = 10\dfrac{7}{8}$

16. $428\dfrac{1}{11} + 22\dfrac{1}{44} = 428\dfrac{4}{44} + 22\dfrac{1}{44} = 450\dfrac{5}{44}$

17. $3\dfrac{2}{5} + 748\dfrac{2}{10} = 3\dfrac{2}{5} + 748\dfrac{1}{5} = 751\dfrac{3}{5}$

18. $4\dfrac{4}{10} - 1\dfrac{1}{5} = 4\dfrac{2}{5} - 1\dfrac{1}{5} = 3\dfrac{1}{5}$

19. $548\dfrac{6}{8} - 31\dfrac{1}{16} = 548\dfrac{12}{16} - 31\dfrac{1}{16} = 517\dfrac{11}{16}$

20. $991\dfrac{1}{3} - 791\dfrac{17}{18} = 991\dfrac{6}{18} - 791\dfrac{17}{18}$

$\qquad = 990\dfrac{24}{18} - 791\dfrac{17}{18} = 199\dfrac{7}{18}$

21.
$$\begin{array}{r} 71.82 \\ \times \ \ 8.01 \\ \hline 7182 \\ 57456 \ \ \\ \hline \mathbf{575.2782} \end{array}$$

22. $\dfrac{936.7}{0.04}$

$$\begin{array}{r} \mathbf{23,417.5} \\ 4\overline{)93,670.0} \\ \underline{8} \ \ \ \ \ \ \ \ \ \\ 13 \ \ \ \ \ \ \\ \underline{12} \ \ \ \ \ \ \\ 1\,6 \ \ \ \ \\ \underline{1\,6} \ \ \ \ \\ 07 \ \ \\ \underline{4} \ \ \\ 30 \ \\ \underline{28} \ \\ 2\,0 \\ \underline{2\,0} \\ 0 \end{array}$$

23. $\dfrac{21}{8} \cdot \dfrac{4}{14} \div \dfrac{9}{2} = \dfrac{\overset{3}{\cancel{21}}}{\cancel{8}} \cdot \dfrac{\overset{1}{\cancel{4}}}{\cancel{14}} \cdot \dfrac{2}{9} = \dfrac{\cancel{3}}{\cancel{2}} \cdot \dfrac{1}{\cancel{2}} \cdot \dfrac{\overset{1}{\cancel{2}}}{\underset{3}{\cancel{9}}}$

$\qquad = \dfrac{1}{6}$

24. $250.025 = 250\dfrac{25}{1000} = 250\dfrac{1}{40}$

25. $xy + yz - z = (1)(7) + (7)(2) - (2)$

$\qquad\qquad\qquad = 7 + 14 - 2$

$\qquad\qquad\qquad = 19$

26. $xyz - xy = (2)(3)(3) - (2)(3) = 18 - 6 = 12$

27. $10 \, \cancel{mi} \times \dfrac{5280 \, \cancel{ft}}{1 \, \cancel{mi}} \times \dfrac{12 \text{ in.}}{1 \, \cancel{ft}}$

$\qquad = \mathbf{10(5280)12 \text{ in.} = 633,600 \text{ in.}}$

28.

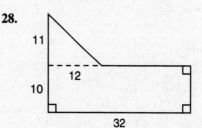

$A = A_{\text{Rectangle}} + A_{\text{Triangle}}$

$\quad = \left[32(10) + \dfrac{11(12)}{2} \right] \text{yd}^2$

$\quad = [320 + 66] \text{ yd}^2 = \mathbf{386 \text{ yd}^2}$

29. $\dfrac{11}{13}$

$$13\overline{)11.000} \quad \begin{array}{r} 0.846 \\ \underline{10\,4} \\ 60 \\ \underline{52} \\ 80 \\ \underline{78} \\ 2 \end{array}$$

The rounded number is **0.85**.

30. $\dfrac{102}{170} = \dfrac{102 \div 2}{170 \div 2} = \dfrac{51 \div 17}{85 \div 17} = \dfrac{3}{5}$

PROBLEM SET 38

1. $\dfrac{\textbf{56 dollars}}{\textbf{7 hours}}, \dfrac{\textbf{7 hours}}{\textbf{56 dollars}}$

$40 \text{ hours} \times \dfrac{\$56}{7 \text{ hours}} = \$320$

2. (a) Range = $9831 - 4016 =$ **5815 tons**

(b) Since the four weights are all different, there is **no mode.**

(c) Median = $\dfrac{6253 + 7132}{2} =$ **6692.5 tons**

(d) Mean = $\dfrac{4016 + 7132 + 9831 + 6253}{4}$

$=$ **6808 tons**

3.

$$\begin{array}{r} 462 \\ \times\ 942 \\ \hline 924 \\ 1848 \\ 4158 \\ \hline \textbf{435,204 points} \end{array}$$

4. $\dfrac{5}{17} \times 34 = \dfrac{170}{17} = \textbf{10}$

5. $3x = 12$

$3(4) = 12$

The solution is **4.**

6. $\dfrac{x}{2} = 57$

$\dfrac{114}{2} = 57$

The solution is **114.**

7. $\dfrac{2}{5} + x = \dfrac{3}{5}$

$\dfrac{2}{5} + \dfrac{1}{5} = \dfrac{3}{5}$

The solution is $\dfrac{\textbf{1}}{\textbf{5}}$.

8. $\dfrac{428}{17} = \textbf{25}\dfrac{\textbf{3}}{\textbf{17}}$

$$17\overline{)428} \quad \begin{array}{r} 25 \\ \underline{34} \\ 88 \\ \underline{85} \\ 3 \end{array}$$

9. $\dfrac{521}{3} = \textbf{173}\dfrac{\textbf{2}}{\textbf{3}}$

$$3\overline{)521} \quad \begin{array}{r} 173 \\ \underline{3} \\ 22 \\ \underline{21} \\ 11 \\ \underline{9} \\ 2 \end{array}$$

10. $50 = 2 \cdot 5 \cdot 5$

$47 = 47$

$120 = 2 \cdot 2 \cdot 2 \cdot 3 \cdot 5$

$\text{LCM }(50, 47, 120) = 2 \cdot 2 \cdot 2 \cdot 3 \cdot 5 \cdot 5 \cdot 47$

$= \textbf{28,200}$

11. $\dfrac{15}{17} - \dfrac{2}{34} = \dfrac{15}{17} - \dfrac{1}{17} = \dfrac{\textbf{14}}{\textbf{17}}$

12. $\dfrac{3}{8} + 1\dfrac{1}{5} - \dfrac{1}{10} = \dfrac{15}{40} + 1\dfrac{8}{40} - \dfrac{4}{40} = \textbf{1}\dfrac{\textbf{19}}{\textbf{40}}$

13. $4 + 13 - 5 \cdot 2 + 7 = 17 - 10 + 7 = \textbf{14}$

14. $2 \cdot 8 - 3 \cdot 1 + 2 \cdot 4 = 16 - 3 + 8 = \textbf{21}$

15. $3\dfrac{1}{5} + 2\dfrac{1}{8} = 3\dfrac{8}{40} + 2\dfrac{5}{40} = \textbf{5}\dfrac{\textbf{13}}{\textbf{40}}$

16. $376\dfrac{4}{5} + 142\dfrac{3}{10} = 376\dfrac{8}{10} + 142\dfrac{3}{10} = 518\dfrac{11}{10}$

$= \textbf{519}\dfrac{\textbf{1}}{\textbf{10}}$

17. $42\frac{7}{8} - 15\frac{3}{4} = 42\frac{7}{8} - 15\frac{6}{8} = 27\frac{1}{8}$

18. $513\frac{11}{20} - 21\frac{4}{5} = 512\frac{31}{20} - 21\frac{16}{20} = 491\frac{15}{20}$

$= 491\frac{3}{4}$

19. $\frac{13.62}{0.05}$

$$\begin{array}{r} 272.4 \\ 5{\overline{\smash{)}1362.0}} \\ \underline{10} \\ 36 \\ \underline{35} \\ 12 \\ \underline{10} \\ 2\,0 \\ \underline{2\,0} \\ 0 \end{array}$$

20. $\frac{14}{16} \cdot \frac{24}{21} \div \frac{2}{3} = \frac{\overset{7}{\cancel{14}}}{\underset{8}{\cancel{16}}} \cdot \frac{\overset{8}{\cancel{24}}}{\underset{7}{\cancel{21}}} \cdot \frac{3}{2} = \frac{\overset{1}{\cancel{7}}}{\underset{1}{\cancel{8}}} \cdot \frac{\overset{1}{\cancel{8}}}{\underset{1}{\cancel{7}}} \cdot \frac{3}{2} = \frac{3}{2}$

21. 5970 is divisible by **2, 3, 5,** and **10.**

22. $241.36 \,\cancel{m} \times \frac{100 \text{ cm}}{1 \,\cancel{m}} = \mathbf{24{,}136 \text{ cm}}$

23. $xz - yz = (7)(3) - (1)(3) = 21 - 3 = \mathbf{18}$

24. $xyz + yz - y = (6)(3)\left(\frac{1}{3}\right) + (3)\left(\frac{1}{3}\right) - (3)$

$= 6 + 1 - 3$

$= \mathbf{4}$

25. $400 \,\cancel{\text{in.}}^2 \times \frac{1 \text{ ft}}{12 \,\cancel{\text{in.}}} \times \frac{1 \text{ ft}}{12 \,\cancel{\text{in.}}}$

$= \frac{400}{(12)(12)} \text{ ft}^2 \approx \mathbf{2.78 \text{ ft}^2}$

26.

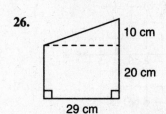

$A = A_{\text{Rectangle}} + A_{\text{Triangle}}$

$= \left[29(20) + \frac{29(10)}{2}\right] \text{cm}^2$

$= [580 + 145] \text{ cm}^2$

$= \mathbf{725 \text{ cm}^2}$

27. $[2(20 + 20) + 2(10 + 10 + 13)] \text{ yd} = \mathbf{146 \text{ yd}}$

28. $92.45 = 92\frac{45}{100} = 92\frac{9}{20}$

29.

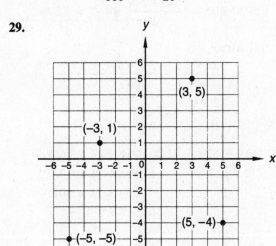

30.

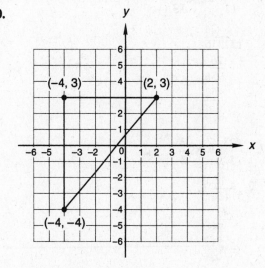

$\text{Area} = \frac{6(7)}{2} = \mathbf{21 \text{ units}^2}$

PROBLEM SET 39

1. $48 \times N = 768$

$$\begin{array}{r} \mathbf{16 \text{ hiding places}} \\ 48{\overline{\smash{)}768}} \\ \underline{48} \\ 288 \\ \underline{288} \\ 0 \end{array}$$

2. $\frac{48 \text{ nuts}}{1 \text{ hour}}, \frac{1 \text{ hour}}{48 \text{ nuts}}$

$20 \,\cancel{\text{hours}} \times \frac{48 \text{ nuts}}{1 \,\cancel{\text{hour}}} = \mathbf{960 \text{ nuts}}$

3. (a) Range = 5.7 − 5.2 = **0.5 m**

(b) Mode = **5.6 m**

(c) Median = $\dfrac{5.4 + 5.6}{2} = \dfrac{11}{2} =$ **5.5 m**

(d) Mean

$= \dfrac{5.6 + 5.4 + 5.3 + 5.7 + 5.2 + 5.6}{6}$

$= \dfrac{32.8}{6} \approx$ **5.47 m**

4. $4165 + 320 + 7142 + 64 + N = 5(2499)$

$11{,}691 + N = 12{,}495$

$N =$ **804**

5. $\dfrac{3}{5} \times 40 = \dfrac{120}{5} =$ **24**

6. $\dfrac{316}{13} = \mathbf{24\dfrac{4}{13}}$

$$\begin{array}{r} 24 \\ 13\overline{)316} \\ 26 \\ \hline 56 \\ 52 \\ \hline 4 \end{array}$$

7. $\dfrac{428}{5} = \mathbf{85\dfrac{3}{5}}$

$$\begin{array}{r} 85 \\ 5\overline{)428} \\ 40 \\ \hline 28 \\ 25 \\ \hline 3 \end{array}$$

8. $8 = 2 \cdot 2 \cdot 2$

$12 = 2 \cdot 2 \cdot 3$

$72 = 2 \cdot 2 \cdot 2 \cdot 3 \cdot 3$

LCM $(8, 12, 72) = 2 \cdot 2 \cdot 2 \cdot 3 \cdot 3 =$ **72**

9. $\dfrac{14}{15} - \dfrac{1}{5} = \dfrac{14}{15} - \dfrac{3}{15} = \mathbf{\dfrac{11}{15}}$

10. $4\dfrac{2}{5} + 3\dfrac{5}{10} = 4\dfrac{4}{10} + 3\dfrac{5}{10} = \mathbf{7\dfrac{9}{10}}$

11. $3 + 2 + 2 \cdot 5 - 3 \cdot 2 = 5 + 10 - 6 =$ **9**

12. $3 - 2 \cdot 1 + 4 \cdot 7 = 3 - 2 + 28 =$ **29**

13. $3\dfrac{2}{7} + 5\dfrac{20}{21} = 3\dfrac{6}{21} + 5\dfrac{20}{21} = 8\dfrac{26}{21} = \mathbf{9\dfrac{5}{21}}$

14. $420\dfrac{3}{5} + 262\dfrac{1}{40} = 420\dfrac{24}{40} + 262\dfrac{1}{40} = \mathbf{682\dfrac{5}{8}}$

15. $43\dfrac{5}{7} - 12\dfrac{2}{14} = 43\dfrac{5}{7} - 12\dfrac{1}{7} = \mathbf{31\dfrac{4}{7}}$

16. $210\dfrac{5}{8} - 17\dfrac{3}{40} = 210\dfrac{25}{40} - 17\dfrac{3}{40} = 193\dfrac{22}{40}$

$= \mathbf{193\dfrac{11}{20}}$

17.

$$\begin{array}{r} 611.2 \\ \times\ 0.0061 \\ \hline 6112 \\ 36672 \\ \hline \mathbf{3.72832} \end{array}$$

18. $\dfrac{0.8136}{0.008}$

$$\begin{array}{r} \mathbf{101.7} \\ 8\overline{)813.6} \\ 8 \\ \hline 013 \\ 8 \\ \hline 5\ 6 \\ 5\ 6 \\ \hline 0 \end{array}$$

19. $\dfrac{18}{20} \cdot \dfrac{24}{16} \div \dfrac{9}{10} = \dfrac{\overset{2}{\cancel{18}}}{\underset{2}{\cancel{20}}} \cdot \dfrac{24}{16} \cdot \dfrac{\overset{1}{\cancel{10}}}{\underset{1}{\cancel{9}}} = \dfrac{\overset{1}{\cancel{2}}}{\underset{1}{\cancel{2}}} \cdot \dfrac{\overset{3}{\cancel{24}}}{\underset{2}{\cancel{16}}}$

$= \mathbf{\dfrac{3}{2}}$

20.

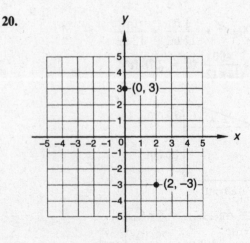

21. $x - 9 = 10$

$x - 9 + 9 = 10 + 9$

$x = \mathbf{19}$

22.
$$x + \frac{4}{5} = \frac{8}{9}$$
$$x + \frac{4}{5} - \frac{4}{5} = \frac{8}{9} - \frac{4}{5}$$
$$x = \frac{40}{45} - \frac{36}{45}$$
$$x = \frac{4}{45}$$

23.
$$x - 3.5 = 4$$
$$x - 3.5 + 3.5 = 4 + 3.5$$
$$x = 7.5$$

24.
$$x + xy + xyz = (1) + (1)(2) + (1)(2)(3)$$
$$= 1 + 2 + 6$$
$$= 9$$

25.
$$xy + x - y = (5)(1) + (5) - (1)$$
$$= 5 + 5 - 1$$
$$= 9$$

26.
$$625.611 \cancel{\text{ cm}} \times \frac{1 \cancel{\text{ m}}}{100 \cancel{\text{ cm}}} \times \frac{1 \text{ km}}{1000 \cancel{\text{ m}}}$$
$$= \frac{625.611}{(100)(1000)} \text{ km} = 0.00625611 \text{ km}$$

27.

(figure: square with triangles, dimensions 20, 20, 15, 9, 11, 18)

$$A = A_{\text{Square}} + A_{\text{Triangle 1}} + A_{\text{Triangle 2}}$$
$$= \left[20(20) + \frac{15(9)}{2} + \frac{18(11)}{2} \right] \text{ft}^2$$
$$= [400 + 67.5 + 99] \text{ ft}^2$$
$$= 566.5 \text{ ft}^2$$

28. $\dfrac{20}{23}$

$$\begin{array}{r} 0.869 \\ 23\overline{)20.000} \\ \underline{18\,4} \\ 1\,60 \\ \underline{1\,38} \\ 220 \\ \underline{207} \\ 13 \end{array}$$

The rounded number is **0.87.**

29. $14.145 = 14\dfrac{145}{1000} = 14\dfrac{29}{200}$

30. (a) **(4, −2)**   (b) **(−4, 4)**

PROBLEM SET 40

1. $\text{Avg} = \dfrac{415 + 478 + 526}{3} = \dfrac{1419}{3}$
$$= \textbf{473 skiers}$$

2. $\dfrac{\textbf{420 cans}}{\textbf{6 minutes}}, \dfrac{\textbf{6 minutes}}{\textbf{420 cans}}$

$$48 \cancel{\text{ hours}} \times \frac{60 \cancel{\text{ minutes}}}{1 \cancel{\text{ hour}}} \times \frac{420 \text{ cans}}{6 \cancel{\text{ minutes}}}$$
$$= \textbf{201,600 cans}$$

3. $32 \times N = 19,440$

$$\begin{array}{r} 607 \\ 32\overline{)19,440} \\ \underline{19\,2} \\ 240 \\ \underline{224} \\ 16 \end{array}$$

Since 16 remain, **608 buses** were needed.

4.

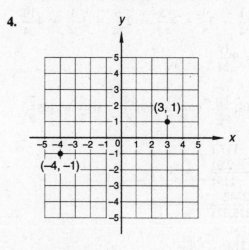

5. $\dfrac{41}{3} = 13\dfrac{2}{3}$

$$\begin{array}{r} 13 \\ 3\overline{)41} \\ \underline{3} \\ 11 \\ \underline{9} \\ 2 \end{array}$$

6. $\dfrac{93}{21} = \dfrac{31}{7} = 4\dfrac{3}{7}$

$$\begin{array}{r} 4 \\ 7\overline{)31} \\ \underline{28} \\ 3 \end{array}$$

7. $27 = 3 \cdot 3 \cdot 3$

$28 = 2 \cdot 2 \cdot 7$

$30 = 2 \cdot 3 \cdot 5$

LCM $(27, 28, 30) = 2 \cdot 2 \cdot 3 \cdot 3 \cdot 3 \cdot 5 \cdot 7$

$\qquad\qquad\qquad = \mathbf{3780}$

8. $\dfrac{5}{16} - \dfrac{1}{8} = \dfrac{5}{16} - \dfrac{2}{16} = \mathbf{\dfrac{3}{16}}$

9. $2\dfrac{1}{5} + 3\dfrac{1}{3} - \dfrac{2}{10} = 2\dfrac{1}{5} + 3\dfrac{1}{3} - \dfrac{1}{5}$

$\qquad = 2 + 3\dfrac{1}{3} = \mathbf{5\dfrac{1}{3}}$

10. $14 - 2 \cdot 3 + 4 \cdot 5 = 14 - 6 + 20 = \mathbf{28}$

11. $2 \cdot 5 - 2 \cdot 2 + 3 = 10 - 4 + 3 = \mathbf{9}$

12. $7\dfrac{1}{8} + 3\dfrac{2}{5} = 7\dfrac{5}{40} + 3\dfrac{16}{40} = \mathbf{10\dfrac{21}{40}}$

13. $674\dfrac{2}{5} - 13\dfrac{7}{10} = 673\dfrac{14}{10} - 13\dfrac{7}{10} = \mathbf{660\dfrac{7}{10}}$

14. $2\dfrac{1}{4} + 3\dfrac{1}{8} + 4\dfrac{5}{12} = 2\dfrac{6}{24} + 3\dfrac{3}{24} + 4\dfrac{10}{24}$

$\qquad = \mathbf{9\dfrac{19}{24}}$

15. $461\dfrac{3}{4} - 65\dfrac{7}{8} = 460\dfrac{14}{8} - 65\dfrac{7}{8} = \mathbf{395\dfrac{7}{8}}$

16.
$$\begin{array}{r} 117.1 \\ \times\ \ 2.01 \\ \hline 1171 \\ 2342\ \ \\ \hline \mathbf{235.371} \end{array}$$

17. $\dfrac{171.6}{0.006}$

$$\begin{array}{r} \mathbf{28,600} \\ 6\overline{)171,600} \\ \underline{12}\ \ \ \ \ \ \\ 51\ \ \ \ \ \\ \underline{48}\ \ \ \ \ \\ 3\,6\ \ \\ \underline{3\,6}\ \ \\ 0\ \ \end{array}$$

18.
$$\begin{array}{r} 6132.810 \\ -\ \ 621.981 \\ \hline \mathbf{5510.829} \end{array}$$

19. $\dfrac{6}{21} \cdot \dfrac{24}{3} \div \dfrac{8}{14} = \dfrac{\overset{2}{\cancel{6}}}{\underset{7}{\cancel{21}}} \cdot \dfrac{\overset{8}{\cancel{24}}}{\underset{1}{\cancel{3}}} \cdot \dfrac{14}{8}$

$\qquad = \dfrac{2}{\underset{1}{\cancel{7}}} \cdot \dfrac{\overset{1}{\cancel{8}}}{1} \cdot \dfrac{\overset{2}{\cancel{14}}}{\underset{1}{\cancel{8}}} = \mathbf{4}$

20. $1\,\cancel{\text{mi}} \times \dfrac{5280\,\cancel{\text{ft}}}{1\,\cancel{\text{mi}}} \times \dfrac{12\,\text{in.}}{1\,\cancel{\text{ft}}}$

$\qquad = 1(5280)12\ \text{in.} = \mathbf{63{,}360\ \text{in.}}$

21. $v - 8 = 9$

$v - 8 + 8 = 9 + 8$

$\qquad v = \mathbf{17}$

22. $x + \dfrac{3}{8} = \dfrac{9}{14}$

$x + \dfrac{3}{8} - \dfrac{3}{8} = \dfrac{9}{14} - \dfrac{3}{8}$

$\qquad\qquad x = \dfrac{36}{56} - \dfrac{21}{56} = \mathbf{\dfrac{15}{56}}$

23. $w - \dfrac{3}{7} = \dfrac{5}{14}$

$w - \dfrac{3}{7} + \dfrac{3}{7} = \dfrac{5}{14} + \dfrac{3}{7}$

$\qquad\qquad w = \dfrac{5}{14} + \dfrac{6}{14} = \mathbf{\dfrac{11}{14}}$

24. $\dfrac{2}{19}x = 6$

$\dfrac{\overset{1}{\cancel{19}}}{\underset{1}{\cancel{2}}} \cdot \dfrac{\overset{1}{\cancel{2}}}{\underset{1}{\cancel{19}}}x = \overset{3}{\cancel{6}} \cdot \dfrac{19}{\underset{1}{\cancel{2}}}$

$\qquad\quad x = \mathbf{57}$

25. $7q = 35$

$\dfrac{7q}{7} = \dfrac{35}{7}$

$\quad q = \mathbf{5}$

26. $8r = \dfrac{1}{9}$

$\dfrac{1}{8} \cdot 8r = \dfrac{1}{9} \cdot \dfrac{1}{8}$

$\qquad r = \mathbf{\dfrac{1}{72}}$

27. $51.785 = 51\dfrac{785}{1000} = \mathbf{51\dfrac{157}{200}}$

28. $xyz + xy + yz - z$

$= \dfrac{1}{3}(12)(2) + \dfrac{1}{3}(12) + 12(2) - 2$

$= 8 + 4 + 24 - 2 = \mathbf{34}$

29.

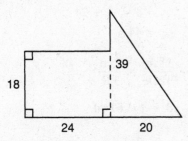

$A = A_{\text{Rectangle}} + A_{\text{Triangle}}$

$= \left[24(18) + \dfrac{20(29)}{2} \right] \text{in.}^2$

$= [432 + 290] \text{ in.}^2$

$= \mathbf{722 \text{ in.}^2}$

30. (a) **41, 43**

(b) **42**

PROBLEM SET 41

1. $1 + 4.5 + 2.9 + 4 + N = 5(3)$

$12.4 + N = 15$

$N = \mathbf{2.6 \text{ in.}}$

2. $\dfrac{\textbf{600 dollars}}{\textbf{30 hanging plants}}, \dfrac{\textbf{30 hanging plants}}{\textbf{600 dollars}}$

70 ~~hanging plants~~ $\times \dfrac{\$600}{30 \text{ ~~hanging plants~~}} = \mathbf{\$1400}$

3. $47 \times N = 2820$

$$
\begin{array}{r}
\mathbf{60 \text{ compartments}} \\
47\overline{)2820} \\
\underline{282} \\
0
\end{array}
$$

4. $\text{Avg} = \dfrac{40 + 20 + 40 + 30 + 20}{5} = \dfrac{150}{5}$

$= \mathbf{30 \text{ in.}}$

5. $\text{Overall average} = \dfrac{(3 \times 38) + (7 \times 31)}{3 + 7}$

$= \dfrac{114 + 217}{10} = \dfrac{331}{10}$

$= \mathbf{33.1 \text{ lb}}$

6. $\text{Overall average} = \dfrac{(12 \times 91) + (4 \times 110)}{12 + 4}$

$= \dfrac{1092 + 440}{16} = \dfrac{1532}{16}$

$= \mathbf{95.75 \text{ cm}}$

7. $\dfrac{82}{5} = \mathbf{16\dfrac{2}{5}}$

$$
\begin{array}{r}
16 \\
5\overline{)82} \\
\underline{5} \\
32 \\
\underline{30} \\
2
\end{array}
$$

8. $\dfrac{121}{15} = \mathbf{8\dfrac{1}{15}}$

$$
\begin{array}{r}
8 \\
15\overline{)121} \\
\underline{120} \\
1
\end{array}
$$

9. $35 = 5 \cdot 7$

$40 = 2 \cdot 2 \cdot 2 \cdot 5$

$120 = 2 \cdot 2 \cdot 2 \cdot 3 \cdot 5$

$\text{LCM } (35, 40, 120) = 2 \cdot 2 \cdot 2 \cdot 3 \cdot 5 \cdot 7$

$= \mathbf{840}$

10. $x + \dfrac{3}{4} = \dfrac{7}{8}$

$x + \dfrac{3}{4} - \dfrac{3}{4} = \dfrac{7}{8} - \dfrac{3}{4}$

$x = \dfrac{7}{8} - \dfrac{6}{8} = \mathbf{\dfrac{1}{8}}$

11. $x - \dfrac{1}{2} = \dfrac{5}{6}$

$x - \dfrac{1}{2} + \dfrac{1}{2} = \dfrac{5}{6} + \dfrac{1}{2}$

$x = \dfrac{5}{6} + \dfrac{3}{6} = \dfrac{8}{6} = \mathbf{\dfrac{4}{3}}$

12. $6x = 18$

$\dfrac{6x}{6} = \dfrac{18}{6}$

$x = \mathbf{3}$

13. $\dfrac{x}{4} = 15$

$4 \cdot \dfrac{x}{4} = 15 \cdot 4$

$x = \mathbf{60}$

14. $5x = 20$

$\dfrac{5x}{5} = \dfrac{20}{5}$

$x = \mathbf{4}$

15. $\dfrac{x}{7} = 5$

$7 \cdot \dfrac{x}{7} = 5 \cdot 7$

$x = \mathbf{35}$

16. $\dfrac{7}{15} - \dfrac{1}{5} = \dfrac{7}{15} - \dfrac{3}{15} = \dfrac{\mathbf{4}}{\mathbf{15}}$

17. $3 \cdot 8 - 2 \cdot 6 + 1 \cdot 7 = 24 - 12 + 7 = \mathbf{19}$

18. $36\dfrac{3}{4} - 21\dfrac{7}{8} = 35\dfrac{14}{8} - 21\dfrac{7}{8} = \mathbf{14}\dfrac{\mathbf{7}}{\mathbf{8}}$

19. $\dfrac{171.6}{0.6}$

$$
\begin{array}{r}
286 \\
6\overline{)1716} \\
\underline{12} \\
51 \\
\underline{48} \\
36 \\
\underline{36} \\
0
\end{array}
$$

20.
$$
\begin{array}{r}
112.4 \\
\times\ 0.071 \\
\hline
1124 \\
7868 \\
\hline
\mathbf{7.9804}
\end{array}
$$

21.
$$
\begin{array}{r}
6781.80 \\
-\ \ 179.89 \\
\hline
\mathbf{6601.91}
\end{array}
$$

22. $\dfrac{16}{18} \cdot \dfrac{24}{36} \div \dfrac{8}{9} = \dfrac{16}{18} \cdot \dfrac{\overset{3}{\cancel{24}}}{\underset{4}{\cancel{36}}} \cdot \dfrac{\overset{1}{\cancel{9}}}{\underset{1}{\cancel{8}}} = \dfrac{\overset{4}{\cancel{16}}}{\underset{6}{\cancel{18}}} \cdot \dfrac{\overset{1}{\cancel{3}}}{\underset{1}{\cancel{4}}} = \dfrac{\overset{2}{\cancel{4}}}{\underset{3}{\cancel{6}}}$

$= \dfrac{\mathbf{2}}{\mathbf{3}}$

23. $5\dfrac{1}{5} + 4\dfrac{4}{9} = 5\dfrac{9}{45} + 4\dfrac{20}{45} = \mathbf{9}\dfrac{\mathbf{29}}{\mathbf{45}}$

24. $x + zx - y = (10) + (2)(10) - (3)$

$= 10 + 20 - 3$

$= \mathbf{27}$

25. $xy + xz + yz = (2)(4) + (2)(6) + (4)(6)$

$= 8 + 12 + 24$

$= \mathbf{44}$

26. $5280 \text{ in.} \times \dfrac{1 \text{ ft}}{12 \text{ in.}} \times \dfrac{1 \text{ yd}}{3 \text{ ft}}$

$= \dfrac{\mathbf{5280}}{\mathbf{(12)(3)}} \text{ yd} \approx \mathbf{146.67 \text{ yd}}$

27.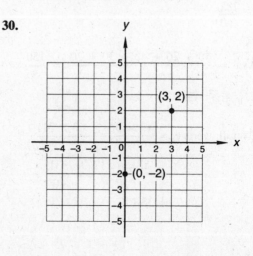

$A = A_{\text{Rectangle}} + A_{\text{Triangle}}$

$= \left[6(31) + \dfrac{15(14)}{2} \right] \text{m}^2$

$= [186 + 105] \text{ m}^2$

$= \mathbf{291 \text{ m}^2}$

28. $[2(30) + 2(10 + 15)] \text{ cm} = \mathbf{110 \text{ cm}}$

29. $\dfrac{19}{24}$

$$
\begin{array}{r}
0.791 \\
24\overline{)19.000} \\
\underline{16\ 8} \\
2\ 20 \\
\underline{2\ 16} \\
40 \\
\underline{24} \\
16
\end{array}
$$

The rounded number is **0.79**.

30.

PROBLEM SET 42

1. Avg = $\dfrac{4(800) + 6(600) + 4(400)}{4 + 6 + 4}$

= $\dfrac{3200 + 3600 + 1600}{14}$

= $\dfrac{8400}{14}$ = **600 units per hour**

2. $\dfrac{89 \text{ points}}{1 \text{ test}}, \dfrac{1 \text{ test}}{89 \text{ points}}$

22 tests $\times \dfrac{89 \text{ points}}{1 \text{ test}}$ = **1958 points**

3. Avg = $\dfrac{(70 \times \$40{,}000) + (30 \times \$30{,}000)}{100}$

= $\dfrac{\$3{,}700{,}000}{100}$ = **\$37,000**

4. (a) $\dfrac{4 \text{ quarts}}{1 \text{ gallon}}, \dfrac{1 \text{ gallon}}{4 \text{ quarts}}$

(b) 1,000,000 gallons $\times \dfrac{4 \text{ quarts}}{1 \text{ gallon}}$

= **4,000,000 quarts**

5.

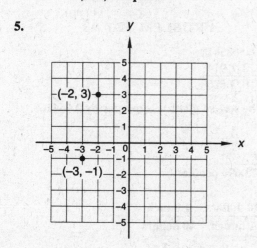

6. $\dfrac{6}{14} \times 126 = \dfrac{756}{14}$ = **54**

7. $\dfrac{94}{7} = 13\dfrac{3}{7}$

$$\begin{array}{r} 13 \\ 7\overline{)94} \\ 7 \\ \hline 24 \\ 21 \\ \hline 3 \end{array}$$

8. $\dfrac{289}{12} = 24\dfrac{1}{12}$

$$\begin{array}{r} 24 \\ 12\overline{)289} \\ 24 \\ \hline 49 \\ 48 \\ \hline 1 \end{array}$$

9. $x + \dfrac{5}{8} = \dfrac{11}{16}$

$x + \dfrac{5}{8} - \dfrac{5}{8} = \dfrac{11}{16} - \dfrac{5}{8}$

$x = \dfrac{11}{16} - \dfrac{10}{16} = \dfrac{1}{16}$

10. $x - \dfrac{1}{4} = \dfrac{5}{12}$

$x - \dfrac{1}{4} + \dfrac{1}{4} = \dfrac{5}{12} + \dfrac{1}{4}$

$x = \dfrac{5}{12} + \dfrac{3}{12} = \dfrac{8}{12} = \dfrac{2}{3}$

11. $14x = 56$

$\dfrac{14x}{14} = \dfrac{56}{14}$

$x = 4$

12. $\dfrac{x}{4} = 9$

$4 \cdot \dfrac{x}{4} = 9 \cdot 4$

$x = 36$

13. $9x = 81$

$\dfrac{9x}{9} = \dfrac{81}{9}$

$x = 9$

14. $\dfrac{x}{13} = 6$

$13 \cdot \dfrac{x}{13} = 6 \cdot 13$

$x = 78$

15. $\dfrac{11}{12} - \dfrac{5}{6} = \dfrac{11}{12} - \dfrac{10}{12} = \dfrac{1}{12}$

16. $19\dfrac{1}{8} - 8\dfrac{3}{4} = 18\dfrac{9}{8} - 8\dfrac{6}{8} = 10\dfrac{3}{8}$

17. $\dfrac{195.8}{1.1}$

$$\begin{array}{r} 178 \\ 11\overline{)1958} \\ \underline{11} \\ 85 \\ \underline{77} \\ 88 \\ \underline{88} \\ 0 \end{array}$$

18. $\dfrac{14}{16} \cdot \dfrac{6}{32} \div \dfrac{3}{4} = \dfrac{14}{16} \cdot \dfrac{\overset{2}{\cancel{6}}}{\underset{8}{\cancel{32}}} \cdot \dfrac{\overset{1}{\cancel{4}}}{\underset{1}{\cancel{3}}} = \dfrac{\overset{7}{\cancel{14}}}{\underset{8}{\cancel{16}}} \cdot \dfrac{\overset{1}{\cancel{2}}}{\underset{4}{\cancel{8}}} = \dfrac{7}{32}$

19.
$$\begin{array}{r} 9876.50 \\ - \ 643.99 \\ \hline 9232.51 \end{array}$$

20.
$$\begin{array}{r} 163.09 \\ \times \ 0.063 \\ \hline 48927 \\ 97854 \\ \hline 10.27467 \end{array}$$

21. $5\dfrac{2}{9} + 3\dfrac{5}{6} = 5\dfrac{4}{18} + 3\dfrac{15}{18} = 8\dfrac{19}{18} = \mathbf{9\dfrac{1}{18}}$

22. $5(4) \div 2 \times (9 - 4) = 20 \div 2 \times 5$
$= 10 \times 5 = \mathbf{50}$

23. $[(15 - 3)(2 + 3) - 15] + 2$
$= [(12)(5) - 15] + 2 = [60 - 15] + 2$
$= 45 + 2 = \mathbf{47}$

24. $6[8 - (10 - 4) \div 3] - 28 \div 4$
$= 6[8 - 6 \div 3] - 28 \div 4$
$= 6[8 - 2] - 28 \div 4 = 6[6] - 28 \div 4$
$= 36 - 7 = \mathbf{29}$

25. $13 + (2 - 1)[5 + (7 - 2)] = 13 + (1)[5 + 5]$
$= 13 + 10 = \mathbf{23}$

26. $mn + zy - y = (4)(3) + (2)(1) - (1)$
$= 12 + 2 - 1$
$= \mathbf{13}$

27. $ax + bx - ab = (2)(2) + (3)(2) - (2)(3)$
$= 4 + 6 - 6$
$= \mathbf{4}$

28. $628 \ \cancel{km} \times \dfrac{1000 \ \cancel{m}}{1 \ \cancel{km}} \times \dfrac{100 \ cm}{1 \ \cancel{m}}$
$= \mathbf{628(1000)(100) \ cm} = \mathbf{62,800,000 \ cm}$

29.

$A = A_{\text{Triangle}} + A_{\text{Rectangle}} + A_{\text{Rectangle}}$

$= \left[\dfrac{54(15)}{2} + 54(5) + 23(20) \right] ft^2$

$= [405 + 270 + 460] \ ft^2$

$= \mathbf{1135 \ ft^2}$

30. $\dfrac{5}{8}$

$$\begin{array}{r} 0.625 \\ 8\overline{)5.000} \\ \underline{4 \ 8} \\ 20 \\ \underline{16} \\ 40 \\ \underline{40} \\ 0 \end{array}$$

PROBLEM SET 43

1.
$$\begin{array}{r} 0.085410 \\ - \ 0.019142 \\ \hline 0.066268 \end{array}$$

The second guess was greater by 0.066268.

2.
$$\begin{array}{r} 1420 \\ \times \ \ 80 \\ \hline \mathbf{113,600} \textbf{ people} \end{array}$$

3. $\dfrac{40 \text{ dollars}}{1 \text{ bunch}} , \dfrac{1 \text{ bunch}}{40 \text{ dollars}}$

$100 \ \cancel{\text{bunches}} \times \dfrac{\$40}{1 \ \cancel{\text{bunch}}} = \mathbf{\$4000}$

4. $42(60) + 20(30) + 10(50)$
$= 2520 + 600 + 500 = \mathbf{3620} \textbf{ steps}$

5. First $= 40$ astronauts
Second $= 2(\text{First}) = 2(40) = 80$ astronauts
Third $= 4(\text{Second}) = 4(80) = 320$ astronauts
Third $- (\text{First} + \text{Second}) = 320 - (40 + 80)$
$= 320 - 120 = \mathbf{200} \textbf{ astronauts}$

6.

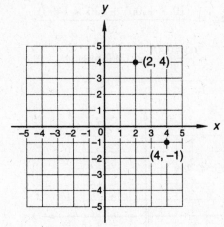

7. $\dfrac{271}{15} = 18\dfrac{1}{15}$

$$\begin{array}{r} 18 \\ 15\overline{)271} \\ \underline{15} \\ 121 \\ \underline{120} \\ 1 \end{array}$$

8. $12 = 2 \cdot 2 \cdot 3$

$22 = 2 \cdot 11$

$40 = 2 \cdot 2 \cdot 2 \cdot 5$

LCM $(12, 22, 40) = 2 \cdot 2 \cdot 2 \cdot 3 \cdot 5 \cdot 11$

$\qquad\qquad\qquad\quad = \mathbf{1320}$

9. $\dfrac{2}{3}x = 3$

$\dfrac{3}{2} \cdot \dfrac{2}{3}x = 3 \cdot \dfrac{3}{2}$

$x = \dfrac{\mathbf{9}}{\mathbf{2}}$

10. $\dfrac{3}{7}x = \dfrac{2}{9}$

$\dfrac{7}{3} \cdot \dfrac{3}{7}x = \dfrac{2}{9} \cdot \dfrac{7}{3}$

$x = \dfrac{\mathbf{14}}{\mathbf{27}}$

11. $\dfrac{x}{5} = 60$

$5 \cdot \dfrac{x}{5} = 60 \cdot 5$

$x = \mathbf{300}$

12. $5x = 60$

$\dfrac{5x}{5} = \dfrac{60}{5}$

$x = \mathbf{12}$

13. $x + \dfrac{1}{8} = \dfrac{1}{2}$

$x + \dfrac{1}{8} - \dfrac{1}{8} = \dfrac{1}{2} - \dfrac{1}{8}$

$x = \dfrac{4}{8} - \dfrac{1}{8} = \dfrac{\mathbf{3}}{\mathbf{8}}$

14. $x - \dfrac{1}{4} = \dfrac{1}{2}$

$x - \dfrac{1}{4} + \dfrac{1}{4} = \dfrac{1}{2} + \dfrac{1}{4}$

$x = \dfrac{2}{4} + \dfrac{1}{4} = \dfrac{\mathbf{3}}{\mathbf{4}}$

15. $61\dfrac{11}{15} - 15\dfrac{3}{5} = 61\dfrac{11}{15} - 15\dfrac{9}{15} = \mathbf{46\dfrac{2}{15}}$

16. $\dfrac{5}{8} + 2\dfrac{1}{4} - \dfrac{1}{10} = \dfrac{25}{40} + 2\dfrac{10}{40} - \dfrac{4}{40} = \mathbf{2\dfrac{31}{40}}$

17. $\begin{array}{r} 2362.80 \\ -\ \ 189.87 \\ \hline \mathbf{2172.93} \end{array}$

18. $\dfrac{612.5}{0.07}$

$$\begin{array}{r} \mathbf{8750} \\ 7\overline{)61{,}250} \\ \underline{56} \\ 5\ 2 \\ \underline{4\ 9} \\ 35 \\ \underline{35} \\ 0 \end{array}$$

19. $\dfrac{12}{16} \cdot \dfrac{12}{21} \div \dfrac{6}{14} = \dfrac{\cancel{12}}{\cancel{16}} \cdot \dfrac{\cancel{12}}{\cancel{21}} \cdot \dfrac{\cancel{14}}{\cancel{6}} = \dfrac{\cancel{3}}{\cancel{4}} \cdot \dfrac{\cancel{2}}{\cancel{3}} \cdot \dfrac{\cancel{2}}{1}$

$= \mathbf{1}$

20. $7 + (6 \cdot 3) \div 2 + 3(11 - 2)$

$= 7 + 18 \div 2 + 3(9) = 7 + 9 + 27 = \mathbf{43}$

21. $6(17 - 5) - (15 - 13 + 4)4 + 6$

$= 6(12) - (6)4 + 6 = 72 - 24 + 6 = \mathbf{54}$

22. $41 - 4[(6 - 2) + 3(5 - 3)]$
$= 41 - 4[4 + 3(2)] = 41 - 4[10]$
$= 41 - 40 = \mathbf{1}$

23. $2\dfrac{1}{3} \times \dfrac{2}{7} = \dfrac{\overset{1}{\cancel{7}}}{3} \times \dfrac{2}{\cancel{7}} = \dfrac{\mathbf{2}}{\mathbf{3}}$

24. $5\dfrac{5}{7} \times 1\dfrac{1}{2} \times 3\dfrac{2}{3} = \dfrac{\overset{20}{\cancel{40}}}{7} \times \dfrac{\overset{1}{\cancel{3}}}{\underset{1}{\cancel{2}}} \times \dfrac{11}{\underset{1}{\cancel{3}}} = \dfrac{\mathbf{220}}{\mathbf{7}}$

25. $\dfrac{9\dfrac{3}{8}}{5\dfrac{3}{4}} = \dfrac{\dfrac{75}{8}}{\dfrac{23}{4}} = \dfrac{75}{\underset{2}{8}} \cdot \dfrac{\overset{1}{\cancel{4}}}{23} = \dfrac{\mathbf{75}}{\mathbf{46}}$

26. $4\dfrac{1}{6} \div 8\dfrac{3}{4} \times 2\dfrac{1}{6} \div 6\dfrac{1}{2}$
$= \dfrac{25}{6} \div \dfrac{35}{4} \times \dfrac{13}{6} \div \dfrac{13}{2}$
$= \dfrac{\overset{5}{\cancel{25}}}{\underset{3}{\cancel{6}}} \times \dfrac{\overset{2}{\cancel{4}}}{\underset{7}{\cancel{35}}} \times \dfrac{\overset{1}{\cancel{13}}}{\underset{3}{\cancel{6}}} \times \dfrac{\overset{1}{\cancel{2}}}{\underset{1}{\cancel{13}}} = \dfrac{\mathbf{10}}{\mathbf{63}}$

27. $xy + yz - z = \left(\dfrac{1}{5}\right)(20) + (20)(3) - (3)$
$= 4 + 60 - 3$
$= \mathbf{61}$

28. $xyz - x = (6)(16)\left(\dfrac{1}{12}\right) - (6) = 8 - 6 = \mathbf{2}$

29. $[2(10 + 10) + 2(0.5)(100)] \text{ cm} = \mathbf{140\ cm}$

30. $59.850 = 59\dfrac{850}{1000} = \mathbf{59\dfrac{17}{20}}$

PROBLEM SET 44

1. $\dfrac{78 \text{ pots}}{312 \text{ dollars}}, \ \dfrac{312 \text{ dollars}}{78 \text{ pots}}$

$400 \ \cancel{\text{pots}} \times \dfrac{\$312}{78 \ \cancel{\text{pots}}} = \mathbf{\$1600}$

2. $\text{Avg} = \dfrac{280 + 220 + 260 + 200}{4} = \dfrac{960}{4}$
$= \mathbf{240\ tons}$

3. $\begin{array}{r} 22{,}000{,}040 \\ -\ 14{,}865{,}932 \\ \hline \mathbf{7{,}134{,}108\ rabbits} \end{array}$

4. $\text{Avg} = \dfrac{(10 \times 1100) + (20 \times 1400)}{30}$

$= \dfrac{39{,}000}{30} = \mathbf{1300\ pills}$

5.

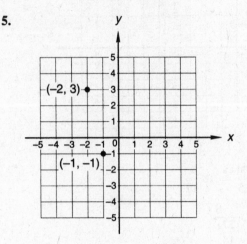

6. $7\dfrac{5}{16} = \dfrac{16 \times 7 + 5}{16} = \dfrac{\mathbf{117}}{\mathbf{16}}$

7. $14 = 2 \cdot 7$
$21 = 3 \cdot 7$
$49 = 7 \cdot 7$
$\text{LCM}(14, 21, 49) = 2 \cdot 3 \cdot 7 \cdot 7 = \mathbf{294}$

8. $\dfrac{7}{8}x = 4$

$\dfrac{8}{7} \cdot \dfrac{7}{8}x = 4 \cdot \dfrac{8}{7}$

$x = \dfrac{\mathbf{32}}{\mathbf{7}}$

9. $4x = 80$
$\dfrac{4x}{4} = \dfrac{80}{4}$
$x = \mathbf{20}$

10. $\dfrac{x}{7} = 84$

$7 \cdot \dfrac{x}{7} = 84 \cdot 7$

$x = \mathbf{588}$

11. $x - \dfrac{3}{7} = \dfrac{9}{14}$

$x - \dfrac{3}{7} + \dfrac{3}{7} = \dfrac{9}{14} + \dfrac{3}{7}$

$x = \dfrac{9}{14} + \dfrac{6}{14} = \dfrac{\mathbf{15}}{\mathbf{14}}$

12.
$$x + \frac{3}{11} = \frac{9}{22}$$
$$x + \frac{3}{11} - \frac{3}{11} = \frac{9}{22} - \frac{3}{11}$$
$$x = \frac{9}{22} - \frac{6}{22} = \mathbf{\frac{3}{22}}$$

13. $3 + (2 \cdot 5)3 + 4(8 \div 2)$
$= 3 + (10)3 + 4(4)$
$= 3 + 30 + 16 = \mathbf{49}$

14. $10 + 2[5(6 - 4) - 3] = 10 + 2[5(2) - 3]$
$= 10 + 2[7] = 10 + 14 = \mathbf{24}$

15. Since $6 \cdot 6 = 36$, $\sqrt{36} = \mathbf{6}$

16. Since $6 \cdot 6 \cdot 6 = 216$, $\sqrt[3]{216} = \mathbf{6}$

17. $5\frac{1}{2} \times \frac{12}{7} = \frac{11}{\underset{1}{\cancel{2}}} \times \frac{\overset{6}{\cancel{12}}}{7} = \mathbf{\frac{66}{7}}$

18. $3\frac{1}{5} \times 1\frac{3}{8} \times 1\frac{1}{11} = \frac{16}{5} \times \frac{\overset{1}{\cancel{11}}}{\underset{1}{\cancel{8}}} \times \frac{\overset{2}{\cancel{12}}}{\underset{1}{\cancel{11}}} = \mathbf{\frac{24}{5}}$

19. $\dfrac{1\frac{2}{5}}{2\frac{1}{7}} = \dfrac{\frac{7}{5}}{\frac{15}{7}} = \frac{7}{5} \cdot \frac{7}{15} = \mathbf{\frac{49}{75}}$

20. $1\frac{1}{6} \div 3 \cdot 6\frac{3}{4} \div 1\frac{1}{2} = \frac{7}{6} \div \frac{3}{1} \cdot \frac{27}{4} \div \frac{3}{2}$
$= \frac{7}{6} \cdot \frac{1}{3} \cdot \frac{\overset{9}{\cancel{27}}}{\underset{2}{\cancel{4}}} \cdot \frac{\overset{1}{\cancel{2}}}{\underset{1}{\cancel{3}}} = \frac{7}{\underset{2}{\cancel{18}}} \cdot \frac{\overset{1}{\cancel{9}}}{2} = \mathbf{\frac{7}{4}}$

21. $\frac{5}{6} + 1\frac{5}{12} - \frac{3}{4} = \frac{10}{12} + 1\frac{5}{12} - \frac{9}{12} = 1\frac{6}{12} = \mathbf{1\frac{1}{2}}$

22. $17\frac{7}{12} - 12\frac{3}{4} = 16\frac{19}{12} - 12\frac{9}{12} = 4\frac{10}{12} = \mathbf{4\frac{5}{6}}$

23.
$$\begin{array}{r} 117.890 \\ -\,112.341 \\ \hline \mathbf{5.549} \end{array}$$

24.
$$\frac{7812}{0.003}$$
$$\begin{array}{r} \mathbf{2,604,000} \\ 3\overline{)7,812,000} \\ \underline{6} \\ 1\,8 \\ \underline{1\,8} \\ 0\,1\,2 \\ \underline{1\,2} \\ 0 \end{array}$$

25. $14 + 3\left[(5 - 1)(8 - 6) - \sqrt[3]{8}\right] + 1$
$= 14 + 3[(4)(2) - 2] + 1 = 14 + 3[6] + 1$
$= 14 + 18 + 1 = \mathbf{33}$

26. $xz + yz - xy = (4)(8) + \left(\frac{1}{2}\right)(8) - (4)\left(\frac{1}{2}\right)$
$= 32 + 4 - 2$
$= \mathbf{34}$

27. $3\,\cancel{\text{mi}} \times \dfrac{5280\,\text{ft}}{1\,\cancel{\text{mi}}} \times \dfrac{12\,\text{in.}}{1\,\cancel{\text{ft}}}$
$= 3(5280)(12)\,\text{in.} = \mathbf{190{,}080\ in.}$

28. $A = A_{\text{Triangle}} + A_{\text{Rectangle}} + A_{\text{Triangle}}$
$= \left[\dfrac{29(7)}{2} + 49(18) + \dfrac{16(18)}{2}\right]\text{ft}^2$
$= [101.5 + 882 + 144]\,\text{ft}^2$
$= \mathbf{1127.5\ ft^2}$

29. $1{,}000{,}000 = 10 \cdot 10 \cdot 10 \cdot 10 \cdot 10 \cdot 10$
We must use 10 as a factor **6 times.**

30. $5\,\cancel{\text{m}^2} \times \dfrac{100\,\text{cm}}{1\,\cancel{\text{m}}} \times \dfrac{100\,\text{cm}}{1\,\cancel{\text{m}}}$
$= 5(100)(100)\,\text{cm}^2 = \mathbf{50{,}000\ cm^2}$

PROBLEM SET 45

1. $\text{Avg} = \dfrac{(40 \times 116) + (60 \times 216)}{100}$
$= \dfrac{17{,}600}{100} = \mathbf{176\ teachers}$

2. $12(12) + 10(2)(12) + 3(12) = 144 + 240 + 36$
$= \mathbf{420\ times}$

3. $\dfrac{1\text{ avocado}}{79\text{ cents}}, \dfrac{79\text{ cents}}{1\text{ avocado}}$

$2370\,\cancel{\text{cents}} \times \dfrac{1\text{ avocado}}{79\,\cancel{\text{cents}}} = \mathbf{30\ avocados}$

4.
$$
\begin{array}{r}
16{,}319.060 \text{ cm} \\
-\quad 309.012 \text{ cm} \\
\hline
\mathbf{16{,}010.048 \text{ cm}}
\end{array}
$$

5.

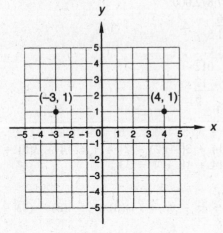

6. $A_{\text{Base}} = \dfrac{5 \text{ ft} \times 4 \text{ ft}}{2} = 10 \text{ ft}^2$

$V = A_{\text{Base}} \times \text{height} = 10 \text{ ft}^2 \times 3 \text{ ft} = \mathbf{30 \text{ ft}^3}$

7. $A_{\text{Base}} = A_{\text{Rectangle}} + A_{\text{Triangle}}$

$\quad = \left[(6 \times 12) + \left(\dfrac{9 \times 8}{2} \right) \right] \text{ft}^2$

$\quad = [72 + 36] \text{ ft}^2$

$\quad = 108 \text{ ft}^2$

$V = A_{\text{Base}} \times \text{height} = 108 \text{ ft}^2 \times 7 \text{ ft} = \mathbf{756 \text{ ft}^3}$

8.
$$\frac{3}{8}x = 1$$
$$\frac{8}{3} \cdot \frac{3}{8}x = 1 \cdot \frac{8}{3}$$
$$x = \frac{8}{3}$$

9.
$$5x = 90$$
$$\frac{5x}{5} = \frac{90}{5}$$
$$x = 18$$

10.
$$\frac{x}{4} = 35$$
$$4 \cdot \frac{x}{4} = 35 \cdot 4$$
$$x = 140$$

11.
$$x - \frac{2}{5} = \frac{1}{100}$$
$$x - \frac{2}{5} + \frac{2}{5} = \frac{1}{100} + \frac{2}{5}$$
$$x = \frac{1}{100} + \frac{40}{100} = \mathbf{\frac{41}{100}}$$

12.
$$x + \frac{3}{13} = \frac{9}{26}$$
$$x + \frac{3}{13} - \frac{3}{13} = \frac{9}{26} - \frac{3}{13}$$
$$x = \frac{9}{26} - \frac{6}{26} = \mathbf{\frac{3}{26}}$$

13. $5 + [(3 \cdot 2)3 - (10 - 2)] = 5 + [(6)3 - 8]$
$= 5 + [18 - 8] = 5 + 10 = \mathbf{15}$

14. $5 + \sqrt[4]{16}(1^2 + 5) + 15 \div 3$
$= 5 + 2(1 + 5) + 15 \div 3$
$= 5 + 2(6) + 15 \div 3 = 5 + 12 + 5 = \mathbf{22}$

15. $4\dfrac{1}{3} \times \dfrac{7}{5} = \dfrac{13}{3} \times \dfrac{7}{5} = \mathbf{\dfrac{91}{15}}$

16. $2\dfrac{1}{4} \times 1\dfrac{2}{5} \times 1\dfrac{3}{14} = \dfrac{9}{4} \times \dfrac{\overset{1}{\cancel{7}}}{5} \times \dfrac{17}{\underset{2}{\cancel{14}}} = \mathbf{\dfrac{153}{40}}$

17. $\dfrac{2\dfrac{1}{3}}{3\dfrac{2}{7}} = \dfrac{\dfrac{7}{3}}{\dfrac{23}{7}} = \dfrac{7}{3} \cdot \dfrac{7}{23} = \mathbf{\dfrac{49}{69}}$

18. $\dfrac{6}{5} \div \dfrac{3}{2} \cdot \dfrac{15}{8} \div \dfrac{9}{8} = \dfrac{6}{5} \cdot \dfrac{2}{3} \cdot \dfrac{\overset{5}{\cancel{15}}}{\underset{1}{\cancel{8}}} \cdot \dfrac{\overset{1}{\cancel{8}}}{\underset{3}{\cancel{9}}}$

$= \dfrac{\overset{2}{\cancel{6}}}{\underset{1}{\cancel{5}}} \cdot \dfrac{2}{3} \cdot \dfrac{\cancel{5}}{1} \cdot \dfrac{1}{3} = \mathbf{\dfrac{4}{3}}$

19. $\dfrac{7}{8} + 1\dfrac{3}{16} - \dfrac{1}{2} = \dfrac{14}{16} + 1\dfrac{3}{16} - \dfrac{8}{16} = \mathbf{1\dfrac{9}{16}}$

20. $14\dfrac{3}{16} - 1\dfrac{1}{2} = 13\dfrac{19}{16} - 1\dfrac{8}{16} = \mathbf{12\dfrac{11}{16}}$

21. $12\dfrac{4}{9} + 10\dfrac{7}{30} = 12\dfrac{40}{90} + 10\dfrac{21}{90} = \mathbf{22\dfrac{61}{90}}$

22.
$$
\begin{array}{r}
23.04 \\
\times\; 0.00012 \\
\hline
4608 \\
2304 \\
\hline
\mathbf{0.0027648}
\end{array}
$$

23.
$$\frac{9663}{0.0006}$$

$$\begin{array}{r} \mathbf{16{,}105{,}000} \\ 6\overline{)96{,}630{,}000} \\ \underline{6} \\ 36 \\ \underline{36} \\ 0\,6 \\ \underline{6} \\ 030 \\ \underline{30} \\ 0 \end{array}$$

24. $km + zm - kz = (3)(6) + \left(\dfrac{1}{3}\right)(6) - (3)\left(\dfrac{1}{3}\right)$

$$= 18 + 2 - 1$$
$$= \mathbf{19}$$

25. $17\,\cancel{\text{km}} \times \dfrac{1000\,\cancel{\text{m}}}{1\,\cancel{\text{km}}} \times \dfrac{100\,\text{cm}}{1\,\cancel{\text{m}}}$

$= 17(1000)(100)\,\text{cm} = \mathbf{1{,}700{,}000\ cm}$

26. $A = A_{\text{Triangle}} + A_{\text{Rectangle}} + A_{\text{Triangle}}$

$= \left[\dfrac{6(24)}{2} + 8(24) + \dfrac{12(16)}{2}\right]\text{m}^2$

$= [72 + 192 + 96]\,\text{m}^2 = \mathbf{360\ m^2}$

27. $100{,}000{,}000 = 10 \cdot 10 \cdot 10 \cdot 10 \cdot 10 \cdot 10$
$\cdot\ 10 \cdot 10$

We must use 10 as a factor **8 times.**

28. (a) **43, 47, 53, 59, 61**

(b) **42, 45, 48, 51, 54, 57, 60, 63, 66**

29. $120\,\cancel{\text{mi}}^2 \times \dfrac{5280\,\text{ft}}{1\,\cancel{\text{mi}}} \times \dfrac{5280\,\text{ft}}{1\,\cancel{\text{mi}}}$

$= \mathbf{120(5280)(5280)\ ft^2 = 3{,}345{,}408{,}000\ ft^2}$

30. $\$20 + \$10 - \$24 = \mathbf{\$6}$

Problem Set 46

1. $\dfrac{\textbf{4 large ones}}{\textbf{40 dollars}}, \dfrac{\textbf{40 dollars}}{\textbf{4 large ones}}$

$120\,\cancel{\text{large ones}} \times \dfrac{\$40}{4\,\cancel{\text{large ones}}} = \mathbf{\$1200}$

2. $4(\$0.50) + 9(\$5.50) + 4(\$7.50)$
$= \$2.00 + \$49.50 + \$30.00 = \mathbf{\$81.50}$

3. $\text{Avg} = \dfrac{(10 \times 26) + (90 \times 30)}{100}$

$= \dfrac{2960}{100} = \mathbf{29.6\ minutes}$

4. $N \times 42 = 5040$

$$\begin{array}{r} \mathbf{120\ items} \\ 42\overline{)5040} \\ \underline{42} \\ 84 \\ \underline{84} \\ 0 \end{array}$$

5. $18 = 2 \cdot 3 \cdot 3$
$42 = 2 \cdot 3 \cdot 7$
$50 = 2 \cdot 5 \cdot 5$
$\text{LCM}\,(18, 42, 50) = 2 \cdot 3 \cdot 3 \cdot 5 \cdot 5 \cdot 7$
$= \mathbf{3150}$

6. $\dfrac{5}{3}x = 20$

$\dfrac{3}{5} \cdot \dfrac{5}{3}x = \overset{4}{\cancel{20}} \cdot \dfrac{3}{\underset{1}{\cancel{5}}}$

$x = \mathbf{12}$

7. $4x = 2$

$\dfrac{4x}{4} = \dfrac{2}{4}$

$x = \dfrac{\mathbf{1}}{\mathbf{2}}$

8. $\dfrac{x}{4} = 7$

$4 \cdot \dfrac{x}{4} = 7 \cdot 4$

$x = \mathbf{28}$

9. $x - 7 = 2$
$x - 7 + 7 = 2 + 7$
$x = \mathbf{9}$

10. $3^3 + 4^4 + 2^5 = 27 + 256 + 32 = \mathbf{315}$

11. $\sqrt[5]{243} = \mathbf{3}$ since $3 \cdot 3 \cdot 3 \cdot 3 \cdot 3 = 243$

12. $\sqrt[3]{125} = \mathbf{5}$ since $5 \cdot 5 \cdot 5 = 125$

13. $\dfrac{3}{4} + 7\dfrac{11}{12} - \dfrac{5}{6} = \dfrac{9}{12} + 7\dfrac{11}{12} - \dfrac{10}{12}$

$= 7\dfrac{10}{12} = \mathbf{7}\dfrac{\mathbf{5}}{\mathbf{6}}$

14. $321\dfrac{7}{12} - 123\dfrac{3}{4} = 320\dfrac{19}{12} - 123\dfrac{9}{12}$

$= 197\dfrac{10}{12} = \mathbf{197}\dfrac{\mathbf{5}}{\mathbf{6}}$

15. $(6)(3) + 4(2 + 12) = 18 + 4(14) = 18 + 56$
$= \mathbf{74}$

16.
$$\begin{array}{r} 111.8 \\ \times\ 0.007 \\ \hline \mathbf{0.7826} \end{array}$$

17. $\dfrac{179.32}{0.004}$

$$\begin{array}{r} \mathbf{44{,}830} \\ 4\overline{)179{,}320} \\ \underline{16} \\ 19 \\ \underline{16} \\ 3\,3 \\ \underline{3\,2} \\ 12 \\ \underline{12} \\ 0 \end{array}$$

18. $7\dfrac{1}{8} \div 2\dfrac{1}{4} \times 3\dfrac{1}{6} = \dfrac{\overset{19}{\cancel{57}}}{\underset{2}{8}} \times \dfrac{\overset{1}{\cancel{4}}}{\underset{3}{\cancel{9}}} \times \dfrac{19}{6} = \dfrac{\mathbf{361}}{\mathbf{36}}$

19. $2\dfrac{1}{4} \times 6\dfrac{3}{4} \div 3\dfrac{1}{8} = \dfrac{9}{\underset{1}{\cancel{4}}} \times \dfrac{27}{\underset{2}{\cancel{4}}} \times \dfrac{\overset{\overset{1}{2}}{\cancel{8}}}{25} = \dfrac{\mathbf{243}}{\mathbf{50}}$

20. $\dfrac{7\frac{2}{3}}{6\frac{5}{6}} = \dfrac{\frac{23}{3}}{\frac{41}{6}} = \dfrac{23}{\underset{1}{\cancel{3}}} \cdot \dfrac{\overset{2}{\cancel{6}}}{41} = \dfrac{\mathbf{46}}{\mathbf{41}}$

21. $\dfrac{4}{5} - \dfrac{1}{2} \div \dfrac{5}{6} = \dfrac{4}{5} - \dfrac{1}{\underset{1}{\cancel{2}}} \cdot \dfrac{\overset{3}{\cancel{6}}}{5} = \dfrac{4}{5} - \dfrac{3}{5} = \dfrac{\mathbf{1}}{\mathbf{5}}$

22. $5(9 - 7 + 4) + 3^2 - \sqrt[3]{27} + 3 \cdot 2$
$= 5(6) + 9 - 3 + 6 = \mathbf{42}$

23. $\dfrac{3}{5}\left(\dfrac{7}{10} - \dfrac{1}{3}\right) = \dfrac{3}{5}\left(\dfrac{21}{30} - \dfrac{10}{30}\right) = \dfrac{\overset{1}{\cancel{3}}}{5} \cdot \dfrac{11}{\underset{10}{\cancel{30}}} = \dfrac{\mathbf{11}}{\mathbf{50}}$

24. $1\dfrac{1}{2} + 7\dfrac{1}{2} \times \dfrac{2}{9} = \dfrac{3}{2} + \dfrac{\overset{5}{\cancel{15}}}{\underset{1}{\cancel{2}}} \times \dfrac{\overset{1}{\cancel{2}}}{\underset{3}{\cancel{9}}} = \dfrac{3}{2} + \dfrac{5}{3}$

$= \dfrac{9}{6} + \dfrac{10}{6} = \dfrac{\mathbf{19}}{\mathbf{6}}$

25. $xyz + y + x + z$
$= (3)(4)(5) + (4) + (3) + (5)$
$= 60 + 4 + 3 + 5$
$= \mathbf{72}$

26. $P = [2(32) + 2(5 + 5 + 10)]$ cm
$= [64 + 40]$ cm
$= 104 \,\cancel{\text{cm}} \times \dfrac{1 \text{ m}}{100 \,\cancel{\text{cm}}} = \mathbf{1.04\ m}$

27.

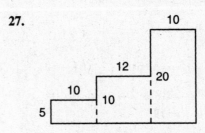

$A_{\text{Base}} = [(5 \times 10) + (10 \times 12)$
$\qquad\quad + (20 \times 10)] \text{ cm}^2 = 370 \text{ cm}^2$

$V = A_{\text{Base}} \times \text{height} = 370 \text{ cm}^2 \times 5 \text{ cm}$
$\quad = \mathbf{1850\ cm^3}$

28.

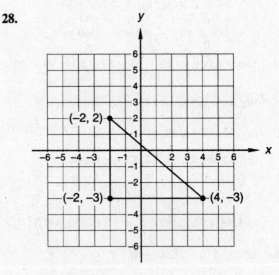

Area $= \dfrac{6(5)}{2} = \mathbf{15\ units^2}$

29. $1{,}000{,}000{,}000 = \mathbf{10 \times 10 \times 10 \times 10 \times 10}$
$\mathbf{\times\ 10 \times 10 \times 10 \times 10}$

30. $5 \,\cancel{\text{yd}^2} \times \dfrac{3 \text{ ft}}{1 \,\cancel{\text{yd}}} \times \dfrac{3 \text{ ft}}{1 \,\cancel{\text{yd}}} = 5(3)(3) \text{ ft}^2 = \mathbf{45\ ft^2}$

PROBLEM SET 47

1. Kings = 12

Castles = 3(Kings) = 3(12) = 36

Princesses = 7(Kings) = 7(12) = 84

Total = 12 + 36 + 84 = **132**

2. 4(6) + 5(7) + N = 10(7)

$$59 + N = 70$$

$$N = \mathbf{11\ lb}$$

3. Blue = 120

Red = Blue + 80 = 120 + 80 = 200

Green = Blue − 20 = 120 − 20 = 100

Total = 120 + 200 + 100 = **420 frogs**

4. $\dfrac{\mathbf{640\ bottles}}{\mathbf{1\ hour}}, \dfrac{\mathbf{1\ hour}}{\mathbf{640\ bottles}}$

$9600 \ \cancel{\text{bottles}} \times \dfrac{1\ \text{hr}}{640 \ \cancel{\text{bottles}}} = \mathbf{15\ hr}$

5.

$A_{\text{Base}} = A_{\text{Rectangle}} + A_{\text{Square}}$

$\quad = [(5)(2) + (2)(2)]\ \text{in.}^2$

$\quad = [10 + 4]\ \text{in.}^2$

$\quad = 14\ \text{in.}^2$

$V = A_{\text{Base}} \times \text{height}$

$\quad = 14\ \text{in.}^2 \times 6\ \text{in.}$

$\quad = \mathbf{84\ in.}^3$

6.

$A_{\text{Base}} = A_{\text{Rectangle}} + A_{\text{Triangle}}$

$\quad = \left[(10)(6) + \dfrac{(6)(5)}{2}\right]\ \text{m}^2$

$\quad = [60 + 16]\ \text{m}^2$

$\quad = 75\ \text{m}^2$

$V = A_{\text{Base}} \times \text{height}$

$\quad = 75\ \text{m}^2 \times 15\ \text{m}$

$\quad = \mathbf{1125\ m}^3$

7. $16 = 2 \cdot 2 \cdot 2 \cdot 2$

$24 = 2 \cdot 2 \cdot 2 \cdot 3$

$36 = 2 \cdot 2 \cdot 3 \cdot 3$

LCM (16, 24, 36) = $2 \cdot 2 \cdot 2 \cdot 2 \cdot 3 \cdot 3$

$\quad = \mathbf{144}$

8. $\dfrac{4}{3}x = 160$

$\dfrac{3}{4} \cdot \dfrac{4}{3}x = 160 \cdot \dfrac{3}{4}$

$x = \mathbf{120}$

9. $x + 6 = 12$

$x + 6 - 6 = 12 - 6$

$x = \mathbf{6}$

10. $x - 5 = 13$

$x - 5 + 5 = 13 + 5$

$x = \mathbf{18}$

11. $2^2 + 3^3 + 4^2 = 4 + 27 + 16 = \mathbf{47}$

12. $\sqrt[4]{16} = \mathbf{2}$ (since $2 \cdot 2 \cdot 2 \cdot 2 = 16$)

13. $7(2 + 6 - 4) + 2^3 - \sqrt[3]{8} = 7(4) + 8 - 2$
$= \mathbf{34}$

14. $\dfrac{4}{5} + 6\dfrac{3}{10} - \dfrac{6}{15} = \dfrac{24}{30} + 6\dfrac{9}{30} - \dfrac{12}{30} = 6\dfrac{21}{30}$

$\quad = \mathbf{6\dfrac{7}{10}}$

15. $615\dfrac{3}{8} - 138\dfrac{3}{4} = 614\dfrac{11}{8} - 138\dfrac{6}{8} = \mathbf{476\dfrac{5}{8}}$

16. $7(8 - 3) + 4 + (6)(2) = 7(5) + 4 + 12 = \mathbf{51}$

17.
$$
\begin{array}{r}
725.890 \\
-\ \ 62.871 \\
\hline
\mathbf{663.019}
\end{array}
$$

18. $\dfrac{381.42}{0.006}$

$$6\overline{)381{,}420} \;\; \mathbf{63{,}570}$$

$$\begin{array}{r} 36 \\ \hline 21 \\ 18 \\ \hline 3\,4 \\ 3\,0 \\ \hline 42 \\ 42 \\ \hline 0 \end{array}$$

19. $7(6-4)+2(4+2)-7\cdot 2$
$=7(2)+2(6)-14=\mathbf{12}$

20. $3\dfrac{3}{5}\times 3\dfrac{1}{2}\div 6\dfrac{3}{4}=\dfrac{\cancel{18}}{5}\times\dfrac{7}{\cancel{2}}\times\dfrac{\cancel{4}}{\cancel{27}}=\dfrac{\mathbf{28}}{\mathbf{15}}$

21. $\dfrac{3\dfrac{4}{5}}{2\dfrac{7}{8}}=\dfrac{\dfrac{19}{5}}{\dfrac{23}{8}}=\dfrac{19}{5}\cdot\dfrac{8}{23}=\dfrac{\mathbf{152}}{\mathbf{115}}$

22. $6\dfrac{2}{9}\div 5\dfrac{1}{9}-\dfrac{7}{8}=\dfrac{\cancel{56}}{\cancel{9}}\cdot\dfrac{\cancel{9}}{\cancel{46}}-\dfrac{7}{8}=\dfrac{28}{23}-\dfrac{7}{8}$

$=\dfrac{224}{184}-\dfrac{161}{184}=\dfrac{\mathbf{63}}{\mathbf{184}}$

23. $4\dfrac{1}{10}-2\dfrac{7}{8}\times 1\dfrac{1}{3}=\dfrac{41}{10}-\dfrac{23}{\cancel{8}}\times\dfrac{\cancel{4}}{3}$

$=\dfrac{41}{10}-\dfrac{23}{6}=\dfrac{123}{30}-\dfrac{115}{30}=\dfrac{8}{30}=\dfrac{\mathbf{4}}{\mathbf{15}}$

24. $xyz+yz-y=\left(\dfrac{1}{6}\right)(3)(2)+(3)(2)-(3)$

$=1+6-3$

$=\mathbf{4}$

25. $6^x=6^3=6\cdot 6\cdot 6=36\cdot 6=\mathbf{216}$

26. $\sqrt[x]{144}=\sqrt{144}=\sqrt{12^2}=\mathbf{12}$

27. $0.5^y=0.5^4=(0.5)(0.5)(0.5)(0.5)$
$=(0.25)(0.25)=\mathbf{0.0625}$

28. $6200\ \cancel{cm}\times\dfrac{1\ \cancel{m}}{100\ \cancel{cm}}\times\dfrac{1\ km}{1000\ \cancel{m}}$

$=\dfrac{6200}{(100)(1000)}\ km=\mathbf{0.062\ km}$

29. $0.0035=\dfrac{35}{10{,}000}=\dfrac{\mathbf{7}}{\mathbf{2000}}$

30. $P=2(30\text{ in.})+2(12\text{ in.})+4(6\text{ in.})$
$=(60+24+24)\text{ in.}$
$=108\ \cancel{\text{in.}}\times\dfrac{1\text{ ft}}{12\ \cancel{\text{in.}}}=\mathbf{9\text{ ft}}$

PROBLEM SET 48

1. $44\times N=1760$

$$44\overline{)1760}\;\; \mathbf{40\ boxes}$$
$$\begin{array}{r}176\\ \hline 0\end{array}$$

2.
$$\begin{array}{r} 44 \\ \times\ 62 \\ \hline 88 \\ 264 \\ \hline \mathbf{2728\ pompons} \end{array}$$

3. $\dfrac{\mathbf{14\ games}}{\mathbf{70\ dollars}},\ \dfrac{\mathbf{70\ dollars}}{\mathbf{14\ games}}$

$200\ \cancel{dollars}\times\dfrac{14\text{ games}}{70\ \cancel{dollars}}=\mathbf{40\ games}$

4. $\text{Avg}=\dfrac{57+55+52}{3}=\dfrac{164}{3}$

$=\mathbf{54\dfrac{2}{3}}\ \text{s}$

5. $\dfrac{2}{5}\cdot A=10$

$\dfrac{5}{2}\cdot\dfrac{2}{5}\cdot A=\cancel{10}^{5}\cdot\dfrac{5}{\cancel{2}_{1}}$

$A=\mathbf{25}$

6. $F\cdot 72=96$

$\dfrac{F\cdot 72}{72}=\dfrac{96}{72}$

$F=\dfrac{\mathbf{4}}{\mathbf{3}}$

7. $\dfrac{6}{7}\cdot 217=B$

$\dfrac{1302}{7}=B$

$\mathbf{186}=B$

8. $B = \frac{6}{\overset{}{\underset{1}{7}}} \cdot \overset{31}{\cancel{217}}$

   $B = \textbf{186}$

9. $\frac{316}{25} = 12\frac{\textbf{16}}{\textbf{25}}$

$$\begin{array}{r} 12 \\ 25\overline{)316} \\ \underline{25} \\ 66 \\ \underline{50} \\ 16 \end{array}$$

10. $8 = 2 \cdot 2 \cdot 2$

    $21 = 3 \cdot 7$

    $24 = 2 \cdot 2 \cdot 2 \cdot 3$

    $\text{LCM}(8, 21, 24) = 2 \cdot 2 \cdot 2 \cdot 3 \cdot 7 = \textbf{168}$

11. $\frac{6}{7}x = 18$

    $\left(\frac{7}{6}\right)\frac{6}{7}x = 18\left(\frac{7}{6}\right)$

    $x = \textbf{21}$

12. $x + \frac{5}{13} = \frac{18}{26}$

    $x + \frac{5}{13} - \frac{5}{13} = \frac{9}{13} - \frac{5}{13}$

    $x = \frac{\textbf{4}}{\textbf{13}}$

13. $7x = 315$

    $\frac{7x}{7} = \frac{315}{7}$

    $x = \textbf{45}$

14. $3[4 - (5 - 4) + 2] + 14$
    $= 3[4 - 1 + 2] + 14 = 3[5] + 14$
    $= 15 + 14 = \textbf{29}$

15. $6(3 - 1) + 2(3 - 1) - 4 = 6(2) + 2(2) - 4$
    $= 12 + 4 - 4 = \textbf{12}$

16. $\frac{4}{9} + \frac{5}{4} \cdot \frac{1}{6} = \frac{4}{9} + \frac{5}{24} = \frac{32}{72} + \frac{15}{72} = \frac{\textbf{47}}{\textbf{72}}$

17. $4\frac{2}{5} \cdot \frac{3}{7} - \frac{4}{7} = \frac{22}{5} \cdot \frac{3}{7} - \frac{4}{7} = \frac{66}{35} - \frac{20}{35}$

    $= \frac{\textbf{46}}{\textbf{35}}$

18. $4^2 + 3^3 - 2^3 = 16 + 27 - 8 = \textbf{35}$

19. $\sqrt[3]{8} + \sqrt[3]{27} = 2 + 3 = \textbf{5}$

20. $\frac{7}{8} + 5\frac{5}{16} - \frac{3}{4} = \frac{14}{16} + 5\frac{5}{16} - \frac{12}{16} = 5\frac{\textbf{7}}{\textbf{16}}$

21. $117\frac{9}{10} - 12\frac{3}{5} = 117\frac{9}{10} - 12\frac{6}{10} = \textbf{105}\frac{\textbf{3}}{\textbf{10}}$

22. $6\frac{3}{8} \times 2\frac{4}{5} \div 2\frac{1}{2} = \frac{51}{\underset{4}{\cancel{8}}} \times \frac{\overset{7}{\cancel{14}}}{5} \times \frac{2}{5}$

    $= \frac{51}{\cancel{4}} \times \frac{7}{5} \times \frac{\overset{1}{\cancel{2}}}{5} = \frac{\textbf{357}}{\textbf{50}}$
    $\ \underset{2}{}$

23. $\dfrac{8\frac{6}{7}}{3\frac{5}{14}} = \dfrac{\frac{62}{7}}{\frac{47}{14}} = \frac{62}{\cancel{7}} \cdot \frac{\overset{2}{\cancel{14}}}{47} = \frac{\textbf{124}}{\textbf{47}}$

24. $4\frac{3}{4} \div 2\frac{1}{3} \times 3\frac{1}{3} \div 1\frac{1}{4}$

    $= \frac{19}{\underset{1}{\cancel{4}}} \times \frac{\overset{1}{\cancel{3}}}{7} \times \frac{\overset{2}{\cancel{10}}}{\underset{1}{\cancel{3}}} \times \frac{\overset{1}{\cancel{4}}}{\underset{1}{\cancel{5}}} = \frac{\textbf{38}}{\textbf{7}}$

25. $\sqrt[x]{243} = \sqrt[5]{243} = \sqrt[5]{3^5} = \textbf{3}$

26. $x^3 = 8^3 = 8 \cdot 8 \cdot 8 = 64 \cdot 8 = \textbf{512}$

27. $y + xyz - z = \left(\frac{1}{4}\right) + (24)\left(\frac{1}{4}\right)(2) - (2)$

    $= \frac{1}{4} + 12 - 2$

    $= \textbf{10}\frac{\textbf{1}}{\textbf{4}}$

28.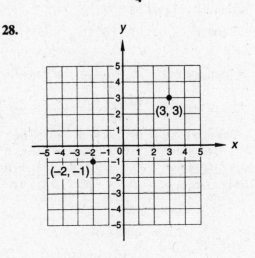

29.

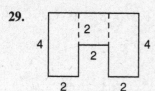

$A_{\text{Base}} = A_{\text{Rectangle 1}} + A_{\text{Square}} + A_{\text{Rectangle 2}}$
$= [(4)(2) + (2)(2) + (4)(2)] \text{ cm}^2$
$= [8 + 4 + 8] \text{ cm}^2$
$= \textbf{20 cm}^2$

$V = A_{\text{Base}} \times \text{height}$
$= 20 \text{ cm}^2 \times 17 \text{ cm}$
$= \textbf{340 cm}^3$

30. $360 \text{ yd} \times \dfrac{3 \text{ ft}}{1 \text{ yd}} \times \dfrac{12 \text{ in.}}{1 \text{ ft}}$
$= 360(3)(12) \text{ in.} = \textbf{12,960 in.}$

PROBLEM SET 49

1. $\dfrac{\textbf{40 pecks}}{\textbf{640 dollars}}, \dfrac{\textbf{640 dollars}}{\textbf{40 pecks}}$

$100 \text{ pecks} \times \dfrac{\$640}{40 \text{ pecks}} = \textbf{\$1600}$

2.
$$
\begin{array}{r}
72.0164200 \text{ in.} \\
- 71.0001403 \text{ in.} \\
\hline
\textbf{1.0162797 in.}
\end{array}
$$

3.
$$
\begin{array}{r}
1305 \\
\times \quad 48 \\
\hline
10440 \\
5220 \quad \\
\hline
\textbf{62,640 flowers}
\end{array}
$$

4. $A_{\text{End}} = \dfrac{(10)(8)}{2} \text{ ft}^2 = 40 \text{ ft}^2$

$V = A_{\text{End}} \times \text{length}$
$= 40 \text{ ft}^2 \times 15 \text{ ft}$
$= \textbf{600 ft}^3$

5. Area of one end $= \dfrac{10 \text{ ft} \times 8 \text{ ft}}{2} = 40 \text{ ft}^2$

Area of one end $= \dfrac{10 \text{ ft} \times 8 \text{ ft}}{2} = 40 \text{ ft}^2$

Area of bottom $= 10 \text{ ft} \times 15 \text{ ft} = 150 \text{ ft}^2$

Area of side $= 15 \text{ ft} \times 9.43 \text{ ft} = 141.45 \text{ ft}^2$

$+$ Area of side $= 15 \text{ ft} \times 9.43 \text{ ft} = 141.45 \text{ ft}^2$

Total surface area $= \textbf{512.9 ft}^2$

6. $\dfrac{5}{12} \cdot A = \dfrac{7}{3}$

$\dfrac{12}{5} \cdot \dfrac{5}{12} \cdot A = \dfrac{7}{\cancel{3}_1} \cdot \dfrac{\cancel{12}^4}{5}$

$A = \dfrac{\textbf{28}}{\textbf{5}}$

7. $F \cdot 26 = 65$

$\dfrac{F \cdot 26}{26} = \dfrac{65}{26}$

$F = \dfrac{\textbf{5}}{\textbf{2}}$

8. $B = \dfrac{3}{\cancel{8}_1} \cdot \cancel{256}^{32}$

$B = \textbf{96}$

9. $\dfrac{213}{8} = \textbf{26}\dfrac{\textbf{5}}{\textbf{8}}$

$$
\begin{array}{r}
26 \\
8\overline{)213} \\
16 \\
\hline
53 \\
48 \\
\hline
5
\end{array}
$$

10. $16 = 2 \cdot 2 \cdot 2 \cdot 2$

$24 = 2 \cdot 2 \cdot 2 \cdot 3$

$30 = 2 \cdot 3 \cdot 5$

$\text{LCM } (16, 24, 30) = 2 \cdot 2 \cdot 2 \cdot 2 \cdot 3 \cdot 5$
$= \textbf{240}$

11. $\dfrac{7}{8}x = 14$

$\dfrac{8}{7} \cdot \dfrac{7}{8}x = 14 \cdot \dfrac{8}{7}$

$x = \textbf{16}$

12. $x - \dfrac{1}{8} = \dfrac{3}{16}$

$x - \dfrac{1}{8} + \dfrac{1}{8} = \dfrac{3}{16} + \dfrac{1}{8}$

$x = \dfrac{3}{16} + \dfrac{2}{16} = \dfrac{\textbf{5}}{\textbf{16}}$

13.
$$x + \frac{5}{8} = \frac{14}{5}$$
$$x + \frac{5}{8} - \frac{5}{8} = \frac{14}{5} - \frac{5}{8}$$
$$x = \frac{112}{40} - \frac{25}{40} = \mathbf{\frac{87}{40}}$$

14. $6 \cdot 3 + (3 \cdot 2)5 - 4(2 \cdot 2)$
$= 18 + (6)5 - 4(4) = 18 + 30 - 16 = \mathbf{32}$

15. $4(6 - 2 + 1) + 3^2 - \sqrt[3]{8} = 4(5) + 9 - 2$
$= 20 + 9 - 2 = \mathbf{27}$

16. $\frac{7}{20} + \frac{4}{5} \cdot \frac{2}{3} = \frac{7}{20} + \frac{8}{15} = \frac{21}{60} + \frac{32}{60}$
$= \mathbf{\frac{53}{60}}$

17. $3\frac{1}{6} \cdot \frac{1}{8} - \frac{4}{12} = \frac{19}{6} \cdot \frac{1}{8} - \frac{4}{12} = \frac{19}{48} - \frac{16}{48}$
$= \frac{3}{48} = \mathbf{\frac{1}{16}}$

18. $3^2 + 4^2 - 5^2 = 9 + 16 - 25 = \mathbf{0}$

19. $\frac{3}{4} + 7\frac{1}{16} - \frac{5}{8} = \frac{12}{16} + 7\frac{1}{16} - \frac{10}{16} = \mathbf{7\frac{3}{16}}$

20. $113\frac{4}{7} - 32\frac{1}{14} = 113\frac{8}{14} - 32\frac{1}{14} = 81\frac{7}{14}$
$= \mathbf{81\frac{1}{2}}$

21.
$$\begin{array}{r} 31.62 \\ \times\ \ 0.08 \\ \hline \mathbf{2.5296} \end{array}$$

22. $3\frac{2}{3} \times 2\frac{3}{4} \div 3\frac{1}{2} = \frac{11}{3} \times \frac{11}{\underset{2}{4}} \times \frac{\overset{1}{2}}{7} = \mathbf{\frac{121}{42}}$

23. $\dfrac{6\frac{3}{4}}{7\frac{11}{12}} = \dfrac{\frac{27}{4}}{\frac{95}{12}} = \frac{27}{\underset{1}{4}} \cdot \frac{\overset{3}{12}}{95} = \mathbf{\frac{81}{95}}$

24.
$$\begin{array}{r} 89.265 \\ -\ \ 6.898 \\ \hline \mathbf{82.367} \end{array}$$

25. $3^x = 3^5 = 3 \times 3 \times 3 \times 3 \times 3 = \mathbf{243}$

26. $\sqrt[p]{8} = \sqrt[3]{8} = \sqrt[3]{2^3} = \mathbf{2}$

27. $xyz - x + xy = (12)\left(\frac{1}{4}\right)(6) - (12) + (12)\left(\frac{1}{4}\right)$
$= 18 - 12 + 3$
$= \mathbf{9}$

28. $A = A_{\text{Triangle 1}} + A_{\text{Triangle 2}}$
$= \left[\frac{33(29)}{2} + \frac{28(9)}{2}\right] \text{m}^2$
$= [478.5 + 126] \text{ m}^2$
$= \mathbf{604.5 \text{ m}^2}$

29. $250{,}001 \text{ cm} \times \frac{1 \text{ m}}{100 \text{ cm}} \times \frac{1 \text{ km}}{1000 \text{ m}}$
$= \frac{250{,}001}{(100)(1000)} \text{ km} = \mathbf{2.50001 \text{ km}}$

30. Overall average $= \dfrac{4(83) + 3(90)}{4 + 3}$
$= \dfrac{332 + 270}{7}$
$= \dfrac{602}{7} = \mathbf{86}$

PROBLEM SET 50

1. $\dfrac{\$64{,}000}{\$40} = \mathbf{1600 \text{ certificates}}$

2.
$$\begin{array}{r} 0.041015 \\ +\ 0.002700 \\ \hline \mathbf{0.043715} \end{array}$$

3. $\dfrac{40 \text{ good ones}}{12 \text{ dollars}}, \dfrac{12 \text{ dollars}}{40 \text{ good ones}}$

$\$3.40 \times \dfrac{40 \text{ good ones}}{12 \text{ dollars}} = 11\frac{1}{3}$ good ones

Thus, **11 good ones** could be purchased.

4. Avg $= \dfrac{760}{4} = \mathbf{190 \text{ lb}}$

5.
$$\frac{5}{7} \cdot A = \frac{7}{3}$$
$$\frac{7}{5} \cdot \frac{5}{7} \cdot A = \frac{7}{3} \cdot \frac{7}{5}$$
$$A = \mathbf{\frac{49}{15}}$$

6.
$$\frac{3}{5} \cdot A = \frac{3}{7}$$
$$\frac{5}{3} \cdot \frac{3}{5} \cdot A = \frac{3}{7} \cdot \frac{5}{3}$$
$$A = \frac{5}{7}$$

7. $B = \frac{4}{7} \cdot \overset{30}{\cancel{210}}$
$$\phantom{B = \frac{4}{7} \cdot} 1$$
$$B = \mathbf{120}$$

8. $A_{\text{Base}} = 3\,\text{ft} \times 4\,\text{ft} = 12\,\text{ft}^2$
$$V = 12\,\text{ft}^2 \times 6\,\text{ft} = \mathbf{72\,ft^3}$$

9. (a) $4{,}203{,}742 = \mathbf{4.203742 \times 10^6}$

   (b) $0.0056305 = \mathbf{5.6305 \times 10^{-3}}$

10. (a) $9.277 \times 10^5 = \mathbf{927{,}700}$

   (b) $2.132 \times 10^{-6} = \mathbf{0.000002132}$

11.
$$\frac{8}{9}x = 16$$
$$\frac{9}{8} \cdot \frac{8}{9}x = \overset{2}{\cancel{16}} \cdot \frac{9}{\cancel{8}}$$
$$\phantom{\frac{9}{8} \cdot \frac{8}{9}x =} 1$$
$$x = \mathbf{18}$$

12.
$$x - 3\frac{1}{4} = \frac{1}{2}$$
$$x - 3\frac{1}{4} + 3\frac{1}{4} = \frac{2}{4} + 3\frac{1}{4}$$
$$x = \mathbf{3\frac{3}{4}}$$

13. $7 \cdot 2 + (3 \cdot 2)4 - 3(2 \cdot 1)$
$$= 14 + (6)4 - 3(2) = 14 + 24 - 6 = \mathbf{32}$$

14. $6(3 - 2 + 1) + 3^2 - \sqrt[3]{8} = 6(2) + 9 - 2$
$$= 12 + 9 - 2 = \mathbf{19}$$

15. $\frac{3}{20} + \frac{4}{5} \cdot \frac{3}{4} = \frac{3}{20} + \frac{12}{20} = \frac{15}{20} = \mathbf{\frac{3}{4}}$

16. $4\frac{2}{3} \cdot \frac{3}{4} - \frac{7}{12} = \frac{14}{3} \cdot \frac{3}{4} - \frac{7}{12} = \frac{42}{12} - \frac{7}{12}$
$$= \frac{35}{12} = \mathbf{2\frac{11}{12}}$$

17. $137\frac{5}{7} - 16\frac{1}{14} = 137\frac{10}{14} - 16\frac{1}{14} = \mathbf{121\frac{9}{14}}$

18. $\frac{4}{5} + 3\frac{1}{10} - \frac{14}{25} = \frac{40}{50} + 3\frac{5}{50} - \frac{28}{50} = \mathbf{3\frac{17}{50}}$

19. $3^2 - \sqrt{9} + \sqrt[3]{8} = 9 - 3 + 2 = \mathbf{8}$

20.
$$\begin{array}{r} 21.32 \\ \times\ 0.06 \\ \hline \mathbf{1.2792} \end{array}$$

21. $\dfrac{6111.2}{0.005}$

$$\begin{array}{r} \mathbf{1{,}222{,}240} \\ 5\overline{)6{,}111{,}200} \\ 5\phantom{{,}111{,}200} \\ \hline 1\ 1\phantom{11{,}200} \\ 1\ 0\phantom{11{,}200} \\ \hline 11\phantom{1{,}200} \\ 10\phantom{1{,}200} \\ \hline 11\phantom{{,}200} \\ 10\phantom{{,}200} \\ \hline 1\ 2 \\ 1\ 0 \\ \hline 20 \\ 20 \\ \hline 0 \end{array}$$

22. $\dfrac{6\frac{4}{5}}{5\frac{3}{4}} = \dfrac{\frac{34}{5}}{\frac{23}{4}} = \frac{34}{5} \cdot \frac{4}{23} = \mathbf{\frac{136}{115}}$

23. $xyz + yz - xy$
$$= (20)\left(\frac{1}{5}\right)(5) + \left(\frac{1}{5}\right)(5) - (20)\left(\frac{1}{5}\right)$$
$$= 20 + 1 - 4$$
$$= \mathbf{17}$$

24. $y^2 = 3^2 = 3 \times 3 = \mathbf{9}$

25. $\sqrt[p]{27} = \sqrt[3]{27} = \sqrt[3]{3^3} = \mathbf{3}$

26. $A_{\text{Base}} = 20\,\text{ft} \times 30\,\text{ft} = 600\,\text{ft}^2$
$$V = A_{\text{Base}} \times \text{height}$$
$$= 600\,\text{ft}^2 \times 10\,\text{ft}$$
$$= \mathbf{6000\,ft^3}$$

27.

Area of front = $10\,\text{ft} \times 20\,\text{ft}$ =	$200\,\text{ft}^2$
Area of back = $10\,\text{ft} \times 20\,\text{ft}$ =	$200\,\text{ft}^2$
Area of top = $20\,\text{ft} \times 30\,\text{ft}$ =	$600\,\text{ft}^2$
Area of bottom = $20\,\text{ft} \times 30\,\text{ft}$ =	$600\,\text{ft}^2$
Area of side = $10\,\text{ft} \times 30\,\text{ft}$ =	$300\,\text{ft}^2$
+ Area of side = $10\,\text{ft} \times 30\,\text{ft}$ =	$300\,\text{ft}^2$
Total surface area	= $\mathbf{2200\,ft^2}$

28. $3{,}059{,}000 \, \text{mi} \times \dfrac{5280 \, \text{ft}}{1 \, \text{mi}} \times \dfrac{12 \, \text{in.}}{1 \, \text{ft}}$

$= 3{,}059{,}000(5280)(12) \, \text{in.} = \mathbf{193{,}818{,}240{,}000 \, in.}$

29. $3\dfrac{1}{18} + 9\dfrac{7}{20} + 12\dfrac{5}{24} = 3\dfrac{20}{360} + 9\dfrac{126}{360} + 12\dfrac{75}{360}$

$= \mathbf{24\dfrac{221}{360}}$

30.

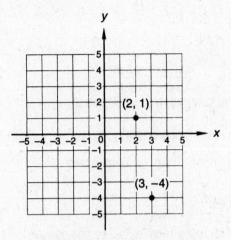

PROBLEM SET 51

1.
$$\begin{array}{r} 228{,}014 \\ -\ 142{,}763 \\ \hline \mathbf{85{,}251} \end{array}$$

2. $\dfrac{\$28{,}000}{140 \text{ items}}, \dfrac{140 \text{ items}}{\$28{,}000}$

$200 \text{ items} \times \dfrac{\$28{,}000}{140 \text{ items}} = \mathbf{\$40{,}000}$

3. $\text{Avg} = \dfrac{260 + 200 + 300 + 240}{4}$

$= \dfrac{1000}{4} = \mathbf{250 \ tons}$

4. $10.1 + 10.6 + 10.3 + N = 4(10.2)$

$31.0 + N = 40.8$

$N = \mathbf{9.8 \ s}$

5.

Area of front	$= 5\,\text{m} \times 30\,\text{m}$	$=$	$150\,\text{m}^2$
Area of back	$= 5\,\text{m} \times 30\,\text{m}$	$=$	$150\,\text{m}^2$
Area of top	$= 30\,\text{m} \times 15\,\text{m}$	$=$	$450\,\text{m}^2$
Area of bottom	$= 30\,\text{m} \times 15\,\text{m}$	$=$	$450\,\text{m}^2$
Area of side	$= 15\,\text{m} \times 5\,\text{m}$	$=$	$75\,\text{m}^2$
+ Area of side	$= 15\,\text{m} \times 5\,\text{m}$	$=$	$75\,\text{m}^2$
Total surface area		$=$	$\mathbf{1350\,m^2}$

6. $A_{\text{End}} = \dfrac{(20)(10)}{2} \, \text{cm}^2 = 100 \, \text{cm}^2$

$V = A_{\text{End}} \times \text{length}$

$= 100 \, \text{cm}^2 \times 41 \, \text{cm}$

$= \mathbf{4100 \ cm^3}$

7. $B = \dfrac{5}{\overset{}{\underset{1}{8}}} \times \overset{21}{\cancel{168}}$

$B = \mathbf{105}$

8. $\dfrac{4}{5} \cdot A = \dfrac{7}{2}$

$\dfrac{5}{4} \cdot \dfrac{4}{5} \cdot A = \dfrac{7}{2} \cdot \dfrac{5}{4}$

$A = \mathbf{\dfrac{35}{8}}$

9. $\dfrac{5}{6} \cdot A = 4\dfrac{1}{3}$

$\dfrac{6}{5} \cdot \dfrac{5}{6} \cdot A = \dfrac{13}{3} \cdot \dfrac{6}{5}$

$A = \mathbf{\dfrac{26}{5}}$

10. $0.4 \times A = 5$

$\dfrac{0.4A}{0.4} = \dfrac{5}{0.4}$

$A = \mathbf{12.5}$

11. $0.44 \times 75 = B$

$B = \mathbf{33}$

12. $D \cdot 33 = 0.0066$

$\dfrac{D \cdot 33}{33} = \dfrac{0.0066}{33}$

$D = \mathbf{0.0002}$

13. $\dfrac{167}{7} = \mathbf{23\dfrac{6}{7}}$

$$\begin{array}{r} 23 \\ 7\overline{)167} \\ \underline{14} \\ 27 \\ \underline{21} \\ 6 \end{array}$$

14.
$$\frac{4}{9}x = 16$$

$$\frac{\cancel{9}}{\cancel{4}} \cdot \frac{\cancel{4}}{\cancel{9}}x = \frac{\cancel{16}}{1} \cdot \frac{9}{\cancel{4}}$$

$$x = 36$$

15.
$$x - \frac{3}{5} = 2\frac{3}{10}$$

$$x - \frac{3}{5} + \frac{3}{5} = 2\frac{3}{10} + \frac{3}{5}$$

$$x = 2\frac{3}{10} + \frac{6}{10}$$

$$x = 2\frac{9}{10}$$

16. $6 \cdot 2 + (6 \cdot 3)7 - 4(3 \cdot 1)$
$= 12 + (18)7 - 4(3) = 12 + 126 - 12 = \mathbf{126}$

17. $5(6 - 3 + 1) + 6^2 - \sqrt{100} = 5(4) + 36 - 10$
$= 20 + 36 - 10 = \mathbf{46}$

18. $\frac{11}{14} + \frac{3}{7} \cdot \frac{3}{4} = \frac{11}{14} + \frac{9}{28} = \frac{22}{28} + \frac{9}{28}$
$= \frac{\mathbf{31}}{\mathbf{28}}$

19. $6\frac{2}{3} \cdot \frac{1}{4} - \frac{3}{4} = \frac{20}{3} \cdot \frac{1}{4} - \frac{3}{4} = \frac{20}{12} - \frac{3}{4}$
$= \frac{20}{12} - \frac{9}{12} = \frac{\mathbf{11}}{\mathbf{12}}$

20. $117\frac{3}{8} - 14\frac{7}{16} = 116\frac{11}{8} - 14\frac{7}{16}$
$= 116\frac{22}{16} - 14\frac{7}{16} = \mathbf{102\frac{15}{16}}$

21. $\frac{3}{5} + 3\frac{4}{15} - \frac{7}{30} = \frac{18}{30} + 3\frac{8}{30} - \frac{7}{30} = \mathbf{3\frac{19}{30}}$

22. $3^2 + \sqrt[3]{27} - \sqrt[3]{8} = 9 + 3 - 2 = \mathbf{10}$

23.
$$\begin{array}{r} 118.321 \\ -\ \ 81.340 \\ \hline \mathbf{36.981} \end{array}$$

24. $4\frac{2}{3} \cdot 2\frac{3}{4} \div 1\frac{5}{12} = \frac{14}{3} \cdot \frac{11}{\cancel{4}} \cdot \frac{\cancel{12}^{\,3}}{17}$

$= \frac{14}{\cancel{3}} \cdot \frac{11}{1} \cdot \frac{\cancel{3}^{\,1}}{17} = \frac{\mathbf{154}}{\mathbf{17}}$

25. $\dfrac{7\frac{2}{3}}{2\frac{5}{6}} = \dfrac{\frac{23}{3}}{\frac{17}{6}} = \frac{23}{\cancel{3}} \cdot \frac{\cancel{6}^{\,2}}{17} = \frac{\mathbf{46}}{\mathbf{17}}$

26. $x + xy + xyz = (1) + (1)(10) + (1)(10)\left(\dfrac{1}{5}\right)$

$= 1 + 10 + 2$

$= \mathbf{13}$

27. $p^3 = (2)^3 = 2 \cdot 2 \cdot 2 = \mathbf{8}$

28. $6 \,\cancel{\text{mi}} \times \dfrac{5280\,\cancel{\text{ft}}}{1\,\cancel{\text{mi}}} \times \dfrac{12\text{ in.}}{1\,\cancel{\text{ft}}}$
$= 6(5280)(12)\text{ in.} = \mathbf{380,160 \text{ in.}}$

29. (a) $98,213,000 = \mathbf{9.8213 \times 10^7}$

(b) $3.207 \times 10^{-5} = \mathbf{0.00003207}$

30. $12\,\cancel{\text{km}}^2 \times \dfrac{1000\text{ m}}{1\,\cancel{\text{km}}} \times \dfrac{1000\text{ m}}{1\,\cancel{\text{km}}}$
$= \mathbf{12(1000)(1000)\text{ m}^2 = 12,000,000\text{ m}^2}$

PROBLEM SET 52

1.
$$\begin{array}{r} 53,007 \\ -\ 47,364 \\ \hline \mathbf{5643 \text{ attendees}} \end{array}$$

2. $\dfrac{\mathbf{1 \text{ bin}}}{\mathbf{14 \text{ uniforms}}}, \dfrac{\mathbf{14 \text{ uniforms}}}{\mathbf{1 \text{ bin}}}$

$140\,\cancel{\text{bins}} \times \dfrac{14\text{ uniforms}}{1\,\cancel{\text{bin}}} = \mathbf{1960 \text{ uniforms}}$

3. $\dfrac{\mathbf{15}}{\mathbf{\$315}}, \dfrac{\mathbf{\$315}}{\mathbf{15}}$

$140 \times \dfrac{\$315}{15} = \mathbf{\$2940}$

4. (a) Range $= 2396$ lb $- 1432$ lb $= \mathbf{964 \text{ lb}}$

(b) Since the five weights are all different, there is **no mode.**

(c) Median $= \mathbf{1840 \text{ lb}}$

(d) Mean
$$= \frac{1432 + 1491 + 1840 + 2153 + 2396}{5}$$
$$= \frac{9312}{5} = \mathbf{1862.4 \text{ lb}}$$

5. $A_{End} = \dfrac{(4)(6)}{2}\text{ ft}^2 = 12\text{ ft}^2$

$V = A_{End} \times \text{length}$
$= 12\text{ ft}^2 \times 23\text{ ft}$
$= \mathbf{276\text{ ft}^3}$

6. $S.A. = 2A_{End} + 2A_{Side} + A_{Bottom}$

$= 2\left[\dfrac{(6)(4)}{2}\right]\text{ ft}^2 + 2[(23)(5)]\text{ ft}^2$
$\qquad + [(6)(23)]\text{ ft}^2$
$= 24\text{ ft}^2 + 230\text{ ft}^2 + 138\text{ ft}^2$
$= \mathbf{392\text{ ft}^2}$

7. $2.15 = 2\dfrac{15}{100} = \mathbf{2\dfrac{3}{20}}$

8. $\dfrac{5}{8} \cdot A = 100$

$\dfrac{8}{5} \cdot \dfrac{5}{8} \cdot A = 100 \cdot \dfrac{8}{5}$

$A = \mathbf{160}$

9. $F \times 64 = 56$

$\dfrac{F \times 64}{64} = \dfrac{56}{64}$

$F = \mathbf{\dfrac{7}{8}}$

10. $D \cdot 507 = 441.09$

$\dfrac{D \cdot 507}{507} = \dfrac{441.09}{507}$

$D = \mathbf{0.87}$

11. $0.52 \times A = 104$

$\dfrac{0.52 \times A}{0.52} = \dfrac{104}{0.52}$

$A = \mathbf{200}$

12. $\dfrac{4}{7}x = 112$

$\dfrac{7}{4} \cdot \dfrac{4}{7}x = 112 \cdot \dfrac{7}{4}$

$x = \mathbf{196}$

13. $x + \dfrac{7}{12} = 3\dfrac{5}{12}$

$x + \dfrac{7}{12} - \dfrac{7}{12} = 3\dfrac{5}{12} - \dfrac{7}{12}$

$x = 2\dfrac{17}{12} - \dfrac{7}{12} = 2\dfrac{10}{12} = \mathbf{2\dfrac{5}{6}}$

14. $28 - 2[(6 - 2) \div (10 - 9) + 3]$
$= 28 - 2[4 \div 1 + 3] = 28 - 2[7] = 28 - 14$
$= \mathbf{14}$

15. $4[(3 - 1)(3 + 2) - 1] + 25$
$= 4[2(5) - 1] + 25 = 4[9] + 25 = 36 + 25$
$= \mathbf{61}$

16. $7\dfrac{3}{4} \cdot \dfrac{2}{3} - \dfrac{5}{12} = \dfrac{31}{4} \cdot \dfrac{2}{3} - \dfrac{5}{12} = \dfrac{62}{12} - \dfrac{5}{12}$

$= \dfrac{57}{12} = \mathbf{\dfrac{19}{4}}$

17. $\dfrac{13}{14} + \dfrac{2}{7} \cdot \dfrac{3}{4} = \dfrac{13}{14} + \dfrac{3}{14} = \dfrac{16}{14} = \mathbf{\dfrac{8}{7}}$

18. $\begin{array}{r} 62.891 \\ - 18.812 \\ \hline \mathbf{44.079} \end{array}$

19. $2^3 + 3^3 - \sqrt[3]{8} = 8 + 27 - 2 = \mathbf{33}$

20. $36\dfrac{6}{7} - 14\dfrac{3}{14} = 36\dfrac{12}{14} - 14\dfrac{3}{14} = \mathbf{22\dfrac{9}{14}}$

21. $\dfrac{7}{9}\left(3\dfrac{1}{6} + \dfrac{1}{3}\right) + \dfrac{5}{6} = \dfrac{7}{9}\left(\dfrac{19}{6} + \dfrac{2}{6}\right) + \dfrac{5}{6}$

$= \dfrac{7}{\underset{3}{\cancel{9}}} \cdot \dfrac{\overset{7}{\cancel{21}}}{6} + \dfrac{5}{6} = \dfrac{49}{18} + \dfrac{5}{6} = \dfrac{49}{18} + \dfrac{15}{18}$

$= \dfrac{64}{18} = \mathbf{\dfrac{32}{9}}$

22. $\dfrac{11}{20} - \left[\left(\dfrac{1}{25} + \dfrac{7}{10}\right) - \dfrac{3}{5}\right]$

$= \dfrac{11}{20} - \left[\left(\dfrac{2}{50} + \dfrac{35}{50}\right) - \dfrac{3}{5}\right]$

$= \dfrac{11}{20} - \left[\dfrac{37}{50} - \dfrac{30}{50}\right] = \dfrac{11}{20} - \dfrac{7}{50}$

$= \dfrac{55}{100} - \dfrac{14}{100} = \mathbf{\dfrac{41}{100}}$

23. $\dfrac{4}{5} + 2\dfrac{7}{10} - \dfrac{3}{20} = \dfrac{16}{20} + 2\dfrac{14}{20} - \dfrac{3}{20} = 2\dfrac{27}{20}$

$= \mathbf{3\dfrac{7}{20}}$

24. $\dfrac{17.025}{0.003}$

$$\begin{array}{r} 5675 \\ 3\overline{)17.025} \\ \underline{15} \\ 20 \\ \underline{18} \\ 22 \\ \underline{21} \\ 15 \\ \underline{15} \\ 0 \end{array}$$

25. $4\dfrac{1}{8} \div 2\dfrac{1}{4} \times 3\dfrac{1}{2} \div 1\dfrac{1}{16} = \dfrac{\overset{11}{\cancel{33}}}{8} \times \dfrac{4}{\underset{3}{\cancel{9}}} \times \dfrac{7}{2} \times \dfrac{\overset{2}{\cancel{16}}}{17}$

$= \dfrac{11}{1} \times \dfrac{4}{3} \times \dfrac{7}{\underset{1}{\cancel{2}}} \times \dfrac{\overset{1}{\cancel{2}}}{17} = \dfrac{308}{51}$

26. $\dfrac{8\dfrac{1}{2}}{4\dfrac{1}{7}} = \dfrac{\dfrac{17}{2}}{\dfrac{29}{7}} = \dfrac{17}{2} \cdot \dfrac{7}{29} = \dfrac{119}{58}$

27. $xyz + yz - x = (16)\left(\dfrac{1}{8}\right)(24) + \left(\dfrac{1}{8}\right)(24) - (16)$

$= 48 + 3 - 16$

$= 35$

28. $p^3 + p + q = (2)^3 + (2) + (3)$

$= 8 + 2 + 3$

$= 13$

29. $2162.18 \, \cancel{cg} \times \dfrac{1\,g}{100 \, \cancel{cg}} \times \dfrac{1 \, kg}{1000 \, g}$

$= \dfrac{2162.18}{(100)(1000)} \, kg = 0.0216218 \, kg$

30. (a) $6,657,000,000 = \mathbf{6.657 \times 10^9}$

(b) $3.209 \times 10^{-7} = \mathbf{0.0000003209}$

PROBLEM SET 53

1. $\begin{array}{r} 0.010300 \text{ in.} \\ -\ 0.001475 \text{ in.} \\ \hline \mathbf{0.008825 \text{ in.}} \end{array}$

2. $\dfrac{\textbf{1900 oscillators}}{\textbf{\$38,000}}, \ \dfrac{\textbf{\$38,000}}{\textbf{1900 oscillators}}$

$5000 \, \cancel{\text{oscillators}} \times \dfrac{\$38,000}{1900 \, \cancel{\text{oscillators}}}$

$= \mathbf{\$100,000}$

3. $\text{Avg} = \dfrac{2(86) + 8(92) + 8(80) + 2(70)}{2 + 8 + 8 + 2}$

$= \dfrac{1688}{20} = \mathbf{84.4 \text{ points}}$

4. (a) $6.3 \times 10^9 = \mathbf{6,300,000,000}$

(b) $0.00000047 = \mathbf{4.7 \times 10^{-7}}$

5. $A_{\text{Base}} = (20)(21) \text{ yd}^2 = 420 \text{ yd}^2$

$V = A_{\text{Base}} \times \text{height}$

$= 420 \text{ yd}^2 \times 10 \text{ yd}$

$= \mathbf{4200 \text{ yd}^3}$

6. $S.A. = 2A_{\text{Front}} + 2A_{\text{Top}} + 2A_{\text{Side}}$

$= 2[21(10)] \text{ yd}^2 + 2[20(21)] \text{ yd}^2$

$+ 2[20(10)] \text{ yd}^2$

$= 420 \text{ yd}^2 + 840 \text{ yd}^2 + 400 \text{ yd}^2$

$= \mathbf{1660 \text{ yd}^2}$

7. $\dfrac{3}{4} \cdot A = 51$

$\dfrac{4}{3} \cdot \dfrac{3}{4} \cdot A = 51 \cdot \dfrac{4}{3}$

$A = \mathbf{68}$

8. $\dfrac{3}{7} \times 91 = B$

$\mathbf{39} = B$

9. $0.003 \times A = 18$

$A = 18 \div 0.003$

$A = \mathbf{6000}$

10. $D \cdot 94 = 68.62$

$D = 68.62 \div 94$

$D = \mathbf{0.73}$

11. (a) $1 \text{ percent of } 24 = \dfrac{24}{100} = \mathbf{0.24}$

(b) $56 \text{ percent of } 24 = 56 \cdot 0.24 = \mathbf{13.44}$

12. (a) $1 \text{ percent of } 3200 = \dfrac{3200}{100} = \mathbf{32}$

(b) $101 \text{ percent of } 3200 = 101 \cdot 32 = \mathbf{3232}$

13. $7\dfrac{3}{22} = \dfrac{22 \times 7 + 3}{22} = \dfrac{\mathbf{157}}{\mathbf{22}}$

14. $\frac{5}{4}x = 120$

$\left(\frac{4}{5}\right)\frac{5}{4}x = \left(\frac{4}{5}\right)120$

$x = \mathbf{96}$

15. $x + \frac{3}{11} = \frac{51}{22}$

$x + \frac{3}{11} - \frac{3}{11} = \frac{51}{22} - \frac{3}{11}$

$x = \frac{51}{22} - \frac{6}{22} = \mathbf{\frac{45}{22}}$

16. $17 - 2[(6-4)(7-3)-7]$
$= 17 - 2[2(4)-7] = 17 - 2[1] = \mathbf{15}$

17. $3^2 + 2[(7+1)(7-5)-12]$
$= 9 + 2[8(2)-12] = 9 + 2[4] = 9 + 8$
$= \mathbf{17}$

18. $6\frac{2}{3} \cdot \frac{1}{4} - \frac{5}{12} = \frac{20}{3} \cdot \frac{1}{4} - \frac{5}{12} = \frac{20}{12} - \frac{5}{12}$

$= \frac{15}{12} = \frac{5}{4} = \mathbf{1\frac{1}{4}}$

19. $\frac{5}{6} + 1\frac{7}{12} - \frac{2}{3} = \frac{10}{12} + 1\frac{7}{12} - \frac{8}{12} = 1\frac{9}{12}$

$= \mathbf{1\frac{3}{4}}$

20. $\left(2\frac{1}{7} - \frac{5}{9}\right)\frac{3}{8} - \frac{25}{42} = \left(\frac{135}{63} - \frac{35}{63}\right)\frac{3}{8} - \frac{25}{42}$

$= \frac{\overset{25}{\cancel{100}}}{\underset{21}{\cancel{63}}} \cdot \frac{\overset{1}{\cancel{3}}}{\underset{2}{8}} - \frac{25}{42} = \frac{25}{42} - \frac{25}{42} = \mathbf{0}$

21. $\left[\frac{1}{8} + \frac{1}{2}\left(\frac{7}{9} - \frac{1}{4}\right)\right] + \frac{2}{3}$

$= \left[\frac{1}{8} + \frac{1}{2}\left(\frac{28}{36} - \frac{9}{36}\right)\right] + \frac{2}{3}$

$= \left[\frac{1}{8} + \frac{1}{2} \cdot \frac{19}{36}\right] + \frac{2}{3} = \left[\frac{9}{72} + \frac{19}{72}\right] + \frac{2}{3}$

$= \frac{28}{72} + \frac{48}{72} = \frac{76}{72} = \mathbf{\frac{19}{18}}$

22. $\sqrt[3]{27} - \sqrt{9} + 3 = 3 - 3 + 3 = \mathbf{3}$

23. $16\frac{3}{5} - 5\frac{1}{10} = 16\frac{6}{10} - 5\frac{1}{10} = 11\frac{5}{10} = \mathbf{11\frac{1}{2}}$

24. $\frac{181.02}{0.006}$

$\begin{array}{r} \mathbf{30{,}170} \\ 6\overline{)181{,}020} \\ \underline{18} \\ 01\ 0 \\ \underline{6} \\ 42 \\ \underline{42} \\ 0 \end{array}$

25. $7\frac{1}{4} \div 3\frac{1}{3} \times 1\frac{2}{3} \times 1\frac{1}{6}$

$= \frac{29}{4} \times \frac{\overset{1}{\cancel{3}}}{\underset{2}{\cancel{10}}} \times \frac{\overset{1}{\cancel{5}}}{\underset{1}{\cancel{3}}} \times \frac{7}{6} = \mathbf{\frac{203}{48}}$

26. $\frac{3\frac{2}{7}}{2\frac{1}{14}} = \frac{\frac{23}{7}}{\frac{29}{14}} = \frac{23}{\cancel{7}} \cdot \frac{\overset{2}{\cancel{14}}}{29} = \mathbf{\frac{46}{29}}$

27. $xyzt + xyz + yzt$

$= \left(\frac{1}{6}\right)(3)(4)(5) + \left(\frac{1}{6}\right)(3)(4) + (3)(4)(5)$

$= 10 + 2 + 60$

$= \mathbf{72}$

28. $\sqrt[x]{8} + y = \sqrt[3]{8} + (8) = 2 + 8 = \mathbf{10}$

29. $625\,\cancel{yd^2} \times \frac{3\,\cancel{ft}}{1\,\cancel{yd}} \times \frac{3\,\cancel{ft}}{1\,\cancel{yd}} \times \frac{12\text{ in.}}{1\,\cancel{ft}} \times \frac{12\text{ in.}}{1\,\cancel{ft}}$

$= 625(3)(3)(12)(12)\text{ in.}^2 = \mathbf{810{,}000\text{ in.}^2}$

30. (a) $\mathbf{(-6, -4)}$

(b) $\mathbf{(2, 6)}$

(c) $\mathbf{(5, -3)}$

PROBLEM SET 54

1. $\begin{array}{r} 140{,}000{,}014 \\ +\ \ 15{,}982{,}000 \\ \hline \mathbf{155{,}982{,}014} \end{array}$

2. $\frac{\mathbf{80\ horses}}{\mathbf{320\ guineas}}, \frac{\mathbf{320\ guineas}}{\mathbf{80\ horses}}$

$320\ \cancel{horses} \times \frac{320\ guineas}{80\ \cancel{horses}} = \mathbf{1280\ guineas}$

3. $7942 - 242 = \mathbf{7700\ troops}$

4. $0.6 \times A = 72$

$$\frac{0.6 \times A}{0.6} = \frac{72}{0.6}$$

$$A = \mathbf{120}$$

5. $D \times 360 = 60$

$$\frac{D \times 360}{360} = \frac{60}{360}$$

$$D = \mathbf{0.1\overline{6}}$$

6. $1000 \, \cancel{cm}^2 \times \dfrac{1 \, m}{100 \, \cancel{cm}} \times \dfrac{1 \, m}{100 \, \cancel{cm}} = \mathbf{0.1 \, m^2}$

7. $A_{End} = \dfrac{10(12)}{2} \, m^2 = 60 \, m^2$

$V = A_{End} \times$ length

$= 60 \, m^2 \times 11 \, m$

$= \mathbf{660 \, m^3}$

8. $S.A. = 2A_{End} + 2A_{Side} + A_{Bottom}$

$= 2\left[\dfrac{10(12)}{2}\right] m^2 + 2[(13)(11)] \, m^2$

$+ (10)(11) \, m^2$

$= 120 \, m^2 + 286 \, m^2 + 110 \, m^2$

$= \mathbf{516 \, m^2}$

9. $F \times 56 = 21$

$$\frac{F \times 56}{56} = \frac{21}{56}$$

$$F = \mathbf{\frac{3}{8}}$$

10. $\dfrac{8}{13} \cdot A = 16$

$$\frac{13}{8} \cdot \frac{8}{13} \cdot A = 16 \cdot \frac{13}{8}$$

$$A = \mathbf{26}$$

11. (a) 1% of $77.2 = \dfrac{77.2}{100} = \mathbf{0.772}$

(b) 500% of $77.2 = 500 \times 0.772 = \mathbf{386}$

12. (a) 1% of $90 = \dfrac{90}{100} = \mathbf{0.9}$

(b) 77% of $90 = 77 \times 0.9 = \mathbf{69.3}$

13. $\dfrac{8}{7}x = 104$

$$\left(\frac{7}{8}\right)\frac{8}{7}x = \left(\frac{7}{8}\right)104$$

$$x = \mathbf{91}$$

14. $x - \dfrac{5}{12} = \dfrac{15}{4}$

$$x - \frac{5}{12} + \frac{5}{12} = \frac{15}{4} + \frac{5}{12}$$

$$x = \frac{45}{12} + \frac{5}{12} = \frac{50}{12} = \mathbf{\frac{25}{6}}$$

15. $\dfrac{t}{11} = \dfrac{9}{44}$

$44t = 99$

$$\frac{44t}{44} = \frac{99}{44}$$

$$t = \mathbf{\frac{9}{4}}$$

16. $\dfrac{5}{6} = \dfrac{2}{y}$

$5y = 12$

$$y = \mathbf{\frac{12}{5}}$$

17. $24 + 2[6(3 - 1) \div (7 - 3) + 6]$

$= 24 + 2[6(2) \div 4 + 6] = 24 + 2[3 + 6]$

$= 24 + 2[9] = \mathbf{42}$

18. $2^3 + 2[(8 + 1)(6 - 5) - 3]$

$= 8 + 2[9(1) - 3]$

$= 8 + 2[6] = 8 + 12 = \mathbf{20}$

19. $5\dfrac{6}{7} \cdot \dfrac{3}{2} - 1\dfrac{1}{14} = \dfrac{41}{7} \cdot \dfrac{3}{2} - \dfrac{15}{14}$

$= \dfrac{123}{14} - \dfrac{15}{14} = \dfrac{108}{14} = \dfrac{54}{7} = \mathbf{7\dfrac{5}{7}}$

20. $\dfrac{5}{6} + 1\dfrac{5}{12} - 1\dfrac{1}{3} = \dfrac{10}{12} + 1\dfrac{5}{12} - 1\dfrac{4}{12} = \mathbf{\dfrac{11}{12}}$

21. $\sqrt[3]{8} + \sqrt[3]{27} - 5 = 2 + 3 - 5 = \mathbf{0}$

22. $23\dfrac{4}{5} - 6\dfrac{1}{15} = 23\dfrac{12}{15} - 6\dfrac{1}{15} = \mathbf{17\dfrac{11}{15}}$

23. $\dfrac{182.101}{0.0006}$

$$
\begin{array}{r}
303{,}501.\overline{6} \\
6\overline{)1{,}821{,}010.0} \\
\underline{18} \\
021 \\
\underline{18} \\
3\,0 \\
\underline{3\,0} \\
010 \\
\underline{6} \\
4\,0 \\
\underline{3\,6} \\
4
\end{array}
$$

24. $8\dfrac{3}{4} \div 1\dfrac{1}{3} \times 1\dfrac{3}{4} \div \dfrac{1}{12}$

$$= \dfrac{\overset{}{35}}{\underset{1}{\cancel{4}}} \times \dfrac{3}{4} \times \dfrac{7}{4} \times \dfrac{\overset{3}{\cancel{12}}}{1} = \dfrac{\mathbf{2205}}{\mathbf{16}}$$

25. $\dfrac{5}{69}\left[\dfrac{1}{7}(4 + 3) + \dfrac{11}{2}\right] = \dfrac{5}{69}\left[\dfrac{7}{7} + \dfrac{11}{2}\right]$

$$= \dfrac{5}{69}\left[\dfrac{13}{2}\right] = \dfrac{\mathbf{65}}{\mathbf{138}}$$

26. $\dfrac{1}{6}\left(2\dfrac{1}{3} + \dfrac{1}{2}\right) + \dfrac{2}{3} = \dfrac{1}{6}\left(\dfrac{7}{3} + \dfrac{1}{2}\right) + \dfrac{2}{3}$

$$= \dfrac{1}{6}\left(\dfrac{14}{6} + \dfrac{3}{6}\right) + \dfrac{2}{3} = \dfrac{1}{6} \cdot \dfrac{17}{6} + \dfrac{2}{3}$$

$$= \dfrac{17}{36} + \dfrac{24}{36} = \dfrac{\mathbf{41}}{\mathbf{36}}$$

27. $m^n = (6^3) = 6 \cdot 6 \cdot 6 = \mathbf{216}$

28. $xyt - yt + x = (6)\left(\dfrac{1}{3}\right)(12) - \left(\dfrac{1}{3}\right)(12) + (6)$

$$= 24 - 4 + 6$$
$$= \mathbf{26}$$

29. $\sqrt[x]{125} = \sqrt[3]{125} = \sqrt[3]{5^3} = \mathbf{5}$

30. (a) $100{,}000 = \mathbf{1.0 \times 10^5}$

(b) $0.0072 = \mathbf{7.2 \times 10^{-3}}$

(c) $3.2 \times 10^5 = \mathbf{320{,}000}$

(d) $3.2 \times 10^{-5} = \mathbf{0.000032}$

PROBLEM SET 55

1. $\dfrac{\textbf{70 items}}{\textbf{\$3500}}, \dfrac{\textbf{\$3500}}{\textbf{70 items}}$

$720 \, \cancel{\text{items}} \times \dfrac{\$3500}{70 \, \cancel{\text{items}}} = \mathbf{\$36{,}000}$

2. $\text{Avg} = \dfrac{(40 \times 10) + (59 \times 30) + (1 \times 300)}{40 + 59 + 1}$

$$= \dfrac{2470}{100} = \textbf{24.7 shillings}$$

3. $\begin{array}{r} 900{,}062 \\ -\ 202{,}020 \\ \hline \mathbf{698{,}042} \textbf{ crones and curmudgeons} \end{array}$

4. $0.3 \times A = 36$

$$\dfrac{0.3 \times A}{0.3} = \dfrac{36}{0.3}$$
$$A = \mathbf{120}$$

5. $D \times 480 = 60$

$$\dfrac{D \times 480}{480} = \dfrac{60}{480}$$
$$D = \mathbf{0.125}$$

6. (a) $0.000387 = \mathbf{3.87 \times 10^{-4}}$

(b) $8.69 \times 10^{11} = \mathbf{869{,}000{,}000{,}000}$

7. (a) 1 percent of $37 = \dfrac{37}{100} = \mathbf{0.37}$

(b) 132 percent of $37 = 132 \cdot 0.37 = \mathbf{48.84}$

8. $\dfrac{3}{40} = \dfrac{75}{1000} = 0.075$

$42\% = \dfrac{42}{100} = \dfrac{21}{50}$

FRACTION	DECIMAL	PERCENT
$\dfrac{51}{100}$	0.51	51%
$\dfrac{3}{40}$	(a) **0.075**	(b) **7.5%**
(c) $\dfrac{\mathbf{21}}{\mathbf{50}}$	(d) **0.42**	42%

9.

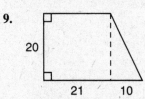

$A_{\text{Base}} = A_{\text{Rectangle}} + A_{\text{Triangle}}$

$$= (20)(21) \text{ cm}^2 + \dfrac{(10)(20)}{2} \text{ cm}^2$$

$$= 420 \text{ cm}^2 + 100 \text{ cm}^2$$

$$= 520 \text{ cm}^2$$

$V = A_{\text{Base}} \times \text{height}$

$$= 520 \text{ cm}^2 \times 10 \text{ cm}$$

$$= \mathbf{5200 \text{ cm}^3}$$

10. $10 = 2 \cdot 5$

$15 = 3 \cdot 5$

$25 = 5 \cdot 5$

$\text{LCM}(10, 15, 25) = 2 \cdot 3 \cdot 5 \cdot 5 = \mathbf{150}$

11. $\dfrac{6}{7} \times 98 = B$

$84 = B$

12. $F \times 64 = 48$

$\dfrac{F \times 64}{64} = \dfrac{48}{64}$

$F = \dfrac{3}{4}$

13. $\dfrac{12}{13}x = 60$

$\left(\dfrac{13}{12}\right)\dfrac{12}{13}x = \left(\dfrac{13}{12}\right)60$

$x = 65$

14. $x + \dfrac{3}{7} = \dfrac{29}{14}$

$x + \dfrac{3}{7} - \dfrac{3}{7} = \dfrac{29}{14} - \dfrac{3}{7}$

$x = \dfrac{29}{14} - \dfrac{6}{14} = \dfrac{23}{14}$

15. $\dfrac{14}{15} = \dfrac{3}{V}$

$14V = 45$

$V = \dfrac{45}{14}$

16. $\dfrac{6}{x} = \dfrac{4}{9}$

$4x = 54$

$\dfrac{4x}{4} = \dfrac{54}{4}$

$x = \dfrac{27}{2}$

17. $\sqrt{144} - 3[(11 - 3) \div (3 - 1) - 4]$

$= 12 - 3[8 \div 2 - 4] = 12 - 3[4 - 4]$

$= 12 - 3 \cdot 0 = 12$

18. $2^2 + 2[2(3 + 1)(3 - 2) - 1]$

$= 4 + 2[2(4)(1) - 1] = 4 + 2[7] = 4 + 14$

$= 18$

19. $4\dfrac{3}{5} \cdot \dfrac{2}{3} - \dfrac{14}{15} = \dfrac{23}{5} \cdot \dfrac{2}{3} - \dfrac{14}{15} = \dfrac{46}{15} - \dfrac{14}{15}$

$= \dfrac{32}{15}$

20. $5\dfrac{2}{3} - 3\dfrac{5}{6} + \dfrac{5}{18} = 5\dfrac{12}{18} - 3\dfrac{15}{18} + \dfrac{5}{18} = 2\dfrac{2}{18}$

$= 2\dfrac{1}{9}$

21. $3^2 + 2^2 - 3 + \sqrt[3]{8} = 9 + 4 - 3 + 2 = 12$

22.
$$\begin{array}{r} 197.3 \\ \times\ 0.013 \\ \hline 5919 \\ 1973 \\ \hline \mathbf{2.5649} \end{array}$$

23.
$$\begin{array}{r} 6211.890 \\ -\ \ \ \ 8.987 \\ \hline \mathbf{6202.903} \end{array}$$

24. $\dfrac{192.03}{0.05}$

$$\begin{array}{r} \mathbf{3840.6} \\ 5\overline{)19{,}203.0} \\ \underline{15} \\ 4\ 2 \\ \underline{4\ 0} \\ 20 \\ \underline{20} \\ 03\ 0 \\ \underline{3\ 0} \\ 0 \end{array}$$

25. $\dfrac{1\dfrac{3}{4}}{2\dfrac{1}{5}} = \dfrac{\dfrac{7}{4}}{\dfrac{11}{5}} = \dfrac{7}{4} \cdot \dfrac{5}{11} = \dfrac{35}{44}$

26. $\dfrac{5}{3}\left(\dfrac{1}{7} + \dfrac{3}{8}\right) - \dfrac{1}{4} = \dfrac{5}{3}\left(\dfrac{8}{56} + \dfrac{21}{56}\right) - \dfrac{1}{4}$

$= \dfrac{5}{3}\left(\dfrac{29}{56}\right) - \dfrac{1}{4} = \dfrac{145}{168} - \dfrac{42}{168} = \dfrac{103}{168}$

27. $\dfrac{1}{4}\left(2\dfrac{1}{4} - \dfrac{1}{8}\right) + \dfrac{3}{16} = \dfrac{1}{4}\left(2\dfrac{2}{8} - \dfrac{1}{8}\right) + \dfrac{3}{16}$

$= \dfrac{1}{4}\left(2\dfrac{1}{8}\right) + \dfrac{3}{16} = \dfrac{17}{32} + \dfrac{6}{32} = \dfrac{23}{32}$

28. $xyz + xz - z = \left(\dfrac{1}{3}\right)(9)(12) + \left(\dfrac{1}{3}\right)(12) - (12)$

$= 36 + 4 - 12$

$= 28$

29. $\sqrt[z]{y} + x^2 = \left(\sqrt[3]{8}\right) + (2)^2 = 2 + 4 = 6$

30.

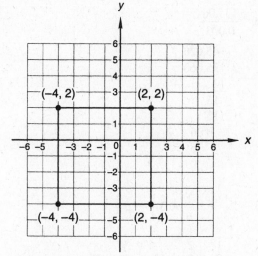

Length of one side = 6 units

Perimeter = 6 + 6 + 6 + 6 = **24 units**

Area = 6 · 6 = **36 units²**

PROBLEM SET 56

1. (a) $\dfrac{\$780}{10\ \text{tires}}$, $\dfrac{10\ \text{tires}}{\$780}$

(b) $58\ \cancel{\text{tires}} \times \dfrac{\$780}{10\ \cancel{\text{tires}}} = \$4524$

(c) $156\ \cancel{\text{dollars}} \times \dfrac{10\ \text{tires}}{780\ \cancel{\text{dollars}}} = \textbf{2 tires}$

2.
$$\begin{array}{r} 9,000,047 \\ -\ 8,793,215 \\ \hline \textbf{206,832} \end{array}$$

3.
$$1200 + 900 + 840 + N = 4(998)$$
$$2940 + N = 3992$$
$$N = \textbf{1052 lb}$$

4. $[60 - (20 + 10)]$ in. = **30 in.**

5. $\dfrac{3}{8} = \dfrac{125}{125} \cdot \dfrac{3}{8} = \dfrac{375}{1000} = 0.375$

$38\% = \dfrac{38}{100} = \dfrac{19}{50}$

FRACTION	DECIMAL	PERCENT
$\dfrac{51}{100}$	0.51	51%
$\dfrac{3}{8}$	(a) **0.375**	(b) **37.5%**
(c) $\dfrac{19}{50}$	(d) **0.38**	38%

6. (a) 1 percent of $52 = \dfrac{52}{100} = \textbf{0.52}$

(b) $0.52 \times 140 = \textbf{72.8}$

7. $\dfrac{3}{4} \cdot A = 60$

$\dfrac{4}{3} \cdot \dfrac{3}{4} \cdot A = 60 \cdot \dfrac{4}{3}$

$A = \textbf{80}$

8. $0.6 \times A = 42$

$\dfrac{0.6 \times A}{0.6} = \dfrac{42}{0.6}$

$A = \textbf{70}$

9. $F \times 57 = 45$

$\dfrac{F \times 57}{57} = \dfrac{45}{57}$

$F = \dfrac{\textbf{15}}{\textbf{19}}$

10. $D \times 280 = 70$

$D = \dfrac{70}{280} = \dfrac{1}{4} = \textbf{0.25}$

11. $\dfrac{4}{9} \times 99 = B$

$\textbf{44} = B$

12. $A_{\text{Base}} = \left[\dfrac{12(16)}{2} + 16(20) + \dfrac{15(16)}{2} \right] \text{cm}^2$

$= [96 + 320 + 120]\ \text{cm}^2$

$= \textbf{536 cm}^2$

Volume = $A_{\text{Base}} \times$ height

$= 536\ \text{cm}^2 \times 5\ \text{cm}$

$= \textbf{2680 cm}^3$

13. $6\dfrac{16}{17} = \dfrac{17 \times 6 + 16}{17} = \dfrac{\textbf{118}}{\textbf{17}}$

14. $\dfrac{x}{9} = \dfrac{5}{6}$

$6x = 45$

$\dfrac{6x}{6} = \dfrac{45}{6}$

$x = \dfrac{\textbf{15}}{\textbf{2}}$

15.
$$\frac{5}{x} = \frac{2}{3}$$
$$2x = 15$$
$$x = \frac{15}{2}$$

16.
$$3\frac{1}{5}x = 8$$
$$\frac{16}{5}x = 8$$
$$\frac{5}{16} \cdot \frac{16}{5}x = \frac{\overset{1}{\cancel{8}}}{1} \cdot \frac{5}{\underset{2}{\cancel{16}}}$$
$$x = \frac{5}{2}$$

17.
$$x + 1\frac{3}{5} = 3\frac{7}{10}$$
$$x + 1\frac{3}{5} - 1\frac{3}{5} = 3\frac{7}{10} - 1\frac{3}{5}$$
$$x = 3\frac{7}{10} - 1\frac{6}{10} = 2\frac{1}{10} = \frac{21}{10}$$

18. $72 - 3[(14 - 4) \div (3 - 1) - 4]$
$= 72 - 3[10 \div 2 - 4] = 72 - 3[5 - 4]$
$= 72 - 3 = \mathbf{69}$

19. $3^3 + 2^2[2(2 + 1)(2 - 1) - 5]$
$= 27 + 4[2(3)(1) - 5] = 27 + 4[1] = \mathbf{31}$

20. $4\frac{4}{5} - 3\frac{2}{3} + \frac{7}{15} = 4\frac{12}{15} - 3\frac{10}{15} + \frac{7}{15}$
$= 1\frac{9}{15} = 1\frac{3}{5}$

21. $2^3 + 3^2 - \sqrt[3]{27} = 8 + 9 - 3 = \mathbf{14}$

22.
$$\begin{array}{r} 132.7 \\ \times\ 0.012 \\ \hline 2654 \\ 1327 \\ \hline \mathbf{1.5924} \end{array}$$

23.
$$\begin{array}{r} 18,251.300 \\ -\quad 62.982 \\ \hline \mathbf{18,188.318} \end{array}$$

24.
$$\frac{135.06}{0.003}$$

$$\begin{array}{r} \mathbf{45,020} \\ 3\overline{)135,060} \\ \underline{12} \\ 15 \\ \underline{15} \\ 0\ 06 \\ \underline{6} \\ 0 \end{array}$$

25. $2\frac{1}{3} \div 1\frac{1}{6} \times 3\frac{1}{4} \div 1\frac{1}{3}$

$$= \frac{\overset{1}{\cancel{7}}}{\underset{1}{\cancel{3}}} \times \frac{\overset{\overset{1}{\cancel{2}}}{\cancel{6}}}{\underset{1}{\cancel{7}}} \times \frac{13}{\underset{2}{\cancel{4}}} \times \frac{3}{4} = \frac{39}{8}$$

26. $\frac{1}{3}\left(\frac{1}{6} + \frac{5}{12}\right) - \frac{1}{36} = \frac{1}{3}\left(\frac{2}{12} + \frac{5}{12}\right) - \frac{1}{36}$

$$\frac{1}{3}\left(\frac{7}{12}\right) - \frac{1}{36} = \frac{7}{36} - \frac{1}{36} = \frac{6}{36} = \frac{1}{6}$$

27. $\frac{1}{4}\left(3\frac{1}{3} - \frac{1}{6}\right) + \frac{1}{12} = \frac{1}{4}\left(\frac{20}{6} - \frac{1}{6}\right) + \frac{1}{12}$

$$= \frac{1}{4}\left(\frac{19}{6}\right) + \frac{1}{12} = \frac{19}{24} + \frac{2}{24} = \frac{21}{24} = \frac{7}{8}$$

28. $xy + x + xyz - z$

$$= (24)\left(\frac{1}{6}\right) + (24) + (24)\left(\frac{1}{6}\right)(3) - (3)$$
$$= 4 + 24 + 12 - 3$$
$$= \mathbf{37}$$

29. $\sqrt[p]{q} + pq = (\sqrt[3]{8}) + (3)(8) = 2 + 24 = \mathbf{26}$

30. (a) $16,000,000,000 = \mathbf{1.6 \times 10^{10}}$

(b) $1.6 \times 10^{-8} = \mathbf{0.000000016}$

PROBLEM SET 57

1. $\frac{80\ \text{trees}}{\$3520}, \frac{\$3520}{80\ \text{trees}}$

$10\ \cancel{\text{trees}} \times \frac{\$3520}{80\ \cancel{\text{trees}}} = \mathbf{\$440}$

2. $\text{Avg} = \frac{(2 \times 43) + (3 \times 58)}{5} = \frac{260}{5}$

$= \mathbf{52\ lb}$

3. $4400 - (1436 + 1892) = 4400 - 3328 = \mathbf{1072}$

4. $5(97) + 642 = 485 + 642 = \mathbf{1127}$

5. $76\% = 0.76 = \dfrac{76}{100} = \dfrac{19}{25}$

$0.6 = 60\% = \dfrac{60}{100} = \dfrac{3}{5}$

Fraction	Decimal	Percent
$\frac{51}{100}$	0.51	51%
(a) $\frac{19}{25}$	(b) **0.76**	76%
(c) $\frac{3}{5}$	0.6	(d) **60%**

6. (a) 1 percent of $3600 = \dfrac{3600}{100} = \mathbf{36}$

(b) $16 \times 36 = \mathbf{576}$

7. $10{,}000 \text{ in.}^2 \times \dfrac{1 \text{ ft}}{12 \text{ in.}} \times \dfrac{1 \text{ ft}}{12 \text{ in.}}$

$= \dfrac{\mathbf{10{,}000}}{\mathbf{(12)(12)}} \text{ ft}^2 \approx \mathbf{69.44 \ ft^2}$

8. $0.8 \times A = 48$

$\dfrac{0.8 \times A}{0.8} = \dfrac{48}{0.8}$

$A = \mathbf{60}$

9. $F \times 60 = 48$

$\dfrac{F \times 60}{60} = \dfrac{48}{60}$

$F = \dfrac{\mathbf{4}}{\mathbf{5}}$

10. $D \times 350 = 70$

$D = \dfrac{70}{350} = \dfrac{1}{5} = \mathbf{0.2}$

11. $F \cdot 2\dfrac{7}{9} = 8\dfrac{1}{3}$

$F \cdot \dfrac{25}{9} = \dfrac{25}{3}$

$F \cdot \dfrac{\overset{1}{\cancel{25}}}{\underset{1}{\cancel{9}}} \cdot \dfrac{\overset{1}{\cancel{9}}}{\underset{1}{\cancel{25}}} = \dfrac{\overset{1}{\cancel{25}}}{\underset{1}{\cancel{3}}} \cdot \dfrac{\overset{3}{\cancel{9}}}{\underset{1}{\cancel{25}}}$

$F = \mathbf{3}$

12. $5\dfrac{5}{8} \cdot A = 4\dfrac{1}{4}$

$\dfrac{45}{8} \cdot A = \dfrac{17}{4}$

$\dfrac{\overset{1}{\cancel{8}}}{\underset{1}{\cancel{45}}} \cdot \dfrac{\overset{1}{\cancel{45}}}{\underset{1}{\cancel{8}}} \cdot A = \dfrac{17}{\underset{1}{\cancel{4}}} \cdot \dfrac{\overset{2}{\cancel{8}}}{45}$

$A = \dfrac{\mathbf{34}}{\mathbf{45}}$

13.

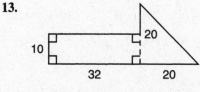

$A_{\text{Base}} = \left[32(10) + \dfrac{20(20)}{2} \right] \text{ft}^2$

$= [430 + 200] \text{ ft}^2$

$= 520 \text{ ft}^2$

$V = A_{\text{Base}} \times \text{height}$

$= 520 \text{ ft}^2 \times 2 \text{ ft}$

$= \mathbf{1040 \ ft^3}$

14. $S.A. = 2A_{\text{End}} + 2A_{\text{Side}} + A_{\text{Bottom}}$

$= 2\left[\dfrac{(12)(7)}{2} \right] \text{m}^2 + 2(16)(9.22) \text{ m}^2$
$+ (12)(16) \text{ m}^2$

$= 84 \text{ m}^2 + 295.04 \text{ m}^2 + 192 \text{ m}^2$

$= \mathbf{571.04 \ m^2}$

15.

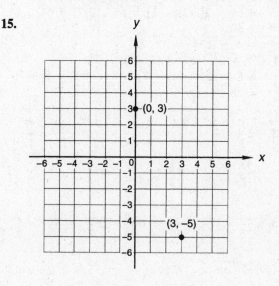

16. $\dfrac{15}{7} = \dfrac{4}{x}$

$15x = 28$

$x = \dfrac{28}{15}$

17. $\dfrac{5}{6} = \dfrac{p}{4}$

$20 = 6p$

$p = \dfrac{20}{6} = \dfrac{10}{3}$

18. $3\dfrac{7}{15}x = 3$

$\dfrac{52}{15}x = 3$

$\dfrac{\cancel{15}^{1}}{\cancel{52}^{1}} \cdot \dfrac{\cancel{52}^{1}}{\cancel{15}^{1}}x = 3 \cdot \dfrac{15}{52}$

$x = \dfrac{45}{52}$

19. $x + 3\dfrac{3}{14} = 4\dfrac{5}{28}$

$x + 3\dfrac{3}{14} - 3\dfrac{3}{14} = 4\dfrac{5}{28} - 3\dfrac{3}{14}$

$x = 3\dfrac{33}{28} - 3\dfrac{6}{28} = \dfrac{27}{28}$

20. $64 - 2[(3-1)(5-2)+1]$
$= 64 - 2[2(3)+1] = 64 - 2[7] = 64 - 14$
$= 50$

21. $2^3 + 2^2[3(2-1)(2+2)-7]$
$= 8 + 4[3(1)(4)-7] = 8 + 4[5] = 8 + 20$
$= 28$

22. $2\dfrac{1}{3} \cdot 1\dfrac{3}{4} - \dfrac{7}{12} = \dfrac{7}{3} \cdot \dfrac{7}{4} - \dfrac{7}{12} = \dfrac{49}{12} - \dfrac{7}{12}$

$\dfrac{42}{12} = \dfrac{7}{2} = 3\dfrac{1}{2}$

23. $6\dfrac{2}{5} - 2\dfrac{1}{4} + \dfrac{3}{40} = 6\dfrac{16}{40} - 2\dfrac{10}{40} + \dfrac{3}{40}$

$= 4\dfrac{9}{40}$

24. $2^4 + 3^3 - 2^3 + \sqrt[3]{8} = 16 + 27 - 8 + 2$
$= 37$

25.
$$\begin{array}{r} 9218.821 \\ -\ \ 61.872 \\ \hline \mathbf{9156.949} \end{array}$$

26. $\dfrac{1}{5}\left(3\dfrac{1}{4} - 2\dfrac{1}{3}\right) + \dfrac{7}{15} = \dfrac{1}{5}\left(\dfrac{39}{12} - \dfrac{28}{12}\right) + \dfrac{7}{15}$

$= \dfrac{1}{5}\left(\dfrac{11}{12}\right) + \dfrac{7}{15} = \dfrac{11}{60} + \dfrac{28}{60} = \dfrac{39}{60} = \dfrac{13}{20}$

27. $\dfrac{2\dfrac{3}{4}}{1\dfrac{7}{8}} = \dfrac{\dfrac{11}{4}}{\dfrac{15}{8}} = \dfrac{11}{\cancel{4}} \cdot \dfrac{\cancel{8}^{2}}{15} = \dfrac{22}{15}$

28. $6\dfrac{2}{3} \div 1\dfrac{1}{6} \times 3\dfrac{1}{3} \div 2\dfrac{4}{5}$

$= \dfrac{\cancel{20}^{10}}{3} \cdot \dfrac{\cancel{6}^{2}}{7} \cdot \dfrac{10}{\cancel{3}_{1}} \cdot \dfrac{5}{\cancel{14}_{7}} = \dfrac{1000}{147}$

29. (a) $1.3 \times 10 = \mathbf{13}$

(b) $0.0392 = \mathbf{3.92 \times 10^{-2}}$

30. (a) $x^y = (3^3) = \mathbf{27}$

(b) $\sqrt[x]{y} = (\sqrt[3]{125}) = \mathbf{5}$

PROBLEM SET 58

1.
$$\begin{array}{r} 69{,}000.00741 \\ -\ 42{,}000.00075 \\ \hline \mathbf{27{,}000.00666} \end{array}$$

2. $143{,}000 - 9614 = \mathbf{133{,}386}$ **more new ideas**

3. $\dfrac{\textbf{53 new ones}}{\textbf{\$742}}, \quad \dfrac{\textbf{\$742}}{\textbf{53 new ones}}$

$25 \ \cancel{\text{new ones}} \times \dfrac{\$742}{53 \ \cancel{\text{new ones}}} = \mathbf{\$350}$

4. $145{,}000 + 5(145{,}000) = 6(145{,}000) = \mathbf{870{,}000}$

5. (a) $\dfrac{195 \text{ mi}}{3 \text{ hr}} = \mathbf{65 \text{ mph}}$

(b) $\dfrac{1 \text{ hr}}{65 \ \cancel{\text{mi}}} \cdot 520 \ \cancel{\text{mi}} = \dfrac{520}{65} \text{ hr} = \mathbf{8 \text{ hr}}$

6. $0.24 = 24\% = \dfrac{24}{100} = \dfrac{6}{25}$

$\dfrac{1}{2} = \dfrac{50}{100} = 50\%$

FRACTION	DECIMAL	PERCENT
$\frac{51}{100}$	0.51	51%
(a) $\mathbf{\frac{6}{25}}$	0.24	(b) **24%**
$\frac{1}{2}$	(c) **0.5**	(d) **50%**

7. (a) $106{,}000{,}000 = \mathbf{1.06 \times 10^8}$

(b) $4.13 \times 10^{-8} = \mathbf{0.0000000413}$

8. $0.7 \times A = 42$

$\dfrac{0.7 \times A}{0.7} = \dfrac{42}{0.7}$

$A = \mathbf{60}$

9. $F \times 72 = 64$

$\dfrac{F \times 72}{72} = \dfrac{64}{72}$

$F = \mathbf{\dfrac{8}{9}}$

10. $D \times 420 = 273$

$\dfrac{D \times 420}{420} = \dfrac{273}{420}$

$D = \mathbf{0.65}$

11. $2\dfrac{2}{7} \cdot 8\dfrac{1}{2} = B$

$\dfrac{\overset{8}{\cancel{16}}}{7} \cdot \dfrac{17}{\underset{1}{\cancel{2}}} = B$

$\dfrac{136}{7} = B$

12. $3\dfrac{1}{3} \cdot A = \dfrac{65}{3}$

$\dfrac{10}{3} \cdot A = \dfrac{65}{3}$

$\dfrac{\overset{1}{\cancel{3}}}{\underset{1}{\cancel{10}}} \cdot \dfrac{\overset{1}{\cancel{10}}}{\underset{1}{\cancel{3}}} \cdot A = \dfrac{\overset{13}{\cancel{65}}}{\underset{1}{\cancel{3}}} \cdot \dfrac{\overset{1}{\cancel{3}}}{\underset{2}{\cancel{10}}}$

$A = \dfrac{13}{2}$

13. $A_{\text{Base}} = \left[17(8) + 10(8) + \dfrac{21(8)}{2} \right] \text{in.}^2$

$= [136 + 80 + 84] \text{ in.}^2$

$= 300 \text{ in.}^2$

Volume $= A_{\text{Base}} \times$ height

$= 300 \text{ in.}^2 \times 3 \text{ in.}$

$= \mathbf{900 \text{ in.}^3}$

14. $\dfrac{16}{5} = \dfrac{3}{x}$

$16x = 15$

$x = \mathbf{\dfrac{15}{16}}$

15. $\dfrac{6}{7} = \dfrac{p}{14}$

$84 = 7p$

$p = \dfrac{84}{7} = \mathbf{12}$

16. $2\dfrac{5}{12}x = 9$

$\dfrac{29}{12}x = \dfrac{9}{1}$

$\dfrac{12}{29} \cdot \dfrac{29}{12}x = \dfrac{9}{1} \cdot \dfrac{12}{29}$

$x = \mathbf{\dfrac{108}{29}}$

17. $x - 2\dfrac{3}{14} = 1\dfrac{1}{21}$

$x - 2\dfrac{3}{14} + 2\dfrac{3}{14} = 1\dfrac{1}{21} + 2\dfrac{3}{14}$

$x = 1\dfrac{2}{42} + 2\dfrac{9}{42} = \mathbf{3\dfrac{11}{42}}$

18. $38 - 2[(6 - 5)(4 - 1) + 1]$
$= 38 - 2[1(3) + 1] = 38 - 2[4] = \mathbf{30}$

19. $2^3 + 3^2[(7 - 2)(3 - 1)3 - 25]$
$= 8 + 9[5(2)(3) - 25] = 8 + 9[5] = 8 + 45$
$= \mathbf{53}$

20. $1\dfrac{1}{3} \cdot 2\dfrac{3}{4} - \dfrac{7}{12} = \dfrac{4}{3} \cdot \dfrac{11}{4} - \dfrac{7}{12}$

$= \dfrac{44}{12} - \dfrac{7}{12} = \dfrac{37}{12} = \mathbf{3\dfrac{1}{12}}$

21. $5\dfrac{3}{5} - 1\dfrac{3}{4} + \dfrac{7}{20} = 5\dfrac{12}{20} - 1\dfrac{15}{20} + \dfrac{7}{20}$

$= 4\dfrac{4}{20} = \mathbf{4\dfrac{1}{5}}$

22. $2^3 + 3^2 - 2^2 + \sqrt{16} = 8 + 9 - 4 + 4 = \mathbf{17}$

23.
$$\begin{array}{r} 16.82 \\ \times\ 0.013 \\ \hline 5046 \\ 1682 \\ \hline \mathbf{0.21866} \end{array}$$

24. $\dfrac{618.21}{0.004}$

$$\begin{array}{r} \mathbf{154{,}552.5} \\ 4\overline{)618{,}210.0} \\ \underline{4} \\ 21 \\ \underline{20} \\ 18 \\ \underline{16} \\ 2\ 2 \\ \underline{2\ 0} \\ 21 \\ \underline{20} \\ 10 \\ \underline{8} \\ 2\ 0 \\ \underline{2\ 0} \\ 0 \end{array}$$

25. $\dfrac{1}{5}\left(\dfrac{2}{3} + \dfrac{1}{2}\right) - \dfrac{2}{15} = \dfrac{1}{5}\left(\dfrac{4}{6} + \dfrac{3}{6}\right) - \dfrac{2}{15}$

$= \dfrac{1}{5}\left(\dfrac{7}{6}\right) - \dfrac{2}{15} = \dfrac{7}{30} - \dfrac{4}{30} = \dfrac{3}{30} = \dfrac{\mathbf{1}}{\mathbf{10}}$

26. $\dfrac{1}{4}\left(2\dfrac{3}{4} - 1\dfrac{1}{8}\right) + \dfrac{3}{16} = \dfrac{1}{4}\left(\dfrac{22}{8} - \dfrac{9}{8}\right) + \dfrac{3}{16}$

$= \dfrac{1}{4}\left(\dfrac{13}{8}\right) + \dfrac{3}{16} = \dfrac{13}{32} + \dfrac{6}{32} = \dfrac{\mathbf{19}}{\mathbf{32}}$

27. $\dfrac{1\frac{4}{7}}{2\frac{3}{4}} = \dfrac{\frac{11}{7}}{\frac{11}{4}} = \dfrac{\overset{1}{\cancel{11}}}{7} \cdot \dfrac{4}{\underset{1}{\cancel{11}}} = \dfrac{\mathbf{4}}{\mathbf{7}}$

28. (a) 1 percent of $19.5 = \dfrac{19.5}{100} = \mathbf{0.195}$

(b) $31 \cdot 0.195 = \mathbf{6.045}$

29. $x^2 + 2xy + y^x = (2)^2 + (2)(2)(3) + (3^2)$
$$= 4 + 12 + 9$$
$$= \mathbf{25}$$

30. $p^q = (2^3) = \mathbf{8}$

---

1. Avg $= \dfrac{(3 \times 42) + (7 \times 12)}{10} = \dfrac{210}{10} = \mathbf{21}$

2. $\dfrac{\textbf{7 big ones}}{\textbf{\$280,000}}, \dfrac{\textbf{\$280,000}}{\textbf{7 big ones}}$

$120{,}000\ \cancel{\text{dollars}} \times \dfrac{7\text{ big ones}}{280{,}000\ \cancel{\text{dollars}}} = \textbf{3 big ones}$

3. $140{,}026 - 132{,}781 = \mathbf{7245\ lb}$

4. (a) $\dfrac{10\text{ mi}}{3\text{ hr}} = 3\dfrac{\mathbf{1}}{\mathbf{3}}$ **mph**

(b) $25\ \cancel{\text{mi}} \cdot \dfrac{3\text{ hr}}{10\ \cancel{\text{mi}}} = \dfrac{75}{10}\text{ hr} = 7\dfrac{\mathbf{1}}{\mathbf{2}}$ **hr**

5. $16\% = 0.16 = \dfrac{16}{100} = \dfrac{4}{25}$

$\dfrac{4}{5} = \dfrac{80}{100} = 80\%$

Fraction	Decimal	Percent
$\frac{51}{100}$	0.51	51%
(a) $\frac{4}{25}$	(b) **0.16**	16%
$\frac{4}{5}$	(c) **0.8**	(d) **80%**

6. (a) $0.093 = \mathbf{9.3 \times 10^{-2}}$

(b) $1.2 \times 10^6 = \mathbf{1{,}200{,}000}$

7. $0.9 \times A = 72$

$\dfrac{0.9 \times A}{0.9} = \dfrac{72}{0.9}$

$A = \mathbf{80}$

8. $F \cdot 6\dfrac{1}{2} = 8\dfrac{1}{4}$

$F \cdot \dfrac{13}{2} = \dfrac{33}{4}$

$F \cdot \dfrac{\overset{1}{\cancel{13}}}{\underset{1}{\cancel{2}}} \cdot \dfrac{\overset{1}{\cancel{2}}}{\underset{1}{\cancel{13}}} = \dfrac{33}{\underset{2}{\cancel{4}}} \cdot \dfrac{\overset{1}{\cancel{2}}}{13}$

$F = \dfrac{\mathbf{33}}{\mathbf{26}}$

9. $D \times 630 = 441$

$\dfrac{D \times 630}{630} = \dfrac{441}{630}$

$D = \mathbf{0.7}$

10.
$$8\frac{1}{4} \cdot A = 7\frac{1}{3}$$

$$\frac{33}{4} \cdot A = \frac{22}{3}$$

$$\frac{\overset{1}{\cancel{4}}}{\underset{1}{\cancel{33}}} \cdot \frac{\overset{1}{\cancel{33}}}{\underset{1}{\cancel{4}}} \cdot A = \frac{\overset{2}{\cancel{22}}}{3} \cdot \frac{4}{\underset{3}{\cancel{33}}}$$

$$A = \frac{8}{9}$$

11.
$$1\frac{1}{4} \cdot 8\frac{2}{3} = B$$

$$\frac{5}{\underset{2}{\cancel{4}}} \cdot \frac{\overset{13}{\cancel{26}}}{3} = B$$

$$\frac{65}{6} = B$$

12.

$A_{\text{Base}} = A_{\text{Rectangle}} + A_{\text{Triangle}}$

$= \left[8(45) + \dfrac{12(38)}{2} \right]$ in.^{2}

$= [360 + 228]$ in.^{2}

$= \mathbf{588}$ **in.**^{2}

Volume $= A_{\text{Base}} \times$ height

$= 588$ in.$^2 \times 4$ in.

$= \mathbf{2352}$ **in.**^{3}

13. (a) 1 percent of $192 = \dfrac{192}{100} = \mathbf{1.92}$

(b) $45 \cdot 1.92 = \mathbf{86.4}$

14. $S.A. = 2A_{\text{Front}} + 2A_{\text{Side}} + 2A_{\text{Top}}$

$= [2(4)(5) + 2(10)(4) + 2(10)(5)]$ m^2

$= \mathbf{220}$ **m**^{2}

15.
$$\frac{\frac{1}{5}}{\frac{1}{4}} = \frac{\frac{9}{10}}{x}$$

$$\frac{1}{5}x = \frac{9}{40}$$

$$\frac{5}{1} \cdot \frac{1}{5}x = \frac{9}{\underset{8}{\cancel{40}}} \cdot \frac{\overset{1}{\cancel{5}}}{1}$$

$$x = \frac{9}{8}$$

16.
$$\frac{\frac{2}{3}}{\frac{2}{5}} = \frac{p}{\frac{7}{12}}$$

$$\frac{2}{5}p = \frac{14}{36}$$

$$\frac{5}{2} \cdot \frac{2}{5}p = \frac{\overset{7}{\cancel{14}}}{36} \cdot \frac{5}{\underset{1}{\cancel{2}}}$$

$$p = \frac{35}{36}$$

17.
$$1\frac{3}{5}x = 6$$

$$\frac{8}{5}x = 6$$

$$\left(\frac{5}{8}\right)\frac{8}{5}x = 6\left(\frac{5}{8}\right)$$

$$x = \frac{15}{4}$$

18.
$$3\frac{1}{4}p = 5$$

$$\frac{13}{4}p = 5$$

$$\left(\frac{4}{13}\right)\frac{13}{4}p = 5\left(\frac{4}{13}\right)$$

$$p = \frac{20}{13}$$

19. $49 - 2\big[(5 - 2^2)(4 + 2) - 5\big]$
$= 49 - 2[1(6) - 5] = 49 - 2[1] = \mathbf{47}$

20. $\sqrt[3]{8} + 2^3\big[2^2(2^3 - 5) - 4\big] = 2 + 8[4(3) - 4]$
$= 2 + 8[8] = 2 + 64 = \mathbf{66}$

21. $2\dfrac{1}{3} \cdot 3\dfrac{1}{4} - \dfrac{11}{12} = \dfrac{7}{3} \cdot \dfrac{13}{4} - \dfrac{11}{12}$

$= \dfrac{91}{12} - \dfrac{11}{12} = \dfrac{80}{12} = \dfrac{20}{3} = \mathbf{6\dfrac{2}{3}}$

22. $14\dfrac{2}{3} - 12\dfrac{7}{8} + \dfrac{11}{48} = 14\dfrac{32}{48} - 12\dfrac{42}{48} + \dfrac{11}{48}$

$= \mathbf{2\dfrac{1}{48}}$

23.
$$\begin{array}{r} 171.6 \\ \times\ 0.007 \\ \hline \mathbf{1.2012} \end{array}$$

24.
$$\begin{array}{r} 1171.610 \\ -\quad 13.321 \\ \hline \mathbf{1158.289} \end{array}$$

25. $\dfrac{611.51}{0.03}$

$$\begin{array}{r} 20{,}383.\overline{6} \\ 3\overline{)61{,}151.0} \\ \underline{6} \\ 01\,1 \\ \underline{9} \\ 25 \\ \underline{24} \\ 11 \\ \underline{9} \\ 2\,0 \\ \underline{1\,8} \\ 2 \end{array}$$

26. $\dfrac{1}{6}\left(\dfrac{1}{3}+\dfrac{1}{2}\right)-\dfrac{5}{36}=\dfrac{1}{6}\left(\dfrac{2}{6}+\dfrac{3}{6}\right)-\dfrac{5}{36}$

$\quad=\dfrac{1}{6}\left(\dfrac{5}{6}\right)-\dfrac{5}{36}=\dfrac{5}{36}-\dfrac{5}{36}=\mathbf{0}$

27. $\dfrac{1}{5}\left(\dfrac{1}{4}-\dfrac{1}{8}\right)+2\dfrac{7}{8}=\dfrac{1}{5}\left(\dfrac{2}{8}-\dfrac{1}{8}\right)+2\dfrac{7}{8}$

$\quad=\dfrac{1}{5}\left(\dfrac{1}{8}\right)+2\dfrac{7}{8}=\dfrac{1}{40}+2\dfrac{35}{40}=2\dfrac{36}{40}=\mathbf{2\dfrac{9}{10}}$

28. $\dfrac{6\dfrac{2}{3}}{2\dfrac{1}{4}}=\dfrac{\dfrac{20}{3}}{\dfrac{9}{4}}=\dfrac{20}{3}\cdot\dfrac{4}{9}=\mathbf{\dfrac{80}{27}}$

29. $6\dfrac{1}{4}\div3\dfrac{2}{3}\times2\dfrac{1}{4}\div\dfrac{1}{8}$

$\quad=\dfrac{25}{\cancelto{1}{4}}\times\dfrac{3}{11}\times\dfrac{9}{\cancelto{2}{4}}\times\dfrac{\cancelto{2}{8}}{1}=\mathbf{\dfrac{675}{22}}$

30. $x^y+3xy^2+3x^2y+y^x$

$\quad=(1^2)+(3)(1)(2)^2+3(1)^2(2)+(2^1)$

$\quad=1+12+6+2$

$\quad=\mathbf{21}$

PROBLEM SET 60

1. $\mathbf{\dfrac{560\ red\ ones}{7\ pesos}},\ \mathbf{\dfrac{7\ pesos}{560\ red\ ones}}$

$60\ \text{pesos}\times\dfrac{560\ \text{red ones}}{7\ \text{pesos}}=\mathbf{4800\ red\ ones}$

2. $\dfrac{56\ \text{mi}}{8\ \text{hr}}\times14\ \text{hr}=\mathbf{98\ mi}$

3. $\dfrac{40\ \text{mi}}{5\ \text{hr}}-\dfrac{60\ \text{mi}}{10\ \text{hr}}=8\ \text{mph}-6\ \text{mph}=\mathbf{2\ mph}$

4. $\text{Avg}=\dfrac{(40+60+40)\ \text{mi}}{(5+10+20)\ \text{hr}}=\dfrac{140\ \text{mi}}{35\ \text{hr}}$

$\quad=\mathbf{4\ mph}$

5. $0.72=\dfrac{72}{100}=\dfrac{18}{25}$

$\dfrac{3}{5}=\dfrac{60}{100}=0.6$

Fraction	Decimal	Percent
$\dfrac{51}{100}$	0.51	51%
(a) $\dfrac{18}{25}$	0.72	(b) **72%**
$\dfrac{3}{5}$	(c) **0.6**	(d) **60%**

6. (a) $d=2r=2\cdot8\ \text{ft}=\mathbf{16\ ft}$

(b) Circumference $=\pi d=\mathbf{16\pi\ ft}\approx\mathbf{50.24\ ft}$

(c) Area $=\pi r^2=\pi(8\ \text{ft})^2$

$\quad=\mathbf{64\pi\ ft^2}\approx\mathbf{200.96\ ft^2}$

7. $0.7\times A=490$

$\dfrac{0.7\times A}{0.7}=\dfrac{490}{0.7}$

$A=\mathbf{700}$

8. $F\cdot3\dfrac{1}{9}=\dfrac{24}{5}$

$F\cdot\dfrac{28}{9}=\dfrac{24}{5}$

$F\cdot\dfrac{9}{28}\cdot\dfrac{28}{9}=\dfrac{\cancelto{6}{24}}{5}\cdot\dfrac{9}{\cancelto{7}{28}}$

$F=\mathbf{\dfrac{54}{35}}$

9. $D\times720=420$

$D=\dfrac{420}{720}=\dfrac{7}{12}=\mathbf{0.58\overline{3}}$

10.
$$3\frac{1}{3} \cdot A = 4\frac{1}{2}$$
$$\frac{10}{3} \cdot A = \frac{9}{2}$$
$$\frac{3}{10} \cdot \frac{10}{3} \cdot A = \frac{9}{2} \cdot \frac{3}{10}$$
$$A = \frac{27}{20}$$

11.
$$2\frac{1}{4} \cdot 2\frac{1}{3} = B$$
$$\frac{\overset{9}{\cancel{9}}}{4} \cdot \frac{7}{\underset{1}{\cancel{3}}} = B$$
$$\frac{21}{4} = B$$

12.

$$A_{Base} = A_{Rectangle} + A_{Triangle}$$
$$= \left[26(11) + \frac{26(9)}{2}\right] \text{ft}^2$$
$$= [286 + 117] \text{ ft}^2$$
$$= 403 \text{ ft}^2$$

$$\begin{aligned}\text{Volume} &= A_{Base} \times \text{height}\\ &= 403 \text{ ft}^2 \times 6 \text{ ft}\\ &= \mathbf{2418 \text{ ft}^3}\end{aligned}$$

13. (a) 1 percent of $5.3 = \dfrac{5.3}{100} = \mathbf{0.053}$

(b) $25 \cdot 0.053 = \mathbf{1.325}$

14. $8000 \, \cancel{\text{km}^2} \times \dfrac{1000 \text{ m}}{1 \, \cancel{\text{km}}} \times \dfrac{1000 \text{ m}}{1 \, \cancel{\text{km}}} = \mathbf{8 \times 10^9 \text{ m}^2}$

15.
$$\frac{\frac{3}{16}}{\frac{3}{5}} = \frac{8}{x}$$
$$\frac{3}{16}x = \frac{24}{5}$$
$$x = \frac{24}{5} \cdot \frac{16}{\underset{1}{\cancel{3}}}$$
$$x = \frac{128}{5}$$

16.
$$\frac{\frac{3}{5}}{\frac{1}{4}} = \frac{\frac{5}{6}}{p}$$
$$\frac{3}{5}p = \frac{5}{24}$$
$$p = \frac{5}{24} \cdot \frac{5}{3}$$
$$p = \frac{25}{72}$$

17.
$$2\frac{4}{5}x = 5$$
$$\frac{14}{5}x = 5$$
$$\left(\frac{5}{14}\right)\frac{14}{5}x = 5\left(\frac{5}{14}\right)$$
$$x = \frac{25}{14}$$

18.
$$4\frac{1}{5}p = 6$$
$$\frac{21}{5}p = 6$$
$$\left(\frac{5}{21}\right)\frac{21}{5}p = 6\left(\frac{5}{21}\right)$$
$$p = \frac{10}{7}$$

19. $54 - 2\left[(6 - 2^2)(3 + 1) - \sqrt{25}\right]$
$= 54 - 2[2(4) - 5] = 54 - 2[3] = 54 - 6$
$= \mathbf{48}$

20. $\sqrt{16} + 2^2\left[2(3^2 - 2^2) - 5\right]$
$= 4 + 4[2(5) - 5] = 4 + 4[5]$
$= 4 + 20 = \mathbf{24}$

21. $3\frac{1}{3} \cdot 2\frac{1}{4} - \frac{5}{12} = \frac{10}{3} \cdot \frac{9}{4} - \frac{5}{12}$
$= \frac{90}{12} - \frac{5}{12} = \mathbf{\frac{85}{12}}$

22. $15\frac{6}{7} - 3\frac{3}{14} + \frac{9}{14} = 15\frac{12}{14} - 3\frac{3}{14} + \frac{9}{14}$
$= 12\frac{18}{14} = 13\frac{4}{14} = \mathbf{13\frac{2}{7}}$

23.
$$\begin{array}{r} 621.8 \\ \times\ 0.018 \\ \hline 49744 \\ 6218 \\ \hline \mathbf{11.1924} \end{array}$$

24.
$$\begin{array}{r} 2612.810 \\ -14.313 \\ \hline 2598.497 \end{array}$$

25. $\dfrac{1821.5}{0.7}$

$$\begin{array}{r} 2602.\overline{142857} \\ 7)\overline{18{,}215.000000} \\ \underline{14} \\ 4\,2 \\ \underline{4\,2} \\ 015 \\ \underline{14} \\ 1\,0 \\ \underline{7} \\ 30 \\ \underline{28} \\ 20 \\ \underline{14} \\ 60 \\ \underline{56} \\ 40 \\ \underline{35} \\ 50 \\ \underline{49} \\ 1 \end{array}$$

26. $\dfrac{1}{5}\left(\dfrac{1}{2} + 2\dfrac{1}{3}\right) - \dfrac{4}{15} = \dfrac{1}{5}\left(\dfrac{3}{6} + \dfrac{14}{6}\right) - \dfrac{4}{15}$

$= \dfrac{1}{5}\left(\dfrac{17}{6}\right) - \dfrac{4}{15} = \dfrac{17}{30} - \dfrac{8}{30} = \dfrac{9}{30} = \mathbf{\dfrac{3}{10}}$

27. $\dfrac{1}{4}\left(\dfrac{1}{6} + 1\dfrac{1}{4}\right) - \dfrac{1}{4} = \dfrac{1}{4}\left(\dfrac{2}{12} + \dfrac{15}{12}\right) - \dfrac{1}{4}$

$= \dfrac{1}{4}\left(\dfrac{17}{12}\right) - \dfrac{1}{4} = \dfrac{17}{48} - \dfrac{12}{48} = \mathbf{\dfrac{5}{48}}$

28. $3\dfrac{1}{2} \times 6\dfrac{1}{3} \div 2\dfrac{1}{3} \times 1\dfrac{1}{3}$

$= \dfrac{\overset{1}{\cancel{7}}}{\underset{1}{\cancel{2}}} \times \dfrac{19}{\underset{1}{\cancel{3}}} \times \dfrac{\overset{1}{\cancel{3}}}{\underset{1}{\cancel{7}}} \times \dfrac{\overset{2}{\cancel{4}}}{3} = \mathbf{\dfrac{38}{3}}$

29. (a) **(−2, −1)**

(b) **(−4, 2)**

(c) **(2, 0)**

30. $xyz + z^z + y^z - y^z$

$= \left(\dfrac{1}{3}\right)(6)(2) + (2^2) + (6^2) - (6^2)$

$= 4 + 4 + 36 - 36$

$= \mathbf{8}$

1. $\dfrac{6900 \text{ ft}}{230 \text{ s}}, \dfrac{230 \text{ s}}{6900 \text{ ft}}$

$100{,}000 \cancel{\text{ ft}} \cdot \dfrac{230 \text{ s}}{6900 \cancel{\text{ ft}}} = 100{,}000\left(\dfrac{1}{30}\right)\text{s}$

$= \mathbf{\dfrac{10{,}000}{3}}\,\text{s}$

2. (a) $\dfrac{480 \text{ mi}}{4 \text{ hr}}, \dfrac{4 \text{ hr}}{480 \text{ mi}}$

(b) $\dfrac{480 \text{ mi}}{4 \text{ hr}} = \mathbf{120 \text{ mph}}$

(c) $\dfrac{1 \text{ hr}}{120 \cancel{\text{ mi}}} \times 1440 \cancel{\text{ mi}} = \mathbf{12 \text{ hr}}$

3. $\dfrac{1 \text{ hr}}{60 \cancel{\text{ mi}}} \times 1200 \cancel{\text{ mi}} = \mathbf{20 \text{ hr}}$

4. $\dfrac{1 \text{ day}}{430 \cancel{\text{ ft}}} \times 7310 \cancel{\text{ ft}} = \mathbf{17 \text{ days}}$

5. $0.22 = \dfrac{22}{100} = \dfrac{11}{50}$

$\dfrac{21}{25} = \dfrac{84}{100} = 84\%$

FRACTION	DECIMAL	PERCENT
$\dfrac{51}{100}$	0.51	51%
(a) $\dfrac{11}{50}$	0.22	(b) **22%**
$\dfrac{21}{25}$	(c) **0.84**	(d) **84%**

6. (a) $0.639 = \mathbf{6.39 \times 10^{-1}}$

(b) $7.01 \times 10^4 = \mathbf{70{,}100}$

7. $0.8 \times A = 96$

$\dfrac{0.8 \times A}{0.8} = \dfrac{96}{0.8}$

$A = \mathbf{120}$

8. $F \times 52 = 30$

$\dfrac{F \times 52}{52} = \dfrac{30}{52}$

$F = \mathbf{\dfrac{15}{26}}$

9. $D \times 700 = 581$

$\dfrac{D \times 700}{700} = \dfrac{581}{700}$

$D = \mathbf{0.83}$

10.
$$3\frac{1}{5} \cdot A = 7\frac{1}{3}$$
$$\frac{16}{5} \cdot A = \frac{22}{3}$$
$$\frac{5}{16} \cdot \frac{16}{5} \cdot A = \frac{22}{3} \cdot \frac{5}{16}$$
$$A = \frac{55}{24}$$

11.
$$2\frac{1}{10} \times 1\frac{3}{4} = B$$
$$\frac{21}{10} \times \frac{7}{4} = B$$
$$B = \frac{147}{40}$$

12.

$$A_{\text{Base}} = A_{\text{Rectangle}} + A_{\text{Triangle}}$$
$$= \left[6(42) + \frac{20(17)}{2} \right] \text{ft}^2$$
$$= [252 + 170] \text{ft}^2$$
$$= 422 \text{ ft}^2$$

$$V = A_{\text{Base}} \times \text{height}$$
$$= 422 \text{ ft}^2 \times 3 \text{ ft}$$
$$= 1266 \text{ ft}^3$$

13. $S.A. = 2A_{\text{End}} + 2A_{\text{Side}} + A_{\text{Bottom}}$
$$= \left[2\left(\frac{4(6)}{2} \right) + 2(5)(10) + (6)(10) \right] \text{in.}^2$$
$$= [24 + 100 + 60] \text{in.}^2$$
$$= 184 \text{ in.}^2$$

14. (a) $r = \dfrac{d}{2} = \dfrac{32 \text{ cm}}{2} = 16 \text{ cm}$

(b) Circumference $= \pi d$
$$= 32\pi \text{ cm} \approx 100.48 \text{ cm}$$

(c) Area $= \pi r^2 = \pi(16 \text{ cm})^2$
$$= 256\pi \text{ cm}^2 \approx 803.84 \text{ cm}^2$$

15.
$$\frac{\frac{5}{2}}{\frac{3}{4}} = \frac{12}{x}$$
$$\frac{5}{2}x = \frac{36}{4}$$
$$\frac{2}{5} \cdot \frac{5}{2}x = \frac{9}{1} \cdot \frac{2}{5}$$
$$x = \frac{18}{5}$$

16.
$$\frac{\frac{4}{7}}{\frac{3}{8}} = \frac{x}{\frac{7}{15}}$$
$$\frac{3}{8}x = \frac{4}{7} \cdot \frac{7}{15}$$
$$\frac{8}{3} \cdot \frac{3}{8}x = \frac{4}{15} \cdot \frac{8}{3}$$
$$x = \frac{32}{45}$$

17.
$$\frac{3}{7}x + 2 = 7\frac{1}{4}$$
$$\frac{3}{7}x + 2 - 2 = \frac{29}{4} - 2$$
$$\frac{3}{7}x = \frac{29}{4} - \frac{8}{4}$$
$$\frac{7}{3} \cdot \frac{3}{7}x = \frac{\overset{7}{\cancel{21}}}{4} \cdot \frac{7}{\underset{1}{\cancel{3}}}$$
$$x = \frac{49}{4}$$

18.
$$5x + 6\frac{3}{14} = 12\frac{2}{7}$$
$$5x + 6\frac{3}{14} - 6\frac{3}{14} = 12\frac{4}{14} - 6\frac{3}{14}$$
$$5x = 6\frac{1}{14}$$
$$\frac{1}{5} \cdot 5x = \frac{\overset{17}{\cancel{85}}}{14} \cdot \frac{1}{\underset{1}{\cancel{5}}}$$
$$x = \frac{17}{14}$$

19.
$$7x - 2 = 11$$
$$7x - 2 + 2 = 11 + 2$$
$$7x = 13$$
$$x = \frac{13}{7}$$

20. $\sqrt{9} + \sqrt{25}[3(3-1)-2] = 3 + 5[3(2)-2]$
$= 3 + 5[4] = 3 + 20 = \textbf{23}$

21.
$$\begin{array}{r} 6111 \\ \times\ 0.0013 \\ \hline 18333 \\ 6111\quad \\ \hline \textbf{7.9443} \end{array}$$

22. $2\dfrac{1}{3} \cdot 3\dfrac{1}{4} - \dfrac{5}{6} = \dfrac{7}{3} \cdot \dfrac{13}{4} - \dfrac{5}{6} = \dfrac{91}{12} - \dfrac{10}{12}$
$= \dfrac{81}{12} = \dfrac{\textbf{27}}{\textbf{4}}$

23. $14\dfrac{4}{5} - 4\dfrac{3}{4} + \dfrac{9}{10} = 14\dfrac{16}{20} - 4\dfrac{15}{20} + \dfrac{18}{20}$
$= 10\dfrac{\textbf{19}}{\textbf{20}}$

24. $\dfrac{1}{3}\left(\dfrac{1}{2} + 2\dfrac{1}{3}\right) - \dfrac{7}{18} = \dfrac{1}{3}\left(\dfrac{3}{6} + \dfrac{14}{6}\right) - \dfrac{7}{18}$
$= \dfrac{1}{3}\left(\dfrac{17}{6}\right) - \dfrac{7}{18} = \dfrac{17}{18} - \dfrac{7}{18} = \dfrac{10}{18} = \dfrac{\textbf{5}}{\textbf{9}}$

25. $2\dfrac{1}{2} \times 3\dfrac{2}{3} \div 1\dfrac{5}{6} \times \dfrac{1}{3}$
$= \dfrac{5}{\cancel{2}_{1}} \times \dfrac{\cancel{11}^{1}}{\cancel{3}_{1}} \times \dfrac{\cancel{6}^{\cancel{3}^{1}}}{\cancel{11}_{1}} \times \dfrac{1}{3} = \dfrac{\textbf{5}}{\textbf{3}}$

26. $\dfrac{1}{4}\left(\dfrac{1}{6} + 3\dfrac{1}{2}\right) - \dfrac{4}{5} = \dfrac{1}{4}\left(\dfrac{1}{6} + \dfrac{21}{6}\right) - \dfrac{4}{5}$
$= \dfrac{1}{4}\left(\dfrac{22}{6}\right) - \dfrac{4}{5} = \dfrac{55}{60} - \dfrac{48}{60} = \dfrac{\textbf{7}}{\textbf{60}}$

27. $\dfrac{4\frac{1}{3}}{5\frac{5}{6}} = \dfrac{\frac{13}{3}}{\frac{35}{6}} = \dfrac{13}{\cancel{3}_{1}} \cdot \dfrac{\cancel{6}^{2}}{35} = \dfrac{\textbf{26}}{\textbf{35}}$

28. $96 - 4\left[(7 - 2^2) \div (1 + 2) + \sqrt{36}\right]$
$= 96 - 4[3 \div 3 + 6] = 96 - 4[7]$
$= 96 - 28 = \textbf{68}$

29. (a) 1 percent of $31.2 = \dfrac{31.2}{100} = \textbf{0.312}$

(b) $50 \cdot 0.312 = \textbf{15.6}$

30. $x^2 + xy + xyz + \sqrt[z]{x}$

$= (9)^2 + (9)\left(\dfrac{1}{3}\right) + (9)\left(\dfrac{1}{3}\right)(2) + (\sqrt[2]{9})$

$= 81 + 3 + 6 + 3$

$= \textbf{93}$

PROBLEM SET 62

1. $\dfrac{14 \text{ big ones}}{9 \text{ crowns}}, \dfrac{9 \text{ crowns}}{14 \text{ big ones}}$

$360 \ \cancel{\text{crowns}} \times \dfrac{14 \text{ big ones}}{9 \ \cancel{\text{crowns}}} = \textbf{560 big ones}$

2. $\dfrac{43 \text{ bottles}}{2 \text{ min}}, \dfrac{2 \text{ min}}{43 \text{ bottles}}$

$860 \ \cancel{\text{bottles}} \times \dfrac{2 \text{ min}}{43 \ \cancel{\text{bottles}}} = \textbf{40 min}$

3. $640 + 160 = 800$ crickets
$F \cdot 800 = 160$
$\dfrac{F \cdot 800}{800} = \dfrac{160}{800}$
$F = \dfrac{\textbf{1}}{\textbf{5}}$

4. $F \cdot 80 = 62$
$\dfrac{F \cdot 80}{80} = \dfrac{62}{80}$
$F = \dfrac{\textbf{31}}{\textbf{40}}$

5. $2\dfrac{1}{3} \cdot 30 = B$
$\dfrac{7}{3} \cdot 30 = B$
$\dfrac{210}{3} = B$
$B = \textbf{70 students}$

6. (a) 1% of $0.031 = \dfrac{0.031}{100} = \textbf{0.00031}$

(b) $92 \cdot 0.00031 = \textbf{0.02852}$

7. $0.12 = \dfrac{12}{100} = \dfrac{3}{25}$

$\dfrac{5}{6} = 0.8\overline{3} = 83.\overline{3}\%$

FRACTION	DECIMAL	PERCENT
$\frac{51}{100}$	0.51	51%
(a) $\frac{3}{25}$	0.12	(b) **12%**
$\frac{5}{6}$	(c) **0.8$\overline{3}$**	(d) **83.$\overline{3}$%**

8. (a) $\dfrac{1\text{ mi}}{8\text{ min}}, \dfrac{8\text{ min}}{1\text{ mi}}$

   (b) $\dfrac{1}{8}$ **mi per min**

   (c) $\dfrac{1\text{ mi}}{8\,\cancel{\text{min}}} \cdot 30\,\cancel{\text{min}} = \dfrac{30}{8}\text{ mi} = \dfrac{15}{4}\text{ mi} = \mathbf{3\dfrac{3}{4}}\text{ mi}$

9. $0.4 \times N = 316$

   $N = \dfrac{316}{0.4} = \mathbf{790}$

10. $F \times \dfrac{45}{6} = 7\dfrac{1}{4}$

    $F \times \dfrac{45}{6} = \dfrac{29}{4}$

    $F = \dfrac{29}{4} \times \dfrac{6}{45} = \dfrac{174}{180} = \mathbf{\dfrac{29}{30}}$

11. $D \times 640 = 560$

    $D = \dfrac{560}{640} = \dfrac{7}{8} = \mathbf{0.875}$

12. $2\dfrac{1}{4} \times N = 6\dfrac{1}{3}$

    $\dfrac{9}{4} \times N = \dfrac{19}{3}$

    $N = \dfrac{19}{3} \times \dfrac{4}{9} = \mathbf{\dfrac{76}{27}}$

13. $3\dfrac{1}{2} \times 1\dfrac{1}{10} = W$

    $\dfrac{7}{2} \times \dfrac{11}{10} = W$

    $\mathbf{\dfrac{77}{20}} = W$

14.

$A_{\text{Base}} = A_{\text{Rectangle}} + A_{\text{Triangle}}$

$= \left[5(16) + \dfrac{(31)(15)}{2} \right] \text{m}^2$

$= [80 + 232.5]\text{ m}^2$

$= 312.5\text{ m}^2$

$V = A_{\text{Base}} \times \text{height}$

$= 312.5\text{ m}^2 \times 2\text{ m}$

$= \mathbf{625\text{ m}^3}$

15. Circumference $= \pi d = 2\pi r = 2\pi(14\text{ ft})$

    $= 28\pi\text{ ft} \approx \mathbf{87.92\text{ ft}}$

    Area $= \pi r^2 = \pi(14\text{ ft})^2$

    $= 196\pi\text{ ft}^2 \approx \mathbf{615.44\text{ ft}^2}$

16. $16{,}000\,\cancel{\text{in.}}^2 \cdot \dfrac{1\text{ ft}}{12\,\cancel{\text{in.}}} \cdot \dfrac{1\text{ ft}}{12\,\cancel{\text{in.}}}$

    $= \dfrac{16{,}000}{(12)(12)}\text{ ft}^2 \approx \mathbf{111.11\text{ ft}^2}$

17. $\dfrac{\frac{1}{3}}{\frac{2}{5}} = \dfrac{6}{x}$

    $\dfrac{1}{3}x = \dfrac{6}{1} \cdot \dfrac{2}{5}$

    $\dfrac{3}{1} \cdot \dfrac{1}{3}x = \dfrac{12}{5} \cdot \dfrac{3}{1}$

    $x = \mathbf{\dfrac{36}{5}}$

18. $\dfrac{\frac{2}{3}}{\frac{3}{4}} = \dfrac{5}{12}{p}$

    $\dfrac{2}{3}p = \dfrac{5}{12} \cdot \dfrac{3}{4}$

    $\dfrac{3}{2} \cdot \dfrac{2}{3}p = \dfrac{5}{16} \cdot \dfrac{3}{2}$

    $p = \mathbf{\dfrac{15}{32}}$

19. $3\dfrac{5}{6}x + 4 = 7\dfrac{1}{2}$

    $\dfrac{23}{6}x + 4 - 4 = \dfrac{15}{2} - \dfrac{8}{2}$

    $\dfrac{23}{6}x = \dfrac{7}{2}$

    $\dfrac{6}{23} \cdot \dfrac{23}{6}x = \dfrac{7}{\cancel{2}_1} \cdot \dfrac{\cancel{6}^3}{23}$

    $x = \mathbf{\dfrac{21}{23}}$

20.
$$3\frac{2}{5}p - \frac{1}{6} = 1\frac{2}{5}$$

$$\frac{17}{5}p - \frac{1}{6} + \frac{1}{6} = \frac{7}{5} + \frac{1}{6}$$

$$\frac{17}{5}p = \frac{42}{30} + \frac{5}{30}$$

$$\frac{5}{17} \cdot \frac{17}{5}p = \frac{47}{\overset{1}{\cancel{30}}} \cdot \frac{\overset{1}{\cancel{5}}}{17}$$
$$\phantom{\frac{5}{17} \cdot \frac{17}{5}p =}\,_{6}$$

$$p = \frac{47}{102}$$

21.
$$7r - 1\frac{3}{7} = 5$$

$$7r - 1\frac{3}{7} + 1\frac{3}{7} = 5 + 1\frac{3}{7}$$

$$7r = 6\frac{3}{7}$$

$$\frac{1}{7} \cdot 7r = \frac{45}{7} \cdot \frac{1}{7}$$

$$r = \frac{45}{49}$$

22. $64 - 3\left[(6 - 2^2)(3^2 - 2^2) + 1\right]$
$= 64 - 3[2(5) + 1] = 64 - 3[11] = 64 - 33$
$= \mathbf{31}$

23. $\sqrt[3]{8} + \sqrt{16}[2(3 - 1) + 7] = 2 + 4[2(2) + 7]$
$= 2 + 4[11] = 2 + 44 = \mathbf{46}$

24. $2\frac{1}{4} \cdot 3\frac{2}{3} - \frac{5}{6} = \frac{9}{4} \cdot \frac{11}{3} - \frac{5}{6} = \frac{99}{12} - \frac{10}{12}$

$= \dfrac{\mathbf{89}}{\mathbf{12}}$

25. $14\frac{3}{4} - 4\frac{1}{5} + \frac{7}{20} = 14\frac{15}{20} - 4\frac{4}{20} + \frac{7}{20}$

$= 10\frac{18}{20} = 10\dfrac{\mathbf{9}}{\mathbf{10}}$

26. $\dfrac{4\frac{2}{3}}{2\frac{1}{6}} = \dfrac{\frac{14}{3}}{\frac{13}{6}} = \dfrac{14}{\overset{}{\cancel{3}}} \cdot \dfrac{\overset{2}{\cancel{6}}}{13} = \dfrac{\mathbf{28}}{\mathbf{13}}$

27. $\frac{1}{4}\left(\frac{1}{2} + 3\frac{1}{4}\right) - \frac{5}{8} = \frac{1}{4}\left(\frac{2}{4} + \frac{13}{4}\right) - \frac{5}{8}$

$= \frac{1}{4}\left(\frac{15}{4}\right) - \frac{10}{16} = \frac{15}{16} - \frac{10}{16} = \dfrac{\mathbf{5}}{\mathbf{16}}$

28. $\frac{1}{3}\left(\frac{1}{2} + 3\frac{1}{3}\right) - \frac{5}{6} = \frac{1}{3}\left(\frac{3}{6} + \frac{20}{6}\right) - \frac{5}{6}$

$= \frac{1}{3}\left(\frac{23}{6}\right) - \frac{5}{6} = \frac{23}{18} - \frac{15}{18} = \frac{8}{18} = \dfrac{\mathbf{4}}{\mathbf{9}}$

29. (a) $3.09 \times 10^{-5} = \mathbf{0.0000309}$

(b) $19,000 = \mathbf{1.9 \times 10^4}$

30. $x^2 + y^2 + 2xy + \sqrt[x]{z}$
$= (3)^2 + (5)^2 + 2(3)(5) + \left(\sqrt[3]{27}\right)$
$= 9 + 25 + 30 + 3$
$= \mathbf{67}$

PROBLEM SET 63

1. $\dfrac{1 \text{ min}}{40 \,\cancel{\text{yd}}} \times 2800 \,\cancel{\text{yd}} = \mathbf{70 \text{ min}}$

2. $\dfrac{20 \text{ min}}{7000 \,\cancel{\text{ft}}} \times 26,950 \,\cancel{\text{ft}} = \mathbf{77 \text{ min}}$

3. Initial rate $= \dfrac{8000 \text{ bottles}}{4 \text{ hr}} = \dfrac{2000 \text{ bottles}}{1 \text{ hr}}$

New rate $= \dfrac{(2 \times 2000) \text{ bottles}}{1 \text{ hr}}$

$= \dfrac{4000 \text{ bottles}}{1 \text{ hr}}$

$\dfrac{1 \text{ hr}}{4000 \,\cancel{\text{bottles}}} \times 40,000 \,\cancel{\text{bottles}} = \mathbf{10 \text{ hr}}$

4. Initial rate $= \dfrac{400 \text{ cm}}{50 \text{ s}} = \dfrac{8 \text{ cm}}{1 \text{ s}}$

New rate $= \dfrac{\left(\frac{1}{2} \times 8\right) \text{ cm}}{1 \text{ s}} = \dfrac{4 \text{ cm}}{1 \text{ s}}$

$\dfrac{4 \text{ cm}}{1 \,\cancel{\text{s}}} \times 20,000 \,\cancel{\text{s}} = \mathbf{80,000 \text{ cm}}$

5.
$$\frac{9}{10} \cdot A = 72$$

$$\frac{10}{9} \cdot \frac{9}{10} \cdot A = \frac{\overset{8}{\cancel{72}}}{1} \cdot \frac{10}{\underset{1}{\cancel{9}}}$$

$$A = \mathbf{80 \text{ e-mail messages}}$$

6. $0.16 = \dfrac{16}{100} = \dfrac{4}{25}$

$\dfrac{2}{5} \cdot \dfrac{20}{20} = \dfrac{40}{100} = 0.4$

Fraction	Decimal	Percent
$\dfrac{51}{100}$	0.51	51%
(a) $\dfrac{4}{25}$	0.16	(b) **16%**
$\dfrac{2}{5}$	(c) **0.4**	(d) **40%**

7. (a) $3.09 \times 10^{-4} = \mathbf{0.000309}$

(b) $6{,}103{,}000 = \mathbf{6.103 \times 10^6}$

8. $D \times 720 = 80$

$D = \dfrac{80}{720} = \dfrac{1}{9} = \mathbf{0.\overline{1}}$

9. $F \times 625 = 75$

$F = \dfrac{75}{625} = \dfrac{\mathbf{3}}{\mathbf{25}}$

10. $6\dfrac{2}{5} \cdot A = 6\dfrac{1}{4}$

$\dfrac{32}{5} \cdot A = \dfrac{25}{4}$

$\dfrac{5}{32} \cdot \dfrac{32}{5} \cdot A = \dfrac{25}{4} \cdot \dfrac{5}{32}$

$A = \dfrac{\mathbf{125}}{\mathbf{128}}$

11. $3\dfrac{1}{5} \times 1\dfrac{1}{2} = B$

$\dfrac{\overset{8}{\cancel{16}}}{5} \times \dfrac{3}{\underset{1}{\cancel{2}}} = B$

$\dfrac{\mathbf{24}}{\mathbf{5}} = B$

12. (a) $C = \pi d = \pi(2 \cdot 10 \text{ cm})$

$= 20\pi \text{ cm} \approx \mathbf{62.8 \text{ cm}}$

(b) $A = \pi r^2 = \pi(10 \text{ cm})^2$

$= 100\pi \text{ cm}^2 \approx \mathbf{314 \text{ cm}^2}$

13.

$A_{\text{Base}} = A_{\text{Rectangle}} + A_{\text{Triangle}}$

$= \left[(30)(11) + \dfrac{(22)(10)}{2} \right] \text{in.}^2$

$= [330 + 110] \text{ in.}^2$

$= 440 \text{ in.}^2$

$V = A_{\text{Base}} \times \text{height}$

$= 440 \text{ in.}^2 \times 5 \text{ in.}$

$= \mathbf{2200 \text{ in.}^3}$

14. $28{,}000 \ \cancel{\text{mi}}^2 \times \dfrac{5280 \text{ ft}}{1 \ \cancel{\text{mi}}} \times \dfrac{5280 \text{ ft}}{1 \ \cancel{\text{mi}}}$

$= (28{,}000)(5280)(5280) \text{ ft}^2$

$= \mathbf{780{,}595{,}200{,}000 \text{ ft}^2}$

15. $\dfrac{\frac{3}{4}}{\frac{2}{5}} = \dfrac{6}{x}$

$\dfrac{3}{4}x = \dfrac{6}{1} \cdot \dfrac{2}{5}$

$\left(\dfrac{4}{3}\right)\dfrac{3}{4}x = \dfrac{12}{5}\left(\dfrac{4}{3}\right)$

$x = \dfrac{\mathbf{16}}{\mathbf{5}}$

16. $\dfrac{\frac{5}{12}}{\frac{5}{3}} = \dfrac{\frac{5}{2}}{k}$

$\dfrac{5}{12}k = \dfrac{5}{2} \cdot \dfrac{5}{3}$

$\left(\dfrac{12}{5}\right)\dfrac{5}{12}k = \dfrac{25}{6}\left(\dfrac{12}{5}\right)$

$k = \mathbf{10}$

17. $6\dfrac{2}{7}x + \dfrac{1}{2} = 2\dfrac{1}{3}$

$\dfrac{44}{7}x + \dfrac{1}{2} - \dfrac{1}{2} = \dfrac{7}{3} - \dfrac{1}{2}$

$\dfrac{44}{7}x = \dfrac{14}{6} - \dfrac{3}{6}$

$\dfrac{7}{44} \cdot \dfrac{44}{7}x = \dfrac{\overset{1}{\cancel{11}}}{6} \cdot \dfrac{7}{\underset{4}{\cancel{44}}}$

$x = \dfrac{7}{24}$

18.
$$\frac{6}{7}x + 3\frac{6}{7} = 7\frac{16}{21}$$
$$\frac{6}{7}x + \frac{27}{7} = \frac{163}{21}$$
$$\frac{6}{7}x + \frac{27}{7} - \frac{27}{7} = \frac{163}{21} - \frac{81}{21}$$
$$\frac{6}{7}x = \frac{82}{21}$$
$$\frac{7}{6} \cdot \frac{6}{7}x = \frac{\overset{41}{\cancel{82}}}{\underset{3}{\cancel{21}}} \cdot \frac{\overset{1}{\cancel{7}}}{\underset{3}{\cancel{6}}}$$
$$x = \frac{41}{9}$$

19.
$$3x + 6 = 28$$
$$3x + 6 - 6 = 28 - 6$$
$$\frac{3x}{3} = \frac{22}{3}$$
$$x = \frac{22}{3}$$

20. $56 - 3[3(3 - 1)(2^2 - 2) - 5]$
$= 56 - 3[3(2)(2) - 5] = 56 - 3[7]$
$= 56 - 21 = \mathbf{35}$

21. $\sqrt{4} + 5[3 \div (2 - 1) + \sqrt{9}]$
$= 2 + 5[3 \div (1) + 3]$
$= 2 + 5[6] = 2 + 30 = \mathbf{32}$

22. $\frac{4}{5} + 2\frac{1}{5} \cdot 1\frac{3}{4} = \frac{4}{5} + \frac{11}{5} \cdot \frac{7}{4} = \frac{16}{20} + \frac{77}{20}$
$= \mathbf{\frac{93}{20}}$

23. $13\frac{7}{10} - 3\frac{2}{5} + \frac{4}{15} = 13\frac{21}{30} - 3\frac{12}{30} + \frac{8}{30}$
$= \mathbf{10\frac{17}{30}}$

24.
$$\begin{array}{r} 621.3 \\ \times\ 0.0014 \\ \hline 24852 \\ 6213 \\ \hline \mathbf{0.86982} \end{array}$$

25.
$$\begin{array}{r} 21.6200 \\ -\ 18.9261 \\ \hline \mathbf{2.6939} \end{array}$$

26. $3\frac{6}{7} \times 2\frac{1}{3} \div 4\frac{1}{2} \div \frac{1}{3}$
$$= \frac{\overset{\overset{1}{\cancel{9}}}{\cancel{27}}}{\cancel{7}} \times \frac{\overset{1}{\cancel{7}}}{\cancel{3}} \times \frac{2}{\underset{1}{\cancel{9}}} \times \frac{3}{1} = \mathbf{6}$$

27. $\frac{1}{2}\left(6\frac{2}{3} + \frac{1}{4}\right) - \frac{5}{24} = \frac{1}{2}\left(\frac{80}{12} + \frac{3}{12}\right) - \frac{5}{24}$
$= \frac{1}{2}\left(\frac{83}{12}\right) - \frac{5}{24} = \frac{83}{24} - \frac{5}{24} = \frac{78}{24} = \mathbf{\frac{13}{4}}$

28. (a) 1 percent of $22 = \frac{22}{100} = \mathbf{0.22}$

(b) $100 \cdot 0.22 = \mathbf{22}$

29. $xyz + x^y + \sqrt[y]{x} - x$
$= (16)(2)\left(\frac{1}{4}\right) + (16^2) + \left(\sqrt[2]{16}\right) - (16)$
$= 8 + 256 + 4 - 16$
$= \mathbf{252}$

30.
$$\downarrow$$
$41,\text{①}90,364$
The rounded number is $41,200,000$
$= \mathbf{4.12 \times 10^7}$.

PROBLEM SET 64

1. $\frac{4 \text{ mi}}{1 \text{ hr}} \times 16 \text{ hr} = \mathbf{64 \text{ mi}}$

2. Initial rate: $\frac{12 \text{ mi}}{4 \text{ hr}} = 3 \text{ mph}$

Faster rate: $3 \text{ mph} + 2 \text{ mph} = 5 \text{ mph}$

$\frac{1 \text{ hr}}{5 \text{ mi}} \cdot (52 - 12) \text{ mi} = \frac{40}{5} \text{ hr} = \mathbf{8 \text{ hr}}$

3. $\left(\frac{5 \text{ mi}}{1 \text{ hr}} \times 12 \text{ hr}\right) + \left(\frac{8 \text{ mi}}{1 \text{ hr}} \times 8 \text{ hr}\right)$
$= 60 \text{ mi} + 64 \text{ mi} = \mathbf{124 \text{ mi}}$

4.
$$\begin{array}{r} 4800 \\ \times\ \ \ 48 \\ \hline 38400 \\ 19200 \\ \hline \mathbf{230,400 \text{ roses}} \end{array}$$

5. $\frac{4}{15} \cdot 7\frac{1}{2} = B$

$\overset{2}{\underset{1}{\cancel{\frac{4}{15}}}} \cdot \overset{1}{\underset{1}{\cancel{\frac{15}{2}}}} = B$

$B = \textbf{2 kielbasas}$

6. $60\% = \frac{60}{100} = \frac{6}{10} = \frac{3}{5}$

Fraction	Decimal	Percent
$\frac{51}{100}$	0.51	51%
(a) $\frac{3}{5}$	(b) **0.6**	60%

7. (a) $10,300 = \textbf{1.03} \times \textbf{10}^4$

(b) $6.019 \times 10^8 = \textbf{601,900,000}$

8. $D \times 240 = 160$

$D = \frac{160}{240} = \frac{2}{3} = \textbf{0.}\overline{\textbf{6}}$

9. $F \times 576 = 36$

$F = \frac{36}{576} = \frac{1}{16}$

10. $5\frac{2}{3} \cdot A = 3\frac{1}{4}$

$\frac{17}{3} \cdot A = \frac{13}{4}$

$\frac{3}{17} \cdot \frac{17}{3} \cdot A = \frac{13}{4} \cdot \frac{3}{17}$

$A = \frac{39}{68}$

11. (a) $C = \pi d = \pi(14 \text{ ft}) = \textbf{14}\pi \textbf{ ft} \approx \textbf{43.96 ft}$

(b) $A = \pi r^2 = \pi(7 \text{ ft})^2 = \textbf{49}\pi \textbf{ ft}^2 \approx \textbf{153.86 ft}^2$

12. (a) Perimeter $= \left(3 + 5 + \frac{\pi \cdot 2 \cdot 2}{2}\right)$ m

$= (8 + 2\pi) \text{ m} \approx \textbf{14.28 m}$

(b)

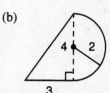

$A_{\text{Total}} = A_{\text{Triangle}} + A_{\text{Semicircle}}$

$= \frac{(3 \text{ m})(4 \text{ m})}{2} + \frac{\pi(2 \text{ m})^2}{2}$

$= 6 \text{ m}^2 + 2\pi \text{ m}^2$

$\approx \textbf{12.28 m}^2$

13.

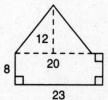

$A_{\text{Base}} = A_{\text{Rectangle}} + A_{\text{Triangle}}$

$= \left[(23 \text{ ft})(8 \text{ ft}) + \frac{(20 \text{ ft})(12 \text{ ft})}{2}\right]$

$= 184 \text{ ft}^2 + 120 \text{ ft}^2$

$= 304 \text{ ft}^2$

$V = A_{\text{Base}} \times \text{height}$

$= 304 \text{ ft}^2 \times 1 \text{ ft}$

$= \textbf{304 ft}^3$

14. $S.A. = 6A_{\text{Side}} = 6(7 \text{ m})^2 = \textbf{294 m}^2$

15. $\dfrac{\frac{1}{5}}{\frac{3}{10}} = \frac{5}{x}$

$\frac{1}{5}x = \frac{5}{1} \cdot \frac{3}{10}$

$\frac{5}{1} \cdot \frac{1}{5}x = \frac{3}{2} \cdot \frac{5}{1}$

$x = \frac{15}{2}$

16. $\dfrac{\frac{7}{12}}{\frac{5}{24}} = \frac{5}{y}$

$\frac{7}{12}y = \frac{5}{1} \cdot \frac{5}{24}$

$\frac{12}{7} \cdot \frac{7}{12}y = \frac{25}{24} \cdot \frac{12}{7}$

$y = \frac{25}{14}$

17. $4\frac{1}{2}x - 1\frac{1}{3} = \frac{17}{12}$

$4\frac{1}{2}x - 1\frac{1}{3} + 1\frac{1}{3} = \frac{17}{12} + 1\frac{1}{3}$

$\frac{9}{2}x = \frac{17}{12} + \frac{16}{12}$

$\left(\frac{2}{9}\right)\frac{9}{2}x = \frac{33}{12}\left(\frac{2}{9}\right)$

$x = \frac{11}{18}$

18.
$$2\frac{1}{2}x - 2\frac{1}{4} = 3\frac{2}{5}$$
$$2\frac{1}{2}x - 2\frac{1}{4} + 2\frac{1}{4} = 3\frac{2}{5} + 2\frac{1}{4}$$
$$\frac{5}{2}x = \frac{68}{20} + \frac{45}{20}$$
$$\left(\frac{2}{5}\right)\frac{5}{2}x = \frac{113}{20}\left(\frac{2}{5}\right)$$
$$x = \frac{113}{50}$$

19.
$$3x + 7 = 25$$
$$3x + 7 - 7 = 25 - 7$$
$$3x = 18$$
$$\frac{3x}{3} = \frac{18}{3}$$
$$x = 6$$

20. $2^2 + 2[2^2(4^2 - 3^2) - 3^3]$
$= 4 + 2[4(7) - 27] = 4 + 2[1] = 6$

21. $\frac{3}{7} + 3\frac{1}{2} \cdot \frac{2}{3} = \frac{3}{7} + \frac{7}{2} \cdot \frac{2}{3} = \frac{3}{7} + \frac{7}{3}$
$= \frac{9}{21} + \frac{49}{21} = \frac{58}{21}$

22. $6\frac{7}{10} - 4\frac{4}{15} + \frac{7}{30} = 6\frac{21}{30} - 4\frac{8}{30} + \frac{7}{30}$
$= 2\frac{20}{30} = 2\frac{2}{3}$

23. $\dfrac{2\frac{1}{3}}{3\frac{1}{5}} = \dfrac{\frac{7}{3}}{\frac{16}{5}} = \frac{7}{3} \cdot \frac{5}{16} = \frac{35}{48}$

24.
$$
\begin{array}{r}
178.220 \\
-\ 19.621 \\
\hline
\mathbf{158.599}
\end{array}
$$

25. $\dfrac{192.61}{0.005}$

$$
\begin{array}{r}
38,522 \\
5\overline{)192,610} \\
\underline{15} \\
42 \\
\underline{40} \\
2\,6 \\
\underline{2\,5} \\
11 \\
\underline{10} \\
10 \\
\underline{10} \\
0
\end{array}
$$

26. $\frac{1}{3}\left(2\frac{1}{3} + 6\frac{1}{2}\right) - \frac{7}{24} = \frac{1}{3}\left(\frac{14}{6} + \frac{39}{6}\right) - \frac{7}{24}$
$= \frac{1}{3}\left(\frac{53}{6}\right) - \frac{7}{24} = \frac{53}{18} - \frac{7}{24} = \frac{212}{72} - \frac{21}{72}$
$= \mathbf{\dfrac{191}{72}}$

27.

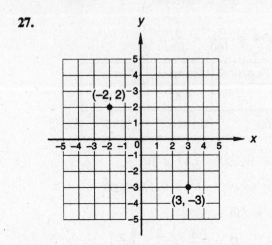

28. (a) 1 percent of 1.34 $= \dfrac{1.34}{100} = \mathbf{0.0134}$

(b) $110 \cdot 0.0134 = \mathbf{1.474}$

29. $xy + \sqrt[y]{z} + xyz + x^y$
$= (3)(3) + \left(\sqrt[3]{1}\right) + (3)(3)(1) + (3^3)$
$= 9 + 1 + 9 + 27$
$= \mathbf{46}$

30.

$$\downarrow$$
$$1\,①9{,}376$$

The rounded number is $110{,}000 = \mathbf{1.1 \times 10^5}$.

PROBLEM SET 65

1.
$$\frac{4}{5} \cdot A = 840$$
$$\frac{5}{4} \cdot \frac{4}{5} \cdot A = 840 \cdot \frac{5}{4}$$
$$A = \mathbf{1050\ pixies}$$

2. Total $= 440 + 360 = 800$
$$F \times 800 = 440$$
$$F = \frac{440}{800} = \mathbf{\frac{11}{20}}$$

3. Initial rate $= \dfrac{400 \text{ jigsaw puzzles}}{8 \text{ hr}}$

$\qquad\qquad = \dfrac{50 \text{ jigsaw puzzles}}{1 \text{ hr}}$

Slower rate $= \dfrac{\left(\dfrac{1}{5} \cdot 50\right) \text{ jigsaw puzzles}}{1 \text{ hr}}$

$\qquad\qquad = \dfrac{10 \text{ jigsaw puzzles}}{1 \text{ hr}}$

$\dfrac{10 \text{ jigsaw puzzles}}{1 \text{ hr}} \cdot 160 \text{ hr}$

$= \textbf{1600 jigsaw puzzles}$

4. Initial rate $= \dfrac{5 \text{ km}}{1\frac{1}{2} \text{ hr}} = \dfrac{5 \text{ km}}{1} \cdot \dfrac{2}{3 \text{ hr}}$

$\qquad\qquad = \dfrac{10}{3} \text{ km/hr}$

Slowed rate $= \dfrac{4 \text{ km}}{2 \text{ hr}} = 2 \text{ km/hr}$

Decrease $= \dfrac{10}{3} - 2 = \dfrac{10}{3} - \dfrac{6}{3} = \dfrac{4}{3}$

$\qquad\qquad = \mathbf{1\frac{1}{3} \text{ km/hr}}$

5. $0.18 = \dfrac{18}{100} = \dfrac{9}{50}$

Fraction	Decimal	Percent
$\frac{51}{100}$	0.51	51%
(a) $\mathbf{\frac{9}{50}}$	0.18	(b) **18%**

6. $\dfrac{6 \text{ mi}}{2 \text{ hr}} \times 9 \text{ hr} = \textbf{27 mi}$

7. $D \times 360 = 300$

$\qquad D = \dfrac{300}{360} = \dfrac{5}{6} = \mathbf{0.8\overline{3}}$

8. $F \times 360 = 300$

$\qquad F = \dfrac{300}{360} = \dfrac{\mathbf{5}}{\mathbf{6}}$

9. $2\dfrac{1}{3} \times 3\dfrac{1}{4} = B$

$\qquad \dfrac{7}{3} \times \dfrac{13}{4} = B$

$\qquad \dfrac{\mathbf{91}}{\mathbf{12}} = B$

10. (a) $C = \pi d = \pi(2 \cdot 3 \text{ in.}) = 6\pi \text{ in.} \approx \textbf{18.84 in.}$

(b) $A = \pi r^2 = \pi(3 \text{ in.})^2 = 9\pi \text{ in.}^2 \approx \textbf{28.26 in.}^2$

11. (a) $P = \left(20 + 12 + 12 + \dfrac{\pi \cdot 20}{2}\right) \text{dkm}$

$\qquad \approx \textbf{75.4 dkm}$

(b) $A_{\text{Figure}} = A_{\text{Rectangle}} + A_{\text{Semicircle}}$

$\qquad = (20 \cdot 12) \text{ dkm}^2$

$\qquad\quad + \left(\dfrac{\pi \cdot 10^2}{2}\right) \text{dkm}^2$

$\qquad \approx 240 \text{ dkm}^2 + 157 \text{ dkm}^2$

$\qquad \approx \textbf{397 dkm}^2$

12.

$A_{\text{Base}} = A_{\text{Rectangle}} + A_{\text{Triangle}}$

$\qquad = \left[(6)(34) + \dfrac{(18)(14)}{2}\right] \text{cm}^2$

$\qquad = [204 + 126] \text{ cm}^2$

$\qquad = 330 \text{ cm}^2$

$V = A_{\text{Base}} \times \text{height}$

$\quad = 330 \text{ cm}^2 \times 20 \text{ cm}$

$\quad = \textbf{6660 cm}^3$

13. $S.A. = 6A_{\text{Side}} = 6(16)(16) \text{ in.}^2 = \textbf{1536 in.}^2$

14. $\dfrac{x}{7} = \dfrac{14}{10} \qquad\qquad \dfrac{y}{5} = \dfrac{14}{10}$

$10x = 98 \qquad\qquad 10y = 70$

$x = \dfrac{98}{10} = \dfrac{49}{5} = \mathbf{9\frac{4}{5}} \qquad y = \dfrac{70}{10} = \mathbf{7}$

15. $\dfrac{2\frac{1}{3}}{\frac{7}{10}} = \dfrac{6}{x}$

$\qquad \dfrac{7}{3}x = \dfrac{42}{10}$

$\qquad \dfrac{3}{7} \cdot \dfrac{7}{3}x = \dfrac{\overset{6}{\cancel{42}}}{10} \cdot \dfrac{3}{\underset{1}{\cancel{7}}}$

$\qquad x = \dfrac{\mathbf{9}}{\mathbf{5}}$

16.
$$\frac{2\frac{6}{7}}{4\frac{12}{21}} = \frac{4\frac{2}{3}}{x}$$

$$\frac{20}{7}x = \overset{2}{\underset{1}{\cancel{\frac{14}{3}}}} \cdot \overset{32}{\underset{3}{\cancel{\frac{96}{21}}}}$$

$$\frac{7}{20} \cdot \frac{20}{7}x = \overset{16}{\cancel{\frac{64}{3}}} \cdot \frac{7}{\underset{5}{\cancel{20}}}$$

$$x = \frac{112}{15}$$

17.
$$3\frac{1}{4}x - 2\frac{1}{3} = \frac{17}{12}$$

$$3\frac{1}{4}x - 2\frac{1}{3} + 2\frac{1}{3} = \frac{17}{12} + 2\frac{1}{3}$$

$$\frac{13}{4}x = \frac{17}{12} + \frac{28}{12}$$

$$\left(\frac{4}{13}\right)\frac{13}{4}x = \frac{45}{12}\left(\frac{4}{13}\right)$$

$$x = \frac{15}{13}$$

18.
$$3\frac{1}{6}x - \frac{3}{8} = 1\frac{17}{24}$$

$$3\frac{1}{6}x - \frac{3}{8} + \frac{3}{8} = 1\frac{17}{24} + \frac{3}{8}$$

$$\frac{19}{6}x = \frac{41}{24} + \frac{9}{24}$$

$$\left(\frac{6}{19}\right)\frac{19}{6}x = \frac{50}{24}\left(\frac{6}{19}\right)$$

$$x = \frac{25}{38}$$

19.
$$17x - 12 = 19$$
$$17x - 12 + 12 = 19 + 12$$
$$17x = 31$$
$$x = \frac{31}{17}$$

20. $3^2 + 2^3[2(3^2 - 4) - 3] = 9 + 8[2(5) - 3]$
$$= 9 + 8[7] = 9 + 56 = \mathbf{65}$$

21. $\frac{2}{7} + 2\frac{1}{3} \cdot \frac{1}{2} = \frac{2}{7} + \frac{7}{3} \cdot \frac{1}{2} = \frac{12}{42} + \frac{49}{42}$
$$= \frac{61}{42}$$

22. $13\frac{2}{3} - 4\frac{1}{2} + 2\frac{5}{6} = 13\frac{4}{6} - 4\frac{3}{6} + 2\frac{5}{6} = \mathbf{12}$

23.
$$\begin{array}{r} 181.2 \\ \times\ 0.013 \\ \hline 5436 \\ 1812 \\ \hline \mathbf{2.3556} \end{array}$$

24.
$$\begin{array}{r} 1921.610 \\ -\ \ 19.897 \\ \hline \mathbf{1901.713} \end{array}$$

25. $\dfrac{175.61}{0.7}$

$$\begin{array}{r} 250.8714285 \\ 7\overline{)1756.1000000} \\ \underline{14} \\ 35 \\ \underline{35} \\ 06\ 1 \\ \underline{5\ 6} \\ 50 \\ \underline{49} \\ 10 \\ \underline{7} \\ 30 \\ \underline{28} \\ 20 \\ \underline{14} \\ 60 \\ \underline{56} \\ 40 \\ \underline{35} \\ 5 \end{array}$$

26. $\frac{1}{4}\left(3\frac{1}{5} + 2\frac{1}{2}\right) - \frac{3}{10} = \frac{1}{4}\left(\frac{32}{10} + \frac{25}{10}\right) - \frac{3}{10}$

$$= \frac{1}{4}\left(\frac{57}{10}\right) - \frac{3}{10} = \frac{57}{40} - \frac{12}{40} = \frac{45}{40} = \mathbf{\frac{9}{8}}$$

27. $3\frac{1}{3} \times 2\frac{1}{2} \div \frac{2}{3} \div 2\frac{5}{6}$

$$= \overset{5}{\underset{1}{\cancel{\frac{10}{3}}}} \times \overset{1}{\underset{1}{\frac{5}{2}}} \times \overset{3}{\underset{1}{\cancel{\frac{3}{2}}}} \times \frac{6}{17} = \mathbf{\frac{75}{17}}$$

28. $xyz + \sqrt{y} + x^2 = (1)(36)(2) + \sqrt{(36)} + (1^2)$
$$= 72 + 6 + 1$$
$$= \mathbf{79}$$

29. (a) 1 percent of $40.3 = \dfrac{40.3}{100} = \mathbf{0.403}$

(b) $120 \cdot 0.403 = \mathbf{48.36}$

30.
$$\downarrow$$
$$61{,}237{,}89\textcircled{9}{,}721.2$$
The rounded number is $\mathbf{6.12379 \times 10^{10}}$.

PROBLEM SET 66

1. $\dfrac{4 \text{ days}}{100 \text{ leagues}} = \dfrac{1 \text{ day}}{25 \text{ leagues}}$

$\dfrac{1 \text{ day}}{(2 \times 25) \text{ leagues}} \times 200 \text{ leagues} = \mathbf{4 \text{ days}}$

2. $\dfrac{36 \text{ mi}}{3 \text{ hr}} - \dfrac{12 \text{ mi}}{2 \text{ hr}} = 12 \text{ mph} - 6 \text{ mph} = \mathbf{6 \text{ mph}}$

3. $\dfrac{3}{5} \times A = 87$

$\dfrac{5}{3} \times \dfrac{3}{5} \times A = 87 \times \dfrac{5}{3}$

$A = \mathbf{145 \text{ responses}}$

4. $\dfrac{104 \text{ packages}}{4 \text{ hours}} = \dfrac{26 \text{ packages}}{1 \text{ hour}}$

$\dfrac{26 \text{ packages}}{1 \text{ hr}} \times 10 \text{ hr} = \mathbf{260 \text{ packages}}$

5. $\dfrac{7}{5} = \dfrac{T}{20}$

$5T = 140$

$T = \dfrac{140}{5} = \mathbf{28 \text{ thespians}}$

6. $\dfrac{3}{2} = \dfrac{15}{S}$

$3S = 30$

$S = \dfrac{30}{3} = \mathbf{10 \text{ sofas}}$

7.

Fraction	Decimal	Percent
$\frac{51}{100}$	0.51	51%
(a) $\mathbf{\frac{73}{100}}$	(b) **0.73**	73%

8. $D \times 450 = 300$

$D = \dfrac{300}{450} = \dfrac{2}{3} = \mathbf{0.\overline{6}}$

9. $F \times 450 = 300$

$F = \dfrac{300}{450} = \mathbf{\dfrac{2}{3}}$

10. $5\dfrac{1}{2} \times 2\dfrac{1}{3} = B$

$\dfrac{11}{2} \times \dfrac{7}{3} = B$

$\mathbf{\dfrac{77}{6}} = B$

11. (a) 1 percent of $2000 = \dfrac{2000}{100} = \mathbf{20}$

(b) $42.1 \cdot 20 = \mathbf{842}$

12. (a) $C = \pi d = \pi(2 \cdot 12 \text{ cm})$
$= \mathbf{24\pi \text{ cm}} \approx \mathbf{75.36 \text{ cm}}$

(b) $A = \pi r^2 = \pi(12 \text{ cm})^2$
$= \mathbf{144\pi \text{ cm}^2} \approx \mathbf{452.16 \text{ cm}^2}$

13. (a) Perimeter
$= \left(12 + 6 + \dfrac{\pi \cdot 2 \cdot 4}{2} + 4 + \dfrac{\pi \cdot 2 \cdot 3}{2}\right) \text{mm}$
$\approx \mathbf{43.98 \text{ mm}}$

(b) $A = A_{\text{Rectangle}} + A_{\text{Big semicircle}}$
$\qquad + A_{\text{Little Semicircle}}$

$= (6 \cdot 12) \text{ mm}^2 + \left(\dfrac{\pi \cdot 4^2}{2}\right) \text{mm}^2$

$\qquad + \left(\dfrac{\pi \cdot 3^2}{2}\right) \text{mm}^2$

$\approx 72 \text{ mm}^2 + 25.12 \text{ mm}^2 + 14.13 \text{ mm}^2$

$\approx \mathbf{111.25 \text{ mm}^2}$

14. $\dfrac{m}{10} = \dfrac{8}{12}$

$12m = 80$

$m = \dfrac{80}{12} = \mathbf{\dfrac{20}{3}}$

$\dfrac{q}{5} = \dfrac{12}{8}$

$8q = 60$

$q = \dfrac{60}{8} = \mathbf{\dfrac{15}{2}}$

15.

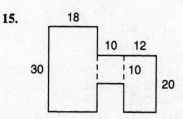

$A_{\text{Base}} = A_{\text{Rectangle 1}} + A_{\text{Square}} + A_{\text{Rectangle 2}}$
$= [(30)(18) + (10)(10) + (12)(20)] \text{ ft}^2$
$= [540 + 100 + 240] \text{ ft}^2$
$= 880 \text{ ft}^2$

$V = A_{\text{Base}} \times \text{height}$
$= 880 \text{ ft}^2 \times 5 \text{ ft}$
$= \mathbf{4400 \text{ ft}^3}$

16. (a) $0.00392 = \mathbf{3.92 \times 10^{-3}}$

(b) $6.03 \times 10^{-9} = \mathbf{0.00000000603}$

17.
$$\frac{\frac{1}{4}}{\frac{4}{5}} = \frac{3}{x}$$

$$\frac{1}{4}x = 3 \cdot \frac{4}{5}$$

$$\frac{4}{1} \cdot \frac{1}{4}x = \frac{12}{5} \cdot \frac{4}{1}$$

$$x = \mathbf{\frac{48}{5}}$$

18.
$$\frac{4\frac{3}{4}}{\frac{8}{9}} = \frac{2\frac{1}{9}}{x}$$

$$\frac{19}{4}x = \frac{8}{9} \cdot \frac{19}{9}$$

$$\frac{4}{19} \cdot \frac{19}{4}x = \frac{\cancel{152}^{\,8}}{81} \cdot \frac{4}{\cancel{19}_{\,1}}$$

$$x = \mathbf{\frac{32}{81}}$$

19.
$$5\frac{1}{3}x - 3\frac{3}{4} = \frac{19}{32}$$

$$5\frac{1}{3}x - 3\frac{3}{4} + 3\frac{3}{4} = \frac{19}{32} + 3\frac{3}{4}$$

$$\frac{16}{3}x = \frac{19}{32} + \frac{120}{32}$$

$$\left(\frac{3}{16}\right)\frac{16}{3}x = \frac{139}{32}\left(\frac{3}{16}\right)$$

$$x = \mathbf{\frac{417}{512}}$$

20.
$$2\frac{1}{2}x - \frac{1}{4} = 2\frac{3}{16}$$

$$2\frac{1}{2}x - \frac{1}{4} + \frac{1}{4} = 2\frac{3}{16} + \frac{1}{4}$$

$$\frac{5}{2}x = \frac{35}{16} + \frac{4}{16}$$

$$\left(\frac{2}{5}\right)\frac{5}{2}x = \frac{39}{16}\left(\frac{2}{5}\right)$$

$$x = \mathbf{\frac{39}{40}}$$

21.
$$22x - 9 = 57$$

$$22x - 9 + 9 = 57 + 9$$

$$22x = 66$$

$$\frac{22x}{22} = \frac{66}{22}$$

$$x = \mathbf{3}$$

22. $15 + 3[(5 - 1)(7 - 2)2 - 4]$
$= 15 + 3[4(5)(2) - 4] = 15 + 3[36]$
$= 15 + 108 = \mathbf{123}$

23. $4^2 + 3^3[2(4^2 - 15) - 2]$
$= 16 + 27[2(1) - 2] = 16 + 27[0] = \mathbf{16}$

24. $\dfrac{1}{5} + 3\dfrac{1}{2} \cdot \dfrac{2}{7} = \dfrac{1}{5} + \dfrac{7}{2} \cdot \dfrac{2}{7} = \dfrac{1}{5} + 1$

$= 1\dfrac{1}{5} = \mathbf{\dfrac{6}{5}}$

25. $11\dfrac{3}{4} - 5\dfrac{1}{5} + 3\dfrac{1}{3} = 11\dfrac{45}{60} - 5\dfrac{12}{60} + 3\dfrac{20}{60}$

$= \mathbf{9\dfrac{53}{60}}$

26. $\dfrac{1}{2}\left(2\dfrac{1}{4} + 1\dfrac{3}{16}\right) - \dfrac{31}{32}$

$= \dfrac{1}{2}\left(\dfrac{36}{16} + \dfrac{19}{16}\right) - \dfrac{31}{32} = \dfrac{1}{2}\left(\dfrac{55}{16}\right) - \dfrac{31}{32}$

$= \dfrac{55}{32} - \dfrac{31}{32} = \dfrac{24}{32} = \mathbf{\dfrac{3}{4}}$

27. $2\dfrac{1}{5} \times 3\dfrac{1}{4} \div 5\dfrac{1}{2} \div 6\dfrac{1}{2}$

$= \dfrac{\cancel{11}^{\,1}}{5} \times \dfrac{\cancel{13}^{\,1}}{\cancel{4}_{\,2}} \times \dfrac{\cancel{2}^{\,1}}{\cancel{11}_{\,1}} \times \dfrac{\cancel{2}^{\,1}}{\cancel{13}_{\,1}} = \mathbf{\dfrac{1}{5}}$

28. $kmx + \sqrt[m]{k} + m^x = (64)(3)(4) + (\sqrt[3]{64}) + (3^4)$
$$= 768 + 4 + 81$$
$$= \mathbf{853}$$

29. $1{,}000{,}000 \text{ in.}^2 \times \dfrac{1 \text{ ft}}{12 \text{ in.}} \times \dfrac{1 \text{ ft}}{12 \text{ in.}}$

$= \mathbf{\dfrac{1{,}000{,}000}{(12)(12)}} \text{ ft}^2 \approx \mathbf{6944.44 \text{ ft}^2}$

30. (a) Range = $21 - 8 = \mathbf{13 \text{ points}}$

(b) Mode = **13 and 21 points**

(c) Median = **13 points**

(d) Mean

$$= \frac{8 + 9 + 12 + 13 + 13 + 15 + 21 + 21}{8}$$

$$= \textbf{14 points}$$

PROBLEM SET 67

1. If $\frac{3}{4}$ did not carry banners, $\frac{1}{4}$ carried banners.

$$\frac{1}{4} \times 1000 = B$$

$$B = \textbf{250 fans}$$

2. $$\frac{7}{16} \cdot A = 210$$

$$\frac{16}{7} \cdot \frac{7}{16} \cdot A = 210 \cdot \frac{16}{7}$$

$$A = \textbf{480 children}$$

3. $F \times 3360 = 480$

$$F = \frac{480}{3360} = \frac{1}{7}$$

4. Initial rate $= \frac{10 \text{ mi}}{5 \text{ hr}} = \frac{2 \text{ mi}}{1 \text{ hr}}$

New rate $= \frac{(3 \times 2) \text{ mi}}{1 \text{ hr}} = \frac{6 \text{ mi}}{1 \text{ hr}}$

$$\frac{1 \text{ hr}}{6 \text{ mi}} \times 24 \text{ mi} = \textbf{4 hr}$$

5. $$\frac{6}{5} = \frac{444}{M}$$

$$6M = 2220$$

$$M = \frac{2220}{6} = \textbf{370 Muslims}$$

6. $\frac{\$4.50}{3 \text{ lb}} = \$1.50 \text{ per lb}$ $\frac{\$14}{8 \text{ lb}} = \$1.75 \text{ per lb}$

The **local market** has the better buy.

7. $\frac{2}{5} = \frac{40}{100} = 0.40 = 40\%$

FRACTION	DECIMAL	PERCENT
$\frac{51}{100}$	0.51	51%
$\frac{2}{5}$	(a) **0.4**	(b) **40%**

8. $$3\frac{3}{5} \times 2\frac{1}{2} = B$$

$$\frac{18}{5} \times \frac{5}{2} = B$$

$$9 = B$$

9. $$0.6 \times A = 144$$

$$\frac{0.6 \times A}{0.6} = \frac{144}{0.6}$$

$$A = \textbf{240}$$

10. $F \times 35 = 21$

$$F = \frac{21}{35} = \frac{3}{5}$$

11. $D \times 810 = 270$

$$D = \frac{270}{810} = \frac{1}{3} = 0.\overline{3}$$

12. (a) $P = \left[20 + 30 + 30 + \frac{\pi(20)}{2} \right] \text{m}$

$$\approx (80 + 31.4) \text{ m}$$

$$\approx \textbf{111.4 m}$$

(b)

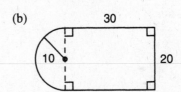

$A = A_{\text{Rectangle}} + A_{\text{Semicircle}}$

$$= \left[30(20) + \frac{\pi(10)^2}{2} \right] \text{m}^2$$

$$\approx [600 + 157] \text{ m}^2$$

$$\approx \textbf{757 m}^2$$

13. $$\frac{s}{5} = \frac{10}{12}$$

$$12s = 50$$

$$s = \frac{50}{12} = \frac{25}{6}$$

$$\frac{t}{13} = \frac{10}{12}$$

$$12t = 130$$

$$t = \frac{130}{12} = \frac{65}{6}$$

14.

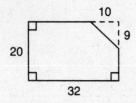

$A_{\text{Base}} = A_{\text{Rectangle}} - A_{\text{Triangle}}$

$= \left[(32)(20) - \dfrac{(10)(9)}{2} \right] \text{cm}^2$

$= [640 - 45] \text{ cm}^2$

$= 595 \text{ cm}^2$

$V = A_{\text{Base}} \times \text{height}$

$= 595 \text{ cm}^2 \times 10 \text{ cm}$

$= \textbf{5950 cm}^3$

15. $12 = 2 \cdot 2 \cdot 3$

$16 = 2 \cdot 2 \cdot 2 \cdot 2$

$30 = 2 \cdot 3 \cdot 5$

$\text{LCM } (12, 16, 30) = 2 \cdot 2 \cdot 2 \cdot 2 \cdot 3 \cdot 5$

$= \textbf{240}$

16. $\dfrac{\frac{1}{4}}{\frac{7}{8}} = \dfrac{6}{x}$

$\dfrac{1}{4}x = 6 \cdot \dfrac{7}{8}$

$\dfrac{4}{1} \cdot \dfrac{1}{4}x = \dfrac{21}{4} \cdot \dfrac{4}{1}$

$x = \textbf{21}$

17. $\dfrac{4\frac{3}{7}}{1\frac{10}{21}} = \dfrac{6}{p}$

$\dfrac{31}{7}p = \dfrac{31}{\cancel{21}} \cdot \dfrac{\cancel{6}^{2}}{1}$
$\phantom{\dfrac{31}{7}p = \dfrac{31}{21}}{}_{7}$

$\dfrac{7}{31} \cdot \dfrac{31}{7}p = \dfrac{\cancel{62}^{2}}{\cancel{7}_{1}} \cdot \dfrac{\cancel{7}^{1}}{\cancel{31}_{1}}$

$p = \textbf{2}$

18. $4\dfrac{1}{7}x - 2\dfrac{2}{3} = 3\dfrac{1}{4}$

$4\dfrac{1}{7}x - 2\dfrac{2}{3} + 2\dfrac{2}{3} = 3\dfrac{1}{4} + 2\dfrac{2}{3}$

$\dfrac{29}{7}x = \dfrac{39}{12} + \dfrac{32}{12}$

$\left(\dfrac{7}{29}\right)\dfrac{29}{7}x = \dfrac{71}{12}\left(\dfrac{7}{29}\right)$

$x = \dfrac{\textbf{497}}{\textbf{348}}$

19. $4\dfrac{1}{3}x - \dfrac{5}{8} = 1\dfrac{5}{24}$

$4\dfrac{1}{3}x - \dfrac{5}{8} + \dfrac{5}{8} = 1\dfrac{5}{24} + \dfrac{5}{8}$

$\dfrac{13}{3}x = \dfrac{29}{24} + \dfrac{15}{24}$

$\left(\dfrac{3}{13}\right)\dfrac{13}{3}x = \dfrac{44}{24}\left(\dfrac{3}{13}\right)$

$x = \dfrac{\textbf{11}}{\textbf{26}}$

20. $18x + 14 = 27$

$18x + 14 - 14 = 27 - 14$

$18x = 13$

$x = \dfrac{\textbf{13}}{\textbf{18}}$

21. $13 + 3[(3 - 1)(6 - 4)2 - 1]$

$= 13 + 3[2(2)(2) - 1] = 13 + 3[7]$

$= 13 + 21 = \textbf{34}$

22. $2^3 + 3\left[3(3^2 - 5) + \sqrt{4}\,\right] = 8 + 3[3(4) + 2]$

$= 8 + 3[14] = 8 + 42 = \textbf{50}$

23. $1\dfrac{2}{7} + 3\dfrac{3}{4} \cdot 2\dfrac{1}{3} = \dfrac{9}{7} + \dfrac{15}{4} \cdot \dfrac{7}{3}$

$= \dfrac{9}{7} + \dfrac{35}{4} = \dfrac{36}{28} + \dfrac{245}{28} = \dfrac{\textbf{281}}{\textbf{28}}$

24. $12\dfrac{4}{5} - 4\dfrac{2}{3} + 11\dfrac{7}{15} = 12\dfrac{12}{15} - 4\dfrac{10}{15} + 11\dfrac{7}{15}$

$= 19\dfrac{9}{15} = \textbf{19}\dfrac{\textbf{3}}{\textbf{5}}$

25. $\begin{array}{r} 9218.980 \\ -\ \ 178.621 \\ \hline \textbf{9040.359} \end{array}$

26. $\dfrac{1}{5}\left(2\dfrac{1}{6} - 1\dfrac{1}{4}\right) - \dfrac{1}{30} = \dfrac{1}{5}\left(\dfrac{26}{12} - \dfrac{15}{12}\right) - \dfrac{1}{30}$

$= \dfrac{1}{5}\left(\dfrac{11}{12}\right) - \dfrac{1}{30} = \dfrac{11}{60} - \dfrac{2}{60} = \dfrac{9}{60} = \dfrac{\textbf{3}}{\textbf{20}}$

27. $4\frac{3}{4} \times 3\frac{7}{12} \div 2\frac{1}{3} \times \frac{1}{6}$

$= \frac{19}{4} \times \frac{43}{\overset{\cancel{12}}{4}} \times \frac{\overset{1}{\cancel{3}}}{7} \times \frac{1}{6} = \frac{817}{672}$

28. $144 \cancel{ft}^2 \times \dfrac{1\ mi}{5280\ \cancel{ft}} \times \dfrac{1\ mi}{5280\ \cancel{ft}}$

$= \dfrac{144}{(5280)(5280)}\ mi^2 \approx 5.17 \times 10^{-6}\ mi^2$

29. (a) $1{,}390{,}000 = \mathbf{1.39 \times 10^6}$

(b) $4.26 \times 10^{17} = \mathbf{426{,}000{,}000{,}000{,}000{,}000}$

30. $xyz + y^2 + x^y - xy$

$= (6)(3)\left(\dfrac{1}{9}\right) + (3)^2 + (6^3) - 6(3)$

$= 2 + 9 + 216 - 18$

$= \mathbf{209}$

PROBLEM SET 68

1. If $\dfrac{4}{5}$ could not read cuneiform, $\dfrac{1}{5}$ could read it.

$\dfrac{1}{5} \times 16{,}000 = B$

$B = \textbf{3200 Sumerians}$

2. $\dfrac{39\ monoliths}{3\ days} = \dfrac{13\ monoliths}{1\ day}$

$\dfrac{1\ day}{(4 \times 13)\ \cancel{monoliths}} \times 208\ \cancel{monoliths} = \textbf{4 days}$

3. $\dfrac{240\ mi}{3\ hr} - \dfrac{240\ mi}{4\ hr} = 80\ mph - 60\ mph$

$= \textbf{20 mph}$

4. $\dfrac{1}{15} = \dfrac{9}{M}$

$M = \textbf{135 theaters}$

5. $\dfrac{7}{9} = \dfrac{210}{C}$

$7C = 1890$

$C = \dfrac{1890}{7} = \textbf{270 comedies}$

6. $\dfrac{26\ lira}{3\ lb} = 8\frac{2}{3}\ lira\ per\ lb$

$\dfrac{33\ lira}{4\ lb} = 8\frac{1}{4}\ lira\ per\ lb$

Mario got the better deal.

7. $a = 0.40 \cdot 2000 = \textbf{800}$

$b = 2000 - 800 = \textbf{1200}$

$c = 100\% - 40\% = \textbf{60\%}$

8. If $\dfrac{3}{4}$ could execute, $\dfrac{1}{4}$ could not execute.

$\dfrac{1}{4} \cdot A = 14$

$\dfrac{4}{1} \cdot \dfrac{1}{4} \cdot A = 14 \cdot \dfrac{4}{1}$

$A = \textbf{56 skateboarders}$

9. $0.68 \cdot 360 = B$

$\mathbf{244.8} = B$

10. $0.06 = \dfrac{6}{100} = \dfrac{3}{50}$

FRACTION	DECIMAL	PERCENT
$\frac{51}{100}$	0.51	51%
(a) $\frac{3}{50}$	0.06	(b) **6%**

11. $D \times 930 = 558$

$D = \dfrac{558}{930} = \dfrac{3}{5} = \textbf{0.6}$

12. $F \times 930 = 558$

$F = \dfrac{558}{930} = \dfrac{3}{5}$

13. $4\frac{3}{5} \times 9\frac{1}{2} = B$

$\dfrac{23}{5} \times \dfrac{19}{2} = B$

$\dfrac{437}{10} = B$

14. (a) $P = \left[\dfrac{\pi(40)}{2} + 60 + 60 + 40\right] ft$

$\approx \textbf{222.8 ft}$

(b)

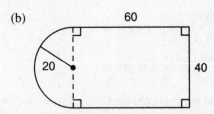

$$A_{\text{Figure}} = A_{\text{Semicircle}} + A_{\text{Rectangle}}$$

$$= \left[\frac{\pi(20)^2}{2} + 60(40) \right] \text{ft}^2$$

$$\approx [628 + 2400] \text{ft}^2$$

$$\approx \textbf{3028 ft}^2$$

15.

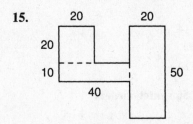

$$A_{\text{Base}} = A_{\text{Square}} + A_{\text{Rectangle 1}} + A_{\text{Rectangle 2}}$$

$$= [(20)(20) + (40)(10) + (50)(20)] \text{ m}^2$$

$$= [400 + 400 + 1000] \text{ m}^2$$

$$= 1800 \text{ m}^2$$

$$V = A_{\text{Base}} \times \text{height}$$

$$= 1800 \text{ m}^2 \times 30 \text{ m}$$

$$= \textbf{54,000 m}^3$$

16. $\dfrac{x}{4} = \dfrac{5}{9}$ $\qquad$ $\dfrac{y}{4} = \dfrac{9}{5}$

$9x = 20$ $\qquad\qquad$ $5y = 36$

$x = \dfrac{20}{9} = \mathbf{2\dfrac{2}{9}}$ $\qquad$ $y = \dfrac{36}{5} = \mathbf{7\dfrac{1}{5}}$

17. $\dfrac{2\frac{2}{3}}{\frac{8}{14}} = \dfrac{\frac{1}{2}}{m}$

$\dfrac{8}{3}m = \dfrac{1}{2} \cdot \dfrac{8}{14}$

$\dfrac{3}{8} \cdot \dfrac{8}{3}m = \dfrac{2}{7} \cdot \dfrac{3}{8}$

$m = \mathbf{\dfrac{3}{28}}$

18. $\dfrac{2\frac{1}{5}}{m} = \dfrac{1\frac{1}{5}}{\frac{1}{3}}$

$\dfrac{11}{5} \cdot \dfrac{1}{3} = \dfrac{6}{5}m$

$\dfrac{5}{6} \cdot \dfrac{6}{5}m = \dfrac{11}{15} \cdot \dfrac{5}{6}$

$m = \mathbf{\dfrac{11}{18}}$

19. $6\dfrac{1}{2}x - \dfrac{1}{4} = 2\dfrac{1}{13}$

$6\dfrac{1}{2}x - \dfrac{1}{4} + \dfrac{1}{4} = 2\dfrac{1}{13} + \dfrac{1}{4}$

$\dfrac{13}{2}x = \dfrac{108}{52} + \dfrac{13}{52}$

$\left(\dfrac{2}{13}\right)\dfrac{13}{2}x = \dfrac{121}{52}\left(\dfrac{2}{13}\right)$

$x = \mathbf{\dfrac{121}{338}}$

20. $3\dfrac{1}{5}x - 7\dfrac{2}{3} = 1\dfrac{3}{4}$

$3\dfrac{1}{5}x - 7\dfrac{2}{3} + 7\dfrac{2}{3} = 1\dfrac{3}{4} + 7\dfrac{2}{3}$

$\dfrac{16}{5}x = \dfrac{21}{12} + \dfrac{92}{12}$

$\left(\dfrac{5}{16}\right)\dfrac{16}{5}x = \dfrac{113}{12}\left(\dfrac{5}{16}\right)$

$x = \mathbf{\dfrac{565}{192}}$

21. $35x - 19 = 51$

$35x - 19 + 19 = 51 + 19$

$35x = 70$

$x = \dfrac{70}{35} = \mathbf{2}$

22. $11 + 3[(5 - 2)(7 - 3)4 - 36]$
$= 11 + 3[3(4)(4) - 36] = 11 + 3[12]$
$= 11 + 36 = \mathbf{47}$

23. $3^2 + 2\left[2(2^3 - 7) + \sqrt{16}\right] = 9 + 2[2(1) + 4]$
$= 9 + 2[6] = 9 + 12 = \mathbf{21}$

24. $1\dfrac{2}{3} + 2\dfrac{1}{2} \cdot 3\dfrac{1}{5} = 1\dfrac{2}{3} + \dfrac{5}{2} \cdot \dfrac{16}{5} = 1\dfrac{2}{3} + 8$
$= 9\dfrac{2}{3} = \mathbf{\dfrac{29}{3}}$

25. $11\frac{2}{3} - 6\frac{1}{2} + 9\frac{1}{12} = 11\frac{8}{12} - 6\frac{6}{12} + 9\frac{1}{12}$

$= 14\frac{3}{12} = \mathbf{14\frac{1}{4}}$

26. $\dfrac{230.95}{0.00025}$

$$
\begin{array}{r}
923,800 \\
25\overline{)23,095,000} \\
22\ 5 \\
\overline{59} \\
50 \\
\overline{95} \\
75 \\
\overline{20\ 0} \\
20\ 0 \\
\overline{0}
\end{array}
$$

27. $\frac{1}{4}\left(3\frac{1}{3} + 2\frac{1}{2}\right) - \frac{11}{12} = \frac{1}{4}\left(\frac{20}{6} + \frac{15}{6}\right) - \frac{11}{12}$

$= \frac{1}{4}\left(\frac{35}{6}\right) - \frac{11}{12} = \frac{35}{24} - \frac{22}{24} = \mathbf{\frac{13}{24}}$

28. $6\frac{1}{2} \times 1\frac{3}{5} \div 1\frac{1}{4} \times \frac{7}{10}$

$= \frac{13}{\underset{1}{\cancel{2}}} \times \frac{\overset{4}{\cancel{8}}}{5} \times \frac{\overset{2}{\cancel{4}}}{5} \times \frac{7}{\underset{5}{\cancel{10}}} = \mathbf{\frac{728}{125}}$

29. $75{,}000\ \cancel{cg} \cdot \dfrac{1\ \cancel{g}}{100\ \cancel{cg}} \cdot \dfrac{1\ \text{kg}}{1000\ \cancel{g}}$

$= \dfrac{75{,}000}{(100)(1000)}\ \text{kg} = \mathbf{0.75\ kg}$

30. $mp + \sqrt[p]{x} + p^m = (2)(4) + (\sqrt[4]{16}) + (4^2)$

$= 8 + 2 + 16$

$= \mathbf{26}$

Problem Set 69

1. $\dfrac{3}{7} = \dfrac{750}{B}$

$3B = 5250$

$B = \mathbf{1750\ beasts}$

2. $\dfrac{13}{2} = \dfrac{52}{F}$

$13F = 104$

$F = \dfrac{104}{13} = \mathbf{8\ students}$

3. $\dfrac{2}{3} \cdot A = 4800$

$\dfrac{3}{2} \cdot \dfrac{2}{3} \cdot A = 4800 \cdot \dfrac{3}{2}$

$A = 7200$ costumes in all

$7200 - 4800 = \mathbf{2400\ costumes}$ were not polychromatic

4. $\dfrac{100\ \text{miles}}{4\ \text{hours}} = \dfrac{25\ \text{miles}}{1\ \text{hour}}$

$\dfrac{1\ \text{hour}}{(4 \times 25)\ \cancel{\text{miles}}} \times 1400\ \cancel{\text{miles}} = \mathbf{14\ hours}$

5. Shanna: $\dfrac{57\ \text{min}}{10\ \text{laps}} = 5.7$ min per lap

Jami: $\dfrac{136\ \text{min}}{17\ \text{laps}} = 8$ min per lap

Shanna swam faster.

6. $0.80 \cdot 14\ \text{oz} = \mathbf{11.2\ oz}$
$14\ \text{oz} - 11.2\ \text{oz} = 2.8\ \text{oz}$

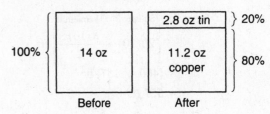

7. $25\% = \dfrac{25}{100} = \dfrac{1}{4}$

Fraction	Decimal	Percent
$\frac{51}{100}$	0.51	51%
(a) $\frac{1}{4}$	(b) **0.25**	25%

8. (a) $0.000913 = \mathbf{9.13 \times 10^{-4}}$

(b) $6.14 \times 10^{-5} = \mathbf{0.0000614}$

9. $0.3 \times A = 123$

$\dfrac{0.3 \times A}{0.3} = \dfrac{123}{0.3}$

$A = \mathbf{410}$

10. $F \times 36 = 32$

$F = \dfrac{32}{36} = \mathbf{\dfrac{8}{9}}$

11. $D \times 360 = 300$

$$D = \frac{300}{360} = \frac{5}{6} = \mathbf{0.8\overline{3}}$$

12. $\quad 3\frac{1}{2} \times A = 4\frac{2}{3}$

$$\frac{7}{2} \times A = \frac{14}{3}$$

$$\frac{2}{7} \times \frac{7}{2} \times A = \frac{14}{3} \times \frac{2}{7}$$

$$A = \mathbf{\frac{4}{3}}$$

13. (a) $P = \left[3(20) + \frac{\pi(20)}{2} \right]$ ft $\approx \mathbf{91.4 \ ft}$

(b)

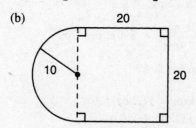

$$A_{\text{Figure}} = A_{\text{Square}} + A_{\text{Semicircle}}$$

$$= \left[20(20) + \frac{\pi(10)^2}{2} \right] \text{ft}^2$$

$$\approx [400 + 157] \text{ft}^2$$

$$\approx \mathbf{557 \ ft^2}$$

14.

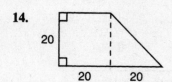

$$A_{\text{Base}} = A_{\text{Square}} + A_{\text{Triangle}}$$

$$= \left[(20)(20) + \frac{(20)(20)}{2} \right] \text{cm}^2$$

$$= [400 + 200] \text{cm}^2$$

$$= 600 \text{ cm}^2$$

height $= 3 \ \cancel{m} \times \dfrac{100 \text{ cm}}{1 \ \cancel{m}} = 300$ cm

$V = A_{\text{Base}} \times$ height

$= 600 \text{ cm}^2 \times 300 \text{ cm}$

$= \mathbf{180{,}000 \ cm^3}$

15.

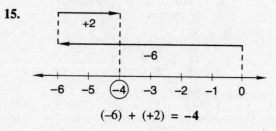

$$(-6) + (+2) = \mathbf{-4}$$

16.

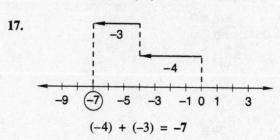

$$(-2) + (+5) = \mathbf{+3}$$

17.

$$(-4) + (-3) = \mathbf{-7}$$

18. $\quad \dfrac{2\frac{3}{4}}{3\frac{7}{8}} = \dfrac{8}{x}$

$$\frac{11}{4}x = \frac{31}{8} \cdot 8$$

$$\frac{4}{11} \cdot \frac{11}{4}x = 31 \cdot \frac{4}{11}$$

$$x = \mathbf{\frac{124}{11}}$$

19. $\quad \dfrac{\frac{6}{7}}{\frac{10}{21}} = \dfrac{9}{x}$

$$\frac{6}{7}x = 9 \cdot \frac{10}{21}$$

$$\frac{7}{6} \cdot \frac{6}{7}x = \frac{30}{7} \cdot \frac{7}{6}$$

$$x = \mathbf{5}$$

20. $\quad 5\frac{1}{2}x + 3\frac{1}{3} = 10\frac{1}{7}$

$$5\frac{1}{2}x + 3\frac{1}{3} - 3\frac{1}{3} = 10\frac{1}{7} - 3\frac{1}{3}$$

$$\frac{11}{2}x = \frac{213}{21} - \frac{70}{21}$$

$$\left(\frac{2}{11}\right)\frac{11}{2}x = \frac{143}{21}\left(\frac{2}{11}\right)$$

$$x = \mathbf{\frac{26}{21}}$$

21. $$4\frac{6}{7}x - \frac{4}{21} = 11\frac{5}{14}$$

$$4\frac{6}{7}x - \frac{4}{21} + \frac{4}{21} = 11\frac{5}{14} + \frac{4}{21}$$

$$\frac{34}{7}x = \frac{159}{14} + \frac{4}{21}$$

$$\frac{34}{7}x = \frac{477}{42} + \frac{8}{42}$$

$$\frac{7}{34} \cdot \frac{34}{7}x = \frac{485}{42} \cdot \frac{7}{34}$$

$$x = \frac{485}{204}$$

22. $$18x + 14 = 72$$

$$18x + 14 - 14 = 72 - 14$$

$$18x = 58$$

$$x = \frac{58}{18} = \frac{29}{9}$$

23. $3^2 + 3[5(3^2 - 2^3) + \sqrt{4}] = 9 + 3[5(1) + 2]$
$= 9 + 3[7] = 9 + 21 = \mathbf{30}$

24. $1\frac{3}{7} + 2\frac{1}{3} \cdot 6\frac{1}{2} = \frac{10}{7} + \frac{7}{3} \cdot \frac{13}{2}$

$= \frac{10}{7} + \frac{91}{6} = \frac{60}{42} + \frac{637}{42} = \mathbf{\frac{697}{42}}$

25. $13\frac{4}{5} - 6\frac{1}{8} + 2\frac{3}{40} = 13\frac{32}{40} - 6\frac{5}{40} + 2\frac{3}{40}$

$= 9\frac{30}{40} = \mathbf{9\frac{3}{4}}$

26. $\frac{1}{10}\left(2\frac{1}{5} + 7\frac{1}{2}\right) - \frac{3}{10}$

$= \frac{1}{10}\left(\frac{22}{10} + \frac{75}{10}\right) - \frac{3}{10} = \frac{1}{10}\left(\frac{97}{10}\right) - \frac{3}{10}$

$= \frac{97}{100} - \frac{30}{100} = \mathbf{\frac{67}{100}}$

27.

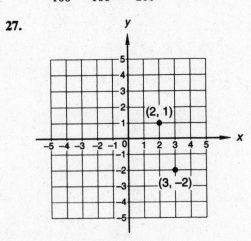

28. (a) 1 percent of $215 = \frac{215}{100} = \mathbf{2.15}$

(b) $14.5 \cdot 2.15 = \mathbf{31.175}$

29. $xyz + x^t + y^t + z^t$
$= (3)(3)(2) + (3^2) + (3^2) + (2^2)$
$= 18 + 9 + 9 + 4$
$= \mathbf{40}$

30. $$\frac{x}{3} = \frac{3}{2}$$

$$2x = 9$$

$$x = \frac{9}{2} = \mathbf{4\frac{1}{2}}$$

$$\frac{y}{5\frac{1}{2}} = \frac{2}{3}$$

$$3y = 2 \cdot \frac{11}{2}$$

$$3y = 11$$

$$y = \frac{11}{3} = \mathbf{3\frac{2}{3}}$$

PROBLEM SET 70

1. Big can $= \frac{\$2.52}{18 \text{ oz}} = \mathbf{\$0.14 \text{ per oz}}$

Small can $= \frac{\$1.08}{12 \text{ oz}} = \mathbf{\$0.09 \text{ per oz}}$

The **small can** was the better buy.

2. $$\frac{3}{5} = \frac{1800}{U}$$

$$3U = 9000$$

$$U = \mathbf{3000}$$

3. $$\frac{7}{8} \cdot A = \$5600$$

$$\frac{8}{7} \cdot \frac{7}{8} \cdot A = \$5600 \cdot \frac{8}{7}$$

$$A = \mathbf{\$6400}$$

4.

FRACTION	DECIMAL	PERCENT
$\frac{51}{100}$	0.51	51%
(a) $\mathbf{\frac{37}{100}}$	0.37	(b) **37%**

5.

$$0.000009\textcircled{1}53$$

The rounded number is $0.0000092 = \mathbf{9.2 \times 10^{-6}}$.

6. $P \cdot 130 = 52$

$$P = \frac{52}{130}$$

$$P = 0.40$$

$$P = \mathbf{40\%}$$

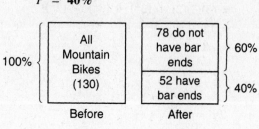

Before   After

7. $0.4 \times A = 216$

$$A = \frac{216}{0.4} = \mathbf{540}$$

8. $F \times 39 = 24$

$$F = \frac{24}{39} = \mathbf{\frac{8}{13}}$$

9. $P \times 480 = 450$

$$P = \frac{450}{480}$$

$$P = \frac{15}{16}$$

$$P = 0.9375$$

$$P = \mathbf{93.75\%}$$

10.

$$4\frac{2}{3} \cdot A = 5\frac{4}{5}$$

$$\frac{14}{3} \cdot A = \frac{29}{5}$$

$$\frac{3}{14} \cdot \frac{14}{3} \cdot A = \frac{29}{5} \cdot \frac{3}{14}$$

$$A = \mathbf{\frac{87}{70}}$$

11. (a) $P = \left[\frac{\pi(20)}{2} + \frac{\pi(20)}{2} + 2(20) \right]$ in.

$$\approx \mathbf{102.8 \text{ in.}}$$

(b)

10

20   10

20

$A_{\text{Figure}} = A_{\text{Square}} + 2A_{\text{Semicircle}}$

$$= \left[20(20) + 2\left(\frac{\pi(10)^2}{2} \right) \right] \text{in.}^2$$

$$\approx \mathbf{714 \text{ in.}^2}$$

12. $V = A_{\text{Base}} \times \text{height}$

$$= 714 \text{ in.}^2 \times 7 \text{ in.}$$

$$= \mathbf{4998 \text{ in.}^3}$$

13. $\dfrac{1 \text{ day}}{(0.5 \times 15) \text{ in.}} \cdot 45 \text{ in.} = \mathbf{6 \text{ days}}$

14. $S.A. = 2A_{\text{End}} + 2A_{\text{Side}} + A_{\text{Bottom}}$

$$= \left[\frac{2(2)(3)}{2} + 2(6)(3.16) + 6(2) \right] \text{cm}^2$$

$$= \mathbf{55.92 \text{ cm}^2}$$

15.

$$\frac{3\frac{1}{5}}{2\frac{1}{3}} = \frac{4}{x}$$

$$\frac{16}{5}x = 4 \cdot \frac{7}{3}$$

$$\frac{5}{16} \cdot \frac{16}{5}x = \frac{28}{3} \cdot \frac{5}{16}$$

$$x = \mathbf{\frac{35}{12}}$$

16.

$$\frac{\frac{3}{4}}{\frac{10}{21}} = \frac{8}{x}$$

$$\frac{3}{4}x = 8 \cdot \frac{10}{21}$$

$$\frac{4}{3} \cdot \frac{3}{4}x = \frac{80}{21} \cdot \frac{4}{3}$$

$$x = \mathbf{\frac{320}{63}}$$

17.

$$6\frac{1}{3}x - 4\frac{7}{8} = 3\frac{3}{4}$$

$$6\frac{1}{3}x - 4\frac{7}{8} + 4\frac{7}{8} = 3\frac{3}{4} + 4\frac{7}{8}$$

$$\frac{19}{3}x = \frac{30}{8} + \frac{39}{8}$$

$$\left(\frac{3}{19} \right)\frac{19}{3}x = \frac{69}{8}\left(\frac{3}{19} \right)$$

$$x = \mathbf{\frac{207}{152}}$$

18.
$$5\frac{6}{7}x - \frac{5}{21} = 11\frac{3}{14}$$
$$5\frac{6}{7}x - \frac{5}{21} + \frac{5}{21} = 11\frac{3}{14} + \frac{5}{21}$$
$$\frac{41}{7}x = \frac{157}{14} + \frac{5}{21}$$
$$\frac{41}{7}x = \frac{471}{42} + \frac{10}{42}$$
$$\left(\frac{7}{41}\right)\frac{41}{7}x = \frac{481}{42}\left(\frac{7}{41}\right)$$
$$x = \frac{481}{246}$$

19. $(1) + (-2) + (1) + (4) + (-6)$
$= (-1) + (1) + (4) + (-6) = (0) + (4) + (-6)$
$= \mathbf{-2}$

20. $(28) + (2) + (-5) + (-10)$
$= (30) + (-5) + (-10) = (25) + (-10)$
$= \mathbf{15}$

21. $(9) + (-9) + (2) + (9) + (-7)$
$= (0) + (2) + (9) + (-7) = (11) + (-7) = \mathbf{4}$

22. $2^2 + 3[2^2(2^4 - 3^2) - \sqrt{4}] = 4 + 3[4(7) - 2]$
$= 4 + 3[26] = 4 + 78 = \mathbf{82}$

23. $2\frac{2}{3} + \frac{1}{6} \cdot 2\frac{1}{4} = \frac{8}{3} + \frac{1}{6} \cdot \frac{9}{4} = \frac{8}{3} + \frac{3}{8}$
$$= \frac{64}{24} + \frac{9}{24} = \mathbf{\frac{73}{24}}$$

24. $16\frac{11}{12} - 3\frac{5}{6} + 2\frac{7}{18} = 16\frac{33}{36} - 3\frac{30}{36} + 2\frac{14}{36}$
$$= \mathbf{15\frac{17}{36}}$$

25.
$$\begin{array}{r} 12.21 \\ \times\ 0.0017 \\ \hline 8547 \\ 1221 \\ \hline \mathbf{0.020757} \end{array}$$

26.
$$\begin{array}{r} 11{,}716.1810 \\ -\ \ \ \ 891.7891 \\ \hline \mathbf{10{,}824.3919} \end{array}$$

27. $\frac{1}{3}\left(2\frac{1}{3} + 3\frac{1}{4}\right) - \frac{13}{36} = \frac{1}{3}\left(\frac{28}{12} + \frac{39}{12}\right) - \frac{13}{36}$
$$= \frac{1}{3}\left(\frac{67}{12}\right) - \frac{13}{36} = \frac{67}{36} - \frac{13}{36} = \frac{54}{36} = \mathbf{\frac{3}{2}}$$

28. $2\frac{1}{3} \times 12\frac{1}{6} \div 1\frac{5}{12} \div \frac{1}{4}$
$$= \frac{7}{\cancel{3}} \times \frac{73}{\cancel{6}} \times \frac{\cancel{12}}{17} \times \frac{\cancel{4}}{1} = \mathbf{\frac{4088}{51}}$$

29. $xy + xyz - x + y^t$
$$= (6)(8) + (6)(8)\left(\frac{1}{24}\right) - (6) + (8^2)$$
$$= 48 + 2 - 6 + 64$$
$$= \mathbf{108}$$

30.
$$\frac{a}{3\frac{1}{2}} = \frac{5}{2} \qquad\qquad \frac{b}{10\frac{2}{3}} = \frac{2}{5}$$
$$2a = 5 \cdot \frac{7}{2} \qquad\qquad 5b = 2 \cdot \frac{32}{3}$$
$$2a = \frac{35}{2} \qquad\qquad 5b = \frac{64}{3}$$
$$a = \frac{35}{2} \cdot \frac{1}{2} \qquad\qquad b = \frac{64}{3} \cdot \frac{1}{5}$$
$$a = \frac{35}{4} \qquad\qquad b = \frac{64}{15}$$
$$a = \mathbf{8\frac{3}{4}} \qquad\qquad b = \mathbf{4\frac{4}{15}}$$

PROBLEM SET 71

1. $\frac{2}{9} = \frac{R}{18}$
$$36 = 9R$$
$$R = \mathbf{4\ students}$$

2. $3\frac{1}{2} \times A = 840$
$$\frac{2}{7} \cdot \frac{7}{2} \cdot A = 840 \cdot \frac{2}{7}$$
$$A = \mathbf{240\ oceanographers}$$

3. $\frac{\$10.90}{12\ disks} = \$10.90$ per dozen
$$\frac{\$80.00}{60\ disks} = \$16.00\ per\ dozen$$
\$10.90 per dozen was the better buy.

4. $\frac{8\,hr}{40\,mi} = \frac{1\,hr}{5\,mi}$
$$\frac{1\,hr}{(2 \times 5)\ \cancel{mi}} \times 20\ \cancel{mi} = \mathbf{2\ hr}$$

5. $P \cdot 275 = 55$

$$P = \frac{55}{275}$$

$$P = 0.20$$

$$P = \mathbf{20\%}$$

100% {	275 flowers in the garden	55 snapdragons } 20%
		220 other flowers } 80%
	Before	After

6.

$$\frac{20}{100} \cdot A = 400$$

$$\frac{100}{20} \cdot \frac{20}{100} \cdot A = 400 \cdot \frac{100}{20}$$

$$A = \mathbf{2000}$$

7.

$$\frac{c}{13\frac{1}{2}} = \frac{5}{8} \qquad\qquad \frac{d}{7} = \frac{8}{5}$$

$$\qquad\qquad\qquad 5d = 56$$

$$8c = 5 \cdot \frac{27}{2} \qquad\qquad \frac{5d}{5} = \frac{56}{5}$$

$$\frac{1}{8} \cdot 8c = \frac{135}{2} \cdot \frac{1}{8} \qquad d = \mathbf{11\frac{1}{5}}$$

$$c = \frac{135}{16}$$

$$c = \mathbf{8\frac{7}{16}}$$

8.

Fraction	Decimal	Percent
$\frac{51}{100}$	0.51	51%
(a) $\frac{3}{10}$	0.3	(b) **30%**

9. $D \times 300 = 120$

$$D = \frac{120}{300}$$

$$D = \mathbf{0.4}$$

10. $2\frac{2}{3} \cdot A = 6\frac{1}{7}$

$$\frac{3}{8} \cdot \frac{8}{3} \cdot A = \frac{43}{7} \cdot \frac{3}{8}$$

$$A = \mathbf{\frac{129}{56}}$$

11. (a) $P = \left[\frac{\pi(20)}{2} + 27 + 33.6 \right] \text{ft} \approx \mathbf{92 \text{ ft}}$

(b) $A_{\text{Figure}} = A_{\text{Semicircle}} + A_{\text{Triangle}}$

$$= \left[\frac{\pi(10)^2}{2} + \frac{27(20)}{2} \right] \text{ft}^2$$

$$\approx [157 + 270] \text{ft}^2$$

$$\approx \mathbf{427 \text{ ft}^2}$$

12. $V = A_{\text{Base}} \times \text{height}$

$$= 427 \text{ ft}^2 \times 2 \text{ ft}$$

$$= \mathbf{854 \text{ ft}^3}$$

13.

$$\frac{4\frac{1}{3}}{2\frac{6}{7}} = \frac{7}{x}$$

$$\frac{13}{3}x = 7 \cdot \frac{20}{7}$$

$$\frac{3}{13} \cdot \frac{13}{3}x = 20 \cdot \frac{3}{13}$$

$$x = \mathbf{\frac{60}{13}}$$

14.

$$\frac{\frac{4}{5}}{\frac{16}{15}} = \frac{x}{3}$$

$$\frac{16}{15}x = \frac{4}{5} \cdot 3$$

$$\frac{15}{16} \cdot \frac{16}{15}x = \frac{12}{5} \cdot \frac{15}{16}$$

$$x = \mathbf{\frac{9}{4}}$$

15.

$$7\frac{1}{8}x + 2\frac{3}{10} = 3\frac{2}{5}$$

$$7\frac{1}{8}x + 2\frac{3}{10} - 2\frac{3}{10} = 3\frac{2}{5} - 2\frac{3}{10}$$

$$\frac{57}{8}x = \frac{34}{10} - \frac{23}{10}$$

$$\left(\frac{8}{57}\right)\frac{57}{8}x = \frac{11}{10}\left(\frac{8}{57}\right)$$

$$x = \mathbf{\frac{44}{285}}$$

16.

$$18x + 7 = 22$$

$$18x + 7 - 7 = 22 - 7$$

$$18x = 15$$

$$x = \frac{15}{18} = \mathbf{\frac{5}{6}}$$

17. $\left(\frac{4}{7}\right)^4 = \frac{4}{7} \cdot \frac{4}{7} \cdot \frac{4}{7} \cdot \frac{4}{7} = \mathbf{\frac{256}{2401}}$

18. $\left(\dfrac{1}{9}\right)^2 = \dfrac{1}{9} \cdot \dfrac{1}{9} = \dfrac{1}{81}$

19. $\sqrt{\dfrac{25}{144}} = \dfrac{\sqrt{25}}{\sqrt{144}} = \dfrac{5}{12}$

20. $\sqrt[3]{\dfrac{27}{125}} = \dfrac{\sqrt[3]{27}}{\sqrt[3]{125}} = \dfrac{3}{5}$

21. $(-8) + (1) + (-3) + (-5)$
$= (-7) + (-3) + (-5) = (-10) + (-5) = \mathbf{-15}$

22. $(2) + (1) + (-4) + (-7) + (8)$
$= (3) + (-4) + (-7) + (8)$
$= (-1) + (-7) + (8) = (-8) + (8) = \mathbf{0}$

23. $(122) + (5) + (-6) + (-11)$
$= (127) + (-6) + (-11)$
$= (121) + (-11) = \mathbf{110}$

24. $(-1) + (6) + (-5) + (-9) + (6)$
$= (5) + (-5) + (-9) + (6)$
$= 0 + (-9) + (6) = \mathbf{-3}$

25. $2^3 + 2^2[3(2^2 - 1) + \sqrt{16}]$
$= 8 + 4[3(3) + 4] = 8 + 4[13]$
$= 8 + 52 = \mathbf{60}$

26. $6\dfrac{1}{3} + \dfrac{2}{3} \cdot 3\dfrac{3}{4} = \dfrac{19}{3} + \dfrac{2}{3} \cdot \dfrac{15}{4} = \dfrac{19}{3} + \dfrac{5}{2}$
$= \dfrac{38}{6} + \dfrac{15}{6} = \dfrac{\mathbf{53}}{\mathbf{6}}$

27. $17\dfrac{5}{12} - 4\dfrac{5}{6} + 1\dfrac{3}{4} = 17\dfrac{10}{24} - 4\dfrac{20}{24} + 1\dfrac{18}{24}$
$= 14\dfrac{8}{24} = \mathbf{14\dfrac{1}{3}}$

28. $\dfrac{1}{4}\left(3\dfrac{1}{3} + 2\dfrac{1}{4}\right) - \dfrac{11}{24}$
$= \dfrac{1}{4}\left(\dfrac{40}{12} + \dfrac{27}{12}\right) - \dfrac{11}{24} = \dfrac{1}{4}\left(\dfrac{67}{12}\right) - \dfrac{11}{24}$
$= \dfrac{67}{48} - \dfrac{22}{48} = \dfrac{45}{48} = \dfrac{\mathbf{15}}{\mathbf{16}}$

29. $3\dfrac{1}{4} \times 4\dfrac{5}{12} \div 3\dfrac{7}{8} \div \dfrac{1}{4} = \dfrac{13}{\cancel{4}} \times \dfrac{53}{\cancel{12}} \times \dfrac{\overset{2}{\cancel{8}}}{31} \times \dfrac{\overset{1}{\cancel{4}}}{1}$
$= \dfrac{\mathbf{1378}}{\mathbf{93}}$

30. $x^y + y^x + \sqrt[y]{z} = (2^3) + (3^2) + (\sqrt[3]{125})$
$= 8 + 9 + 5$
$= \mathbf{22}$

PROBLEM SET 72

1. $\dfrac{\$5}{20 \cancel{\text{dozen}}} \times 1 \cancel{\text{dozen}} = \mathbf{\$0.25}$

   $10 \cancel{\text{dozen}} \times \dfrac{\$0.25}{1 \cancel{\text{dozen}}} = \mathbf{\$2.50}$

2. $\dfrac{3}{5} \times 150 = B$

   $B = \mathbf{90 \text{ clowns}}$

3. $\dfrac{2}{5} = \dfrac{1400}{P}$

   $2P = 7000$

   $P = \mathbf{3500}$

4. $\dfrac{400 \text{ yd}}{20 \cancel{\text{min}}} \times 60 \cancel{\text{min}} = \mathbf{1200 \text{ yd}}$

5. $\dfrac{\$762}{1 \text{ year}} = \mathbf{\$762 \text{ per year}}$

   $\dfrac{\$1850}{3 \text{ years}} = \mathbf{\$616.67 \text{ per year}}$

   $\dfrac{\$3000}{5 \text{ years}} = \mathbf{\$600 \text{ per year}}$

   Five years of coverage for $3000 is the least expensive insurance plan.

6. $0.70 \cdot 420 = B$

   $B = \mathbf{294 \text{ runners}}$

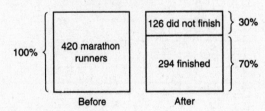

7. $\dfrac{30}{100} \cdot A = 360$

   $\dfrac{100}{30} \cdot \dfrac{30}{100} \cdot A = 360 \cdot \dfrac{100}{30}$

   $A = \mathbf{1200}$

8. $P \cdot 50 = 12$

   $P = \dfrac{12}{50}$

   $P = 0.24$

   $P = \mathbf{24\%}$

9. $x \le 5$

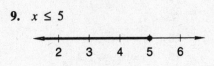

10. $x > 0$

11.

$21\underset{\downarrow}{\textcircled{6}},348,219$

The rounded number is $216,000,000$

$= 2.16 \times 10^8$.

12. $D \times 420 = 147$

$$D = \frac{147}{420} = \frac{7}{20} = \mathbf{0.35}$$

13. (a) $P = \left[\dfrac{\pi(20)}{2} + 25 + 32.02\right]$ in.

$\approx \mathbf{88.42}$ **in.**

(b) $A_{\text{Figure}} = A_{\text{Semicircle}} + A_{\text{Triangle}}$

$= \left[\dfrac{\pi(10)^2}{2} + \dfrac{(25)(20)}{2}\right]$ in.^{2}

$\approx [157 + 250]$ in.^{2}

$\approx \mathbf{407}$ **in.**^{2}

14. $V = A_{\text{Base}} \times$ height

$= 407$ in.$^2 \times 5$ in.

$= \mathbf{2035}$ **in.**^{3}

15.

$$\frac{p}{2} = \frac{\frac{10}{3}}{\frac{8}{7}}$$

$$\frac{8}{7}p = \frac{20}{3}$$

$$\frac{7}{8} \cdot \frac{8}{7}p = \frac{20}{3} \cdot \frac{7}{8}$$

$$p = \frac{35}{6}$$

$$p = 5\frac{5}{6}$$

16.

$$\frac{6\frac{1}{4}}{2\frac{1}{3}} = \frac{5}{x}$$

$$\frac{\frac{25}{4}}{\frac{7}{3}} = \frac{5}{x}$$

$$\frac{25}{4}x = 5 \cdot \frac{7}{3}$$

$$\frac{4}{25} \cdot \frac{25}{4}x = \frac{35}{3} \cdot \frac{4}{25}$$

$$x = \frac{28}{15}$$

17.

$$6\frac{2}{3}x - 4\frac{1}{4} = 3\frac{1}{12}$$

$$6\frac{2}{3}x - 4\frac{1}{4} + 4\frac{1}{4} = 3\frac{1}{12} + 4\frac{1}{4}$$

$$\frac{20}{3}x = \frac{37}{12} + \frac{51}{12}$$

$$\left(\frac{3}{20}\right)\frac{20}{3}x = \frac{88}{12}\left(\frac{3}{20}\right)$$

$$x = \frac{11}{10}$$

18.

$$20x + 18 = 132$$

$$20x + 18 - 18 = 132 - 18$$

$$20x = 114$$

$$\frac{20x}{20} = \frac{114}{20}$$

$$x = \frac{57}{10}$$

19. $\sqrt[4]{\dfrac{16}{81}} = \dfrac{\sqrt[4]{16}}{\sqrt[4]{81}} = \dfrac{\mathbf{2}}{\mathbf{3}}$

20. $\left(\dfrac{9}{11}\right)^2 = \dfrac{9^2}{11^2} = \dfrac{\mathbf{81}}{\mathbf{121}}$

21. $(6) + (-9) + (1) + (-5) = (-3) + (1) + (-5)$
$= (-2) + (-5) = \mathbf{-7}$

22. $(9) + (-1) + (-7) + (7) = (8) + (-7) + (7)$
$= (1) + (7) = \mathbf{8}$

23. $13 + 2[2^3(2^2 - 1)(3 - 1) + 1]$
$= 13 + 2[8(3)(2) + 1] = 13 + 2[49]$
$= 13 + 98 = \mathbf{111}$

24. $2^3 + 2^2[3(2^2 - 1) + \sqrt{16}]$
$= 8 + 4[3(3) + 4] = 8 + 4[13]$
$= 8 + 52 = \mathbf{60}$

25. $2\dfrac{11}{48} - \dfrac{1}{6} \cdot 3\dfrac{3}{8} = \dfrac{107}{48} - \dfrac{1}{6} \cdot \dfrac{27}{8}$

$= \dfrac{107}{48} - \dfrac{27}{48} = \dfrac{80}{48} = \dfrac{\mathbf{5}}{\mathbf{3}}$

26. $18\dfrac{5}{12} - 12\dfrac{1}{6} + 1\dfrac{3}{4} = 18\dfrac{5}{12} - 12\dfrac{2}{12} + 1\dfrac{9}{12}$

$= 7\dfrac{12}{12} = \mathbf{8}$

27. $6\dfrac{1}{4} \times 2\dfrac{1}{3} \div 1\dfrac{5}{12} \div \dfrac{1}{6} = \dfrac{25}{\underset{1}{\cancel{4}}} \times \dfrac{7}{\underset{1}{\cancel{3}}} \times \dfrac{\overset{3}{\cancel{12}}}{17} \times \dfrac{6}{1}$

$= \dfrac{\mathbf{1050}}{\mathbf{17}}$

28. $\dfrac{181.31}{0.005}$

$$5\overline{)181,310}\quad \mathbf{36{,}262}$$

$$\begin{array}{r} 15 \\ \hline 31 \\ 30 \\ \hline 1\,3 \\ 1\,0 \\ \hline 31 \\ 30 \\ \hline 10 \\ 10 \\ \hline 0 \end{array}$$

29. $\dfrac{1}{5}\left(3\dfrac{1}{4} - 2\dfrac{1}{3}\right) + \dfrac{13}{60}$

$= \dfrac{1}{5}\left(\dfrac{39}{12} - \dfrac{28}{12}\right) + \dfrac{13}{60} = \dfrac{1}{5}\left(\dfrac{11}{12}\right) + \dfrac{13}{60}$

$= \dfrac{11}{60} + \dfrac{13}{60} = \dfrac{24}{60} = \mathbf{\dfrac{2}{5}}$

30. $x^x + y^y + x^y + y^x$

$= (3^3) + (2^2) + (3^2) + (2^3)$

$= 27 + 4 + 9 + 8$

$= \mathbf{48}$

PROBLEM SET 73

1. $2\dfrac{1}{2} \cdot A = 5400$

$\dfrac{2}{5} \cdot \dfrac{5}{2} \cdot A = 5400 \cdot \dfrac{2}{5}$

$A = \mathbf{2160}$

2. $\dfrac{15}{2} = \dfrac{3000}{N}$

$15N = 6000$

$N = \mathbf{400\ students}$

3. $\dfrac{7}{2} = \dfrac{7700}{L}$

$7L = 15{,}400$

$L = \mathbf{2200\ biochemists}$

4. $\dfrac{1400\ \text{bottles}}{1\ \text{hour}} \times 8\ \text{hours} = \mathbf{11{,}200\ bottles}$

5. $0.20 \cdot \$85\ \text{million} = E$

$E = \mathbf{\$17\ million}$

6. $\dfrac{80}{100} \times 280 = B$

$\mathbf{224} = B$

7. $x < 6$

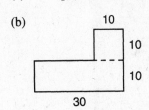

8. $x \geq -22$

9. $1{,}000{,}000\ \cancel{\text{m}}^2 \times \dfrac{100\ \text{cm}}{1\ \cancel{\text{m}}} \times \dfrac{100\ \text{cm}}{1\ \cancel{\text{m}}}$

$= \mathbf{1{,}000{,}000(100)(100)\ cm^2}$

$= \mathbf{10{,}000{,}000{,}000\ cm^2}$

10.

FRACTION	DECIMAL	PERCENT
$\dfrac{51}{100}$	0.51	51%
(a) $\mathbf{\dfrac{23}{100}}$	(b) **0.23**	23%

11. $6.04 \times 10^{-6} = \mathbf{0.00000604}$

12. $D \times 600 = 360$

$D = \dfrac{360}{600} = \dfrac{3}{5} = \mathbf{0.6}$

13. (a) $P = [2(20) + 2(30)]\ \text{m} = \mathbf{100\ m}$

(b)

```
          10
         ┌─────┐
         │     │ 10
  ┌──────┘- - -┘
  │            │ 10
  └────────────┘
       30
```

$A_{\text{Figure}} = A_{\text{Rectangle}} + A_{\text{Square}}$

$= [(10)(30) + (10)(10)]\ \text{m}^2$

$= [300 + 100]\ \text{m}^2$

$= \mathbf{400\ m^2}$

14. height $= 200\ \cancel{\text{cm}} \times \dfrac{1\ \text{m}}{100\ \cancel{\text{cm}}} = 2\ \text{m}$

$V = A_{\text{Base}} \times \text{height}$

$= 400\ \text{m}^2 \times 2\ \text{m}$

$= \mathbf{800\ m^3}$

15. $V = Bh = \pi r^2 h$

$= \pi(2\ \text{in.})^2 \cdot 5\ \text{in.}$

$\approx (3.14)(4)(5)\ \text{in.}^3$

$\approx \mathbf{62.8\ in.^3}$

16. $S.A. = 2B + L.A.$

$\quad = 2\pi r^2 + \pi dh$

$\quad = 2\pi(2 \text{ in.})^2 + \pi(4 \text{ in.})(5 \text{ in.})$

$\quad \approx 25.12 \text{ in.}^2 + 62.8 \text{ in.}^2$

$\quad \approx \mathbf{87.92 \text{ in.}^2}$

17.

$$\frac{5\frac{1}{3}}{3\frac{1}{6}} = \frac{6}{x}$$

$$\frac{\frac{16}{3}}{\frac{19}{6}} = \frac{6}{x}$$

$$\frac{16}{3}x = 6 \cdot \frac{19}{6}$$

$$\frac{3}{16} \cdot \frac{16}{3}x = 19 \cdot \frac{3}{16}$$

$$x = \frac{57}{16} = \mathbf{3\frac{9}{16}}$$

18.

$$\frac{\frac{4}{5}}{\frac{8}{25}} = \frac{x}{5}$$

$$5 \cdot \frac{4}{5} = \frac{8}{25}x$$

$$\frac{25}{8} \cdot \frac{8}{25}x = 4 \cdot \frac{25}{8}$$

$$x = \mathbf{\frac{25}{2}}$$

19.

$$5\frac{1}{3}x - 2\frac{1}{4} = 4\frac{7}{12}$$

$$5\frac{1}{3}x - 2\frac{1}{4} + 2\frac{1}{4} = 4\frac{7}{12} + 2\frac{1}{4}$$

$$\frac{16}{3}x = \frac{55}{12} + \frac{27}{12}$$

$$\left(\frac{3}{16}\right)\frac{16}{3}x = \frac{82}{12}\left(\frac{3}{16}\right)$$

$$x = \mathbf{\frac{41}{32}}$$

20. $\left(\dfrac{3}{8}\right)^3 = \dfrac{3^3}{8^3} = \mathbf{\dfrac{27}{512}}$

21. $\sqrt{\dfrac{16}{144}} = \dfrac{\sqrt{16}}{\sqrt{144}} = \dfrac{4}{12} = \mathbf{\dfrac{1}{3}}$

22. $(-8) + (7) + (4) + (-6) = (-1) + (4) + (-6)$
$= (3) + (-6) = \mathbf{-3}$

23. $(21) + (-19) + (8) + (-11)$
$= (2) + (8) + (-11) = (10) + (-11) = \mathbf{-1}$

24. $\sqrt[3]{8} + \sqrt{4}\left[(2^3 - 4)3 + 1\right]$
$= 2 + 2[4(3) + 1] = 2 + 2[13]$
$= 2 + 26 = \mathbf{28}$

25. $3\dfrac{7}{24} - \dfrac{1}{8} \cdot 2\dfrac{5}{6} = \dfrac{79}{24} - \dfrac{1}{8} \cdot \dfrac{17}{6}$

$= \dfrac{79}{24} - \dfrac{17}{48} = \dfrac{158}{48} - \dfrac{17}{48} = \dfrac{141}{48} = \mathbf{\dfrac{47}{16}}$

26. $19\dfrac{6}{7} - \dfrac{3}{14} + 1\dfrac{4}{21} = 19\dfrac{36}{42} - \dfrac{9}{42} + 1\dfrac{8}{42}$

$= 20\dfrac{35}{42} = \mathbf{20\dfrac{5}{6}}$

27.
$$\begin{array}{r} 18.87 \\ \times\ 0.0032 \\ \hline 3774 \\ 5661 \\ \hline \mathbf{0.060384} \end{array}$$

28. $\dfrac{1}{4}\left(3\dfrac{1}{5} - 2\dfrac{1}{3}\right) + \dfrac{7}{60}$

$= \dfrac{1}{4}\left(\dfrac{48}{15} - \dfrac{35}{15}\right) + \dfrac{7}{60} = \dfrac{1}{4}\left(\dfrac{13}{15}\right) + \dfrac{7}{60}$

$= \dfrac{13}{60} + \dfrac{7}{60} = \dfrac{20}{60} = \mathbf{\dfrac{1}{3}}$

29. $\dfrac{4}{5} \cdot A = 60$

$\dfrac{5}{4} \cdot \dfrac{4}{5} \cdot A = \overset{15}{\cancel{60}} \cdot \dfrac{5}{\underset{1}{\cancel{4}}}$

$A = \mathbf{75 \text{ hound dogs}}$

30. $x^y + y^x + \sqrt[y]{x} = 8^3 + 3^8 + \sqrt[3]{8}$
$\qquad = 512 + 6561 + 2$
$\qquad = \mathbf{7075}$

PROBLEM SET 74

1. $5\dfrac{1}{2} \times 980 = B$

$\dfrac{11}{2} \times 980 = B$

$\mathbf{5390} = B$

2. $\dfrac{32 \text{ mi}}{8 \text{ hr}} + 3\dfrac{\text{mi}}{\text{hr}} = (4 + 3)\dfrac{\text{mi}}{\text{hr}} = 7\dfrac{\text{mi}}{\text{hr}}$

$\dfrac{1 \text{ hr}}{7 \text{ mi}} \times 49 \text{ mi} = \mathbf{7 \text{ hr}}$

3. $\dfrac{7}{2} = \dfrac{W}{4200}$

$2W = 29,400$

$W = \mathbf{14,700 \text{ winners}}$

4. $\frac{3}{4} \cdot A = 42$

$$\frac{4}{3} \cdot \frac{3}{4} \cdot A = 42 \cdot \frac{4}{3}$$

$$A = \textbf{56 bats}$$

5. $P \cdot 175 = 35$

$$P = \frac{35}{175}$$

$$P = 0.2$$

$$P = \textbf{20\%}$$

	35 succeeded } 20%
100% { 175 scuba divers	140 failed } 80%
Before	After

6. $\frac{30}{100} \times N = 123$

$$N = 123 \cdot \frac{100}{30} = \textbf{410}$$

7. $x > -12$

(number line: open circle at -12, shaded right; marks -13, -12, -11, -10, -9)

8. $x \leq 52$

(number line: filled point at 52, shaded left; marks 49, 50, 51, 52, 53)

9.

FRACTION	DECIMAL	PERCENT
$\frac{51}{100}$	0.51	51%
$\frac{37}{100}$	(a) **0.37**	(b) **37%**

10.

0.003 ⓪ 796

The rounded number is $0.0031 = \textbf{3.1} \times \textbf{10}^{-3}$.

11. $P \times 500 = 125$

$$P = \frac{125}{500} = \frac{1}{4} = \textbf{0.25}$$

12. (a) $P = \left[10 + 10 + \frac{\pi(10)}{2} + \frac{\pi(10)}{2} \right]$ m

$$\approx \textbf{51.4 m}$$

(b)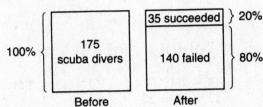

$A_{\text{Figure}} = A_{\text{Square}} + 2A_{\text{Semicircle}}$

$$= \left[(10)(10) + \frac{2\pi(5)^2}{2} \right] \text{m}^2$$

$$\approx [100 + 78.5] \text{ m}^2$$

$$\approx \textbf{178.5 m}^2$$

13. $V = A_{\text{Base}} \times \text{height}$

$$\approx (178.5 \text{ m}^2) \cdot 50 \text{ m}$$

$$\approx \textbf{8925 m}^3$$

14. $V = \pi r^2 h \approx (3.14)(7)^2(14) \text{ dm}^3$

$$\approx \textbf{2154.04 dm}^3$$

$S.A. = 2\pi r^2 + \pi dh$

$$= 2\pi(7 \text{ dm})^2 + \pi(14 \text{ dm})(14 \text{ dm})$$

$$\approx 307.72 \text{ dm}^2 + 615.44 \text{ dm}^2$$

$$\approx \textbf{923.16 dm}^2$$

15.

$$\frac{4\frac{2}{3}}{1\frac{5}{6}} = \frac{8}{x}$$

$$\frac{\frac{14}{3}}{\frac{11}{6}} = \frac{8}{x}$$

$$\frac{14}{3}x = 8 \cdot \frac{11}{6}$$

$$\frac{3}{14} \cdot \frac{14}{3}x = \frac{44}{3} \cdot \frac{3}{14}$$

$$x = \frac{\textbf{22}}{\textbf{7}}$$

16.

$$2\frac{1}{4}x + 4\frac{1}{3} = 5\frac{7}{12}$$

$$2\frac{1}{4}x + 4\frac{1}{3} - 4\frac{1}{3} = 5\frac{7}{12} - 4\frac{1}{3}$$

$$\frac{9}{4}x = \frac{67}{12} - \frac{52}{12}$$

$$\left(\frac{4}{9}\right)\frac{9}{4}x = \frac{15}{12}\left(\frac{4}{9}\right)$$

$$x = \frac{\textbf{5}}{\textbf{9}}$$

17.
$$\frac{x}{4} = \frac{\frac{3}{10}}{\frac{6}{5}}$$

$$\frac{6}{5}x = \frac{12}{10}$$

$$\frac{5}{6} \cdot \frac{6}{5}x = \frac{12}{10} \cdot \frac{5}{6}$$

$$x = \mathbf{1}$$

18.
$$\sqrt[3]{\frac{64}{343}} + \left(\frac{2}{3}\right)^2 = \frac{\sqrt[3]{64}}{\sqrt[3]{343}} + \frac{2^2}{3^2}$$

$$= \frac{4}{7} + \frac{4}{9} = \frac{36}{63} + \frac{28}{63} = \mathbf{\frac{64}{63}}$$

19.
$$\frac{1}{5}\left(4\frac{2}{5} - 2\frac{1}{3}\right) + \frac{4}{15} = \frac{1}{5}\left(\frac{66}{15} - \frac{35}{15}\right) + \frac{4}{15}$$

$$= \frac{1}{5}\left(\frac{31}{15}\right) + \frac{4}{15} = \frac{31}{75} + \frac{20}{75} = \frac{51}{75} = \mathbf{\frac{17}{25}}$$

20. $-4 + 5 + -10 + 2$
$= (-4) + (+5) + (-10) + (+2)$
$= (+1) + (-10) + (+2) = (-9) + (+2) = \mathbf{-7}$

21. $-11 + -2 + 20 + -7$
$= (-11) + (-2) + (+20) + (-7)$
$= (-13) + (+20) + (-7) = (+7) + (-7) = \mathbf{0}$

22. $12 + -7 + -10 + 1$
$= (+12) + (-7) + (-10) + (+1)$
$= (+5) + (-10) + (+1)$
$= (-5) + (+1) = \mathbf{-4}$

23. $(-21) + (+7) + (-2) + (+6)$
$= (-14) + (-2) + (+6)$
$= (-16) + (+6) = \mathbf{-10}$

24. $16\frac{7}{8} - 2\frac{1}{4} \cdot 1\frac{1}{2} + 1\frac{7}{16}$

$$= 16\frac{7}{8} - \frac{9}{4} \cdot \frac{3}{2} + 1\frac{7}{16}$$

$$= 16\frac{7}{8} - \frac{27}{8} + 1\frac{7}{16}$$

$$= 16\frac{14}{16} - 3\frac{6}{16} + 1\frac{7}{16} = 14\frac{15}{16}$$

25.
$$\begin{array}{r} 166.5 \\ \times\ 0.0024 \\ \hline 6660 \\ 3330 \\ \hline \mathbf{0.39960} \end{array}$$

26.
$$\begin{array}{r} 19{,}621.810 \\ -\ \ \ \ 698.971 \\ \hline \mathbf{18{,}922.839} \end{array}$$

27. $\dfrac{213.19}{0.005}$

$$\begin{array}{r} 42{,}638 \\ 5\overline{)213{,}190} \\ \underline{20} \\ 13 \\ \underline{10} \\ 3\,1 \\ \underline{3\,0} \\ 19 \\ \underline{15} \\ 40 \\ \underline{40} \\ 0 \end{array}$$

28. $13 + 2[2^2(3^2 - 2^3)(2^2 + 1) - 15]$
$= 13 + 2[4(9 - 8)(4 + 1) - 15]$
$= 13 + 2[4 \cdot 1 \cdot 5 - 15] = 13 + 2[5]$
$= 13 + 10 = \mathbf{23}$

29.

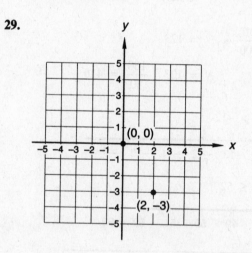

30. $x^y + y^x - \sqrt[x]{z} = (3^2) + (2^3) - (\sqrt[3]{64})$
$$= 9 + 8 - 4$$
$$= \mathbf{13}$$

PROBLEM SET 75

1.
$$\frac{3}{5} \cdot A = \$5200$$

$$\frac{5}{3} \cdot \frac{3}{5} \cdot A = \$5200 \cdot \frac{5}{3}$$

$$A \approx \mathbf{\$8666.67}$$

2.
$$\frac{350 \text{ mi}}{5 \text{ hr}} = \mathbf{70 \text{ mph}}$$

3.
$$\frac{8}{3} = \frac{2400}{D}$$

$$8D = 7200$$

$$D = \mathbf{900 \text{ doubters}}$$

4. If $\frac{4}{5}$ were not in the hive, $\frac{1}{5}$ were in the hive.

$$\frac{1}{5} \cdot A = 1350$$

$$\frac{5}{1} \cdot \frac{1}{5} \cdot A = 1350 \cdot \frac{5}{1}$$

$$A = 6750$$

$6750 - 1350 = $ **5400 bees** were not in the hive

5. $$\frac{30}{2} = \frac{t}{5}$$

$$2t = 150$$

$$t = \textbf{75 s}$$

6. $$\frac{7}{3} = \frac{23}{D}$$

$$7D = 69$$

$$D = \frac{\textbf{69}}{\textbf{7}} \textbf{ drawings}$$

7. $$0.25 \times A = 85$$

$$\frac{0.25A}{0.25} = \frac{85}{0.25}$$

$$A = \textbf{340 apartment complexes}$$

25% {	85 allow pets		
75% {	255 do not allow pets	340 apartment complexes	} 100%
	Parts of Total	Total	

8. $$P \times 325 = 65$$

$$\frac{P \times 325}{325} = \frac{65}{325}$$

$$P = 0.2 = \textbf{20\%}$$

9. $x < -34$

$$\xleftarrow{\hspace{1cm}} \underset{-37}{|} \quad \underset{-36}{|} \quad \underset{-35}{|} \quad \underset{-34}{\oplus} \quad \underset{-33}{|} \xrightarrow{\hspace{1cm}}$$

10. $x \geq 21$

$$\xleftarrow{\hspace{1cm}} \underset{20}{|} \quad \underset{21}{\bullet} \quad \underset{22}{|} \quad \underset{23}{|} \quad \underset{24}{|} \xrightarrow{\hspace{1cm}}$$

11.

FRACTION	DECIMAL	PERCENT
$\frac{51}{100}$	0.51	51%
(a) $\frac{71}{100}$	0.71	(b) **71%**

12. $26{,}900{,}000{,}000 = \textbf{2.69} \times \textbf{10}^{\textbf{10}}$

13. $$D \times 790 = 474$$

$$D = \frac{474}{790} = \frac{3}{5} = \textbf{0.6}$$

14. (a) $P = \left[2(8) + 16 + \frac{\pi(8)}{2} \right] \text{yd} \approx \textbf{44.56 yd}$

(b)

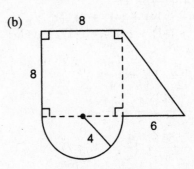

$A = A_{\text{Square}} + A_{\text{Triangle}} + A_{\text{Semicircle}}$

$= (8)(8) + \frac{(6)(8)}{2} + \frac{\pi(4)^2}{2}$

$\approx \textbf{113.12 yd}^{\textbf{2}}$

15. $4 \text{ ft} \times \frac{1 \text{ yd}}{3 \text{ ft}} = \frac{4}{3} \text{ yd}$

$V = A_{\text{Base}} \times \text{height}$

$\approx 113.12 \text{ yd}^2 \times \frac{4}{3} \text{ yd}$

$\approx \textbf{150.83 yd}^{\textbf{3}}$

16. $S.A. = 2\pi r^2 + \pi dh$

$= 2\pi(4 \text{ cm})^2 + \pi(8 \text{ cm})(9 \text{ cm})$

$\approx 100.48 \text{ cm}^2 + 226.08 \text{ cm}^2$

$\approx \textbf{326.56 cm}^{\textbf{2}}$

17. $1{,}000{,}000 \text{ in.} \times \frac{1 \text{ ft}}{12 \text{ in.}} \times \frac{1 \text{ mi}}{5280 \text{ ft}}$

$= \frac{1{,}000{,}000}{(12)(5280)} \text{ mi} \approx \textbf{15.78 mi}$

18. $\left(\frac{2}{3} \right)^3 = \frac{2}{3} \times \frac{2}{3} \times \frac{2}{3} = \frac{\textbf{8}}{\textbf{27}}$

19. $\sqrt[3]{\frac{8}{27}} = \sqrt[3]{\left(\frac{2}{3} \right)^3} = \frac{\textbf{2}}{\textbf{3}}$

20. $\left(\frac{1}{5} \right)^3 = \frac{1}{5} \times \frac{1}{5} \times \frac{1}{5} = \frac{\textbf{1}}{\textbf{125}}$

21. $\sqrt[4]{\frac{81}{256}} = \frac{\sqrt[4]{81}}{\sqrt[4]{256}} = \frac{\textbf{3}}{\textbf{4}}$

22. $10 + 3[3^2(1^5 + 2^3)(3^2 + 1)]$
$= 10 + 3[9(9)(10)] = 10 + 3[810]$
$= 10 + 2430 = \mathbf{2440}$

23. $8 + -9 + 6 + -14$
$= (+8) + (-9) + (+6) + (-14)$
$= (-1) + (+6) + (-14) = (5) + (-14) = \mathbf{-9}$

24. $-30 + 14 + -1 + 17$
$= (-30) + (+14) + (-1) + (+17)$
$= (-16) + (-1) + (+17) = (-17) + (+17) = \mathbf{0}$

25. $11\frac{2}{8} - 5\frac{1}{2} \cdot 1\frac{2}{3} + 1\frac{1}{4}$
$= 11\frac{1}{4} - \frac{11}{2} \cdot \frac{5}{3} + 1\frac{1}{4}$
$= 11\frac{1}{4} - \frac{55}{6} + 1\frac{1}{4}$
$= 11\frac{3}{12} - 9\frac{2}{12} + 1\frac{3}{12} = 3\frac{4}{12} = 3\frac{1}{3} = \mathbf{\frac{10}{3}}$

26. $\frac{1}{4}\left(3\frac{2}{3} - 1\frac{1}{4}\right) + \frac{6}{7} = \frac{1}{4}\left(\frac{44}{12} - \frac{15}{12}\right) + \frac{6}{7}$
$= \frac{1}{4}\left(\frac{29}{12}\right) + \frac{6}{7} = \frac{29}{48} + \frac{6}{7} = \frac{203}{336} + \frac{288}{336}$
$= \mathbf{\frac{491}{336}}$

27. $\frac{657.12}{0.0012}$

$$
\begin{array}{r}
547{,}600 \\
12\overline{)6{,}571{,}200} \\
\underline{6\,0} \\
57 \\
\underline{48} \\
91 \\
\underline{84} \\
7\,2 \\
\underline{7\,2} \\
0
\end{array}
$$

28. $\dfrac{3\frac{1}{2}}{1\frac{2}{5}} = \dfrac{10}{m}$

$\dfrac{\frac{7}{2}}{\frac{7}{5}} = \dfrac{10}{m}$

$\dfrac{7}{2}m = 10 \cdot \dfrac{7}{5}$

$\dfrac{2}{7} \cdot \dfrac{7}{2}m = 14 \cdot \dfrac{2}{7}$

$m = \mathbf{4}$

29. $5\frac{1}{4}y - 2\frac{1}{2} = 4\frac{5}{12}$

$5\frac{1}{4}y - 2\frac{1}{2} + 2\frac{1}{2} = 4\frac{5}{12} + 2\frac{1}{2}$

$\frac{21}{4}y = \frac{53}{12} + \frac{30}{12}$

$\left(\frac{4}{21}\right)\frac{21}{4}y = \frac{83}{12}\left(\frac{4}{21}\right)$

$y = \mathbf{\frac{83}{63}}$

30. $\dfrac{\$57}{3 \text{ dingos}} = \$19$ per dingo

$\dfrac{\$99.50}{5 \text{ dingos}} = \$19.90$ per dingo

3 dingos for \$57 is the better deal.

PROBLEM SET 76

1. $F \times 29{,}000 = 3000$
$F = \dfrac{3000}{29{,}000} = \mathbf{\dfrac{3}{29}}$

2. $\dfrac{7}{8} \times 16{,}000 = B$
$B = \mathbf{14{,}000 \text{ malingerers}}$

3. $\dfrac{40 \text{ mi}}{5 \text{ hr}} + 4\dfrac{\text{mi}}{\text{hr}} = (8 + 4)\dfrac{\text{mi}}{\text{hr}} = 12\dfrac{\text{mi}}{\text{hr}}$
$\dfrac{1 \text{ hr}}{12 \cancel{\text{ mi}}} \times 36 \cancel{\text{ mi}} = \mathbf{3 \text{ hr}}$

4. $\dfrac{7}{5} = \dfrac{112}{N}$
$7N = 560$
$N = \mathbf{80 \text{ onlookers}}$

5. $\dfrac{16}{2} = \dfrac{T}{6}$
$2T = 96$
$T = \mathbf{48 \text{ teams}}$

6. $0.40 \cdot M = 22$
$\dfrac{0.40 \cdot M}{0.40} = \dfrac{22}{0.40}$
$M = \mathbf{55 \text{ marsupials}}$

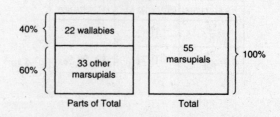

7. $\dfrac{60}{100} \times 480 = B$

$\qquad$ **288** $= B$

8. $x > -4$

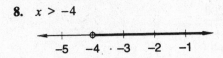

9. $x \le 4$

10.

Fraction	Decimal	Percent
$\dfrac{51}{100}$	0.51	51%
(a) $\dfrac{39}{100}$	0.39	(b) **39%**

11. (a) $\left(\dfrac{2}{3}\right)^4 = \dfrac{2^4}{3^4} = \dfrac{\mathbf{16}}{\mathbf{81}}$

(b) $\sqrt{\dfrac{64}{81}} = \dfrac{\sqrt{64}}{\sqrt{81}} = \dfrac{\mathbf{8}}{\mathbf{9}}$

12. $0.4 \times A = 148$

$\qquad A = \dfrac{148}{0.4} = \mathbf{370}$

13. (a) $P = \left[24 + 2(12) + \dfrac{2\pi(12)}{2}\right]$ ft

$\qquad \approx \mathbf{85.68\ ft}$

(b)

$A_{\text{Figure}} = A_{\text{Rectangle}} + 2A_{\text{Semicircle}}$

$\qquad = \left[(12)(24) + \dfrac{2\pi(6)^2}{2}\right]$ ft^2

$\qquad \approx [288 + 113.04]$ ft^2

$\qquad = \mathbf{401.04\ ft^2}$

14. $1\ \cancel{yd} \times \dfrac{3\ ft}{1\ \cancel{yd}} = \mathbf{3\ ft}$

$\qquad V = A_{\text{Base}} \times \text{height}$

$\qquad\quad \approx 401.04\ \text{ft}^2 \times 3\ \text{ft}$

$\qquad\quad \approx \mathbf{1203.12\ ft^3}$

15. $V = \pi r^2 h \approx (3.14)(10)^2(11)\ \text{m}^3 \approx \mathbf{3454\ m^3}$

16. $\dfrac{4}{x} = \dfrac{6\frac{1}{2}}{7\frac{1}{3}}$

$\qquad \dfrac{13}{2}x = \dfrac{22}{3} \cdot 4$

$\qquad \dfrac{2}{13} \cdot \dfrac{13}{2}x = \dfrac{88}{3} \cdot \dfrac{2}{13}$

$\qquad\qquad x = \dfrac{\mathbf{176}}{\mathbf{39}}$

17. $\dfrac{\frac{3}{7}}{\frac{9}{21}} = \dfrac{z}{3}$

$\qquad 3 \cdot \dfrac{3}{7} = \dfrac{9}{21}z$

$\qquad \dfrac{21}{9} \cdot \dfrac{9}{21}z = \dfrac{9}{7} \cdot \dfrac{21}{9}$

$\qquad\qquad z = \mathbf{3}$

18. $6\frac{1}{2}x + 3\frac{1}{4} = 5\frac{1}{8}$

$\quad 6\frac{1}{2}x + 3\frac{1}{4} - 3\frac{1}{4} = 5\frac{1}{8} - 3\frac{1}{4}$

$\qquad\qquad \dfrac{13}{2}x = \dfrac{41}{8} - \dfrac{26}{8}$

$\qquad \dfrac{2}{13} \cdot \dfrac{13}{2}x = \dfrac{15}{8} \cdot \dfrac{2}{13}$

$\qquad\qquad x = \dfrac{\mathbf{15}}{\mathbf{52}}$

19. $(4 \times 10^2) \times (2 \times 10^8)$
$= (4 \times 2) \times (10^2 \times 10^8) = \mathbf{8 \times 10^{10}}$

20. $(2 \times 10^7) \times (3 \times 10^{-4})$
$= (2 \times 3) \times (10^7 \times 10^{-4}) = \mathbf{6 \times 10^3}$

21. $(7 \times 10^{11}) \times (3 \times 10^{-13})$
$= (7 \times 3) \times (10^{11} \times 10^{-13}) = 21 \times 10^{-2}$
$= \mathbf{2.1 \times 10^{-1}}$

22. $(4.73 \times 10^{-3}) \times (2 \times 10^{13})$
$= (4.73 \times 2) \times (10^{-3} \times 10^{13}) = \mathbf{9.46 \times 10^{10}}$

23. $-6 + 4 + -10 + 2$
$= (-6) + (4) + (-10) + (+2)$
$= (-2) + (-10) + (+2) = (-12) + (+2)$
$= \mathbf{-10}$

24. $-36 + 21 + -6 + -3$
$= (-36) + (+21) + (-6) + (-3)$
$= (-15) + (-6) + (-3) = (-21) + (-3)$
$= \mathbf{-24}$

25. $13\frac{3}{4} - 1\frac{4}{5} \cdot \frac{3}{2} + \frac{7}{20}$

$= \frac{55}{4} - \frac{9}{5} \cdot \frac{3}{2} + \frac{7}{20} = \frac{55}{4} - \frac{27}{10} + \frac{7}{20}$

$= \frac{275}{20} - \frac{54}{20} + \frac{7}{20} = \frac{228}{20} = \mathbf{\frac{57}{5}}$

26.
$$\begin{array}{r} 19,611.62 \\ - \quad 687.91 \\ \hline \mathbf{18,923.71} \end{array}$$

27. $\frac{218.31}{0.006}$

$$\begin{array}{r} \mathbf{36,385} \\ 7\overline{)218,310} \\ \underline{18} \\ 38 \\ \underline{36} \\ 2\,3 \\ \underline{1\,8} \\ 51 \\ \underline{48} \\ 30 \\ \underline{30} \\ 0 \end{array}$$

28. $\frac{1}{6}\left(2\frac{1}{3} \cdot \frac{1}{2} - \frac{3}{4}\right) + \frac{5}{24}$

$= \frac{1}{6}\left(\frac{14}{12} - \frac{9}{12}\right) + \frac{5}{24} = \frac{1}{6}\left(\frac{5}{12}\right) + \frac{5}{24}$

$= \frac{5}{72} + \frac{15}{72} = \frac{20}{72} = \mathbf{\frac{5}{18}}$

29. $4^2 + \sqrt{16}\left[2^2(3^2 - 2^2)(4^2 - 3^2) - 10\right]$
$= 16 + 4[4(5)(7) - 10] = 16 + 4[130]$
$= 16 + 520 = \mathbf{536}$

30. $xy + x^2 + y^2 + x^y = 4(2) + 4^2 + 2^2 + 4^2$
$\qquad\qquad\qquad = 8 + 16 + 4 + 16$
$\qquad\qquad\qquad = \mathbf{44}$

PROBLEM SET 77

1. $\frac{2}{5} \cdot A = 42$

$\frac{5}{2} \cdot \frac{2}{5} \cdot A = 42 \cdot \frac{5}{2}$

$A = \mathbf{105\ laws}$

2. If $\frac{3}{7}$ were neophytes, $\frac{4}{7}$ were not neophytes.

$\frac{4}{7} \times 28{,}000 = B$

$B = \mathbf{16{,}000\ students}$

3. $\frac{60\ mi}{3\ hr} + N = \frac{200\ mi}{5\ hr}$

$\qquad N = 40\ mph - 20\ mph = \mathbf{20\ mph}$

4. $\frac{5}{17} = \frac{200}{U}$

$5U = 3400$

$U = \mathbf{680\ undaunted}$

5. $\frac{5}{2} = \frac{C}{7}$

$2C = 35$

$C = \mathbf{\frac{35}{2}\ cups\ of\ flour}$

6. $B = 1.10 \cdot 50$

$B = \mathbf{55\ kilojoules}$

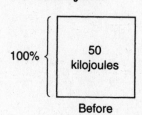

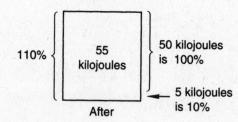

7. $P \cdot 8.2 = 20.5$

$P = \frac{20.5}{8.2}$

$P = 2.5$

$P = \mathbf{250\%}$

8. $x > -1$

9. $x \leq 0$

10. $1.40 \cdot 310 = B$

$\mathbf{434} = B$

11. $\dfrac{26}{50} = \dfrac{52}{100} = 52\% = 0.52$

Fraction	Decimal	Percent
$\dfrac{51}{100}$	0.51	51%
$\dfrac{26}{50}$	(a) **0.52**	(b) **52%**

12. (a) $\left(\dfrac{3}{4}\right)^3 = \dfrac{3^3}{4^3} = \dfrac{\mathbf{27}}{\mathbf{64}}$

    (b) $\sqrt[4]{\dfrac{16}{81}} = \sqrt[4]{\left(\dfrac{2}{3}\right)^4} = \dfrac{\mathbf{2}}{\mathbf{3}}$

13. $0.65 \times A = 780$

    $A = \dfrac{780}{0.65} = \mathbf{1200}$

14. $V = \pi r^2 h$

    $= [\pi(3)^2]\,\text{m}^2 \times 10\,\text{m}$

    $\approx 28.26\,\text{m}^2 \times 10\,\text{m}$

    $\approx \mathbf{282.6\,m^3}$

    $S.A. = 2\pi r^2 + \pi d h$

    $= [2\pi(3^2_\bullet) + \pi(6)(10)]\,\text{m}^2$

    $= 78\pi\,\text{m}^2$

    $\approx \mathbf{244.92\,m^2}$

15.
    $\downarrow$

    $1,0\,\textcircled{3}\,9,296,119.4$

    The rounded number is 1,040,000,000
    $= \mathbf{1.04 \times 10^9}$.

16. $\dfrac{4\frac{1}{4}}{6\frac{4}{5}} = \dfrac{15}{x}$

    $\dfrac{\frac{17}{4}}{\frac{34}{5}} = \dfrac{15}{x}$

    $\dfrac{17}{4}x = 102$

    $\dfrac{4}{17} \cdot \dfrac{17}{4}x = 102 \cdot \dfrac{4}{17}$

    $x = \mathbf{24}$

17. $\dfrac{\frac{5}{8}}{\frac{15}{22}} = \dfrac{x}{2}$

    $\dfrac{5}{4} = \dfrac{15}{22}x$

    $\dfrac{22}{15} \cdot \dfrac{15}{22}x = \dfrac{5}{4} \cdot \dfrac{22}{15}$

    $x = \dfrac{\mathbf{11}}{\mathbf{6}}$

18. $7\frac{1}{5}x + 2\frac{1}{9} = 4\frac{1}{3}$

    $7\frac{1}{5}x + 2\frac{1}{9} - 2\frac{1}{9} = 4\frac{1}{3} - 2\frac{1}{9}$

    $\dfrac{36}{5}x = \dfrac{39}{9} - \dfrac{19}{9}$

    $\dfrac{5}{36} \cdot \dfrac{36}{5}x = \dfrac{20}{9} \cdot \dfrac{5}{36}$

    $x = \dfrac{\mathbf{25}}{\mathbf{81}}$

19. $(4 \times 10^{23}) \times (3 \times 10^{-14})$
    $= (4 \times 3) \times (10^{23} \times 10^{-14})$
    $= 12 \times 10^9 = \mathbf{1.2 \times 10^{10}}$

20. $(1.8 \times 10^{-3}) \times (2 \times 10^5) = \mathbf{3.6 \times 10^2}$

21. $(2.5 \times 10^{60}) \times (3 \times 10^{15}) = \mathbf{7.5 \times 10^{75}}$

22. $\dfrac{1}{5}\left(3\frac{1}{2} \cdot \dfrac{4}{7} - \dfrac{11}{14}\right) + \dfrac{3}{28}$

    $= \dfrac{1}{5}\left(\dfrac{28}{14} - \dfrac{11}{14}\right) + \dfrac{3}{28} = \dfrac{1}{5}\left(\dfrac{17}{14}\right) + \dfrac{3}{28}$

    $= \dfrac{34}{140} + \dfrac{15}{140} = \dfrac{49}{140} = \dfrac{\mathbf{7}}{\mathbf{20}}$

23. $-15 + -3 + 14 + 5$
    $= (-15) + (-3) + (+14) + (+5)$
    $= (-18) + (+14) + (+5) = (-4) + (+5) = \mathbf{1}$

24. $-29 + 11 + -5 + -3$
    $= (-29) + (+11) + (-5) + (-3)$
    $= (-18) + (-5) + (-3) = (-23) + (-3) = \mathbf{-26}$

25. $10\frac{1}{6} - 2\frac{5}{6} \cdot \dfrac{5}{4} + \dfrac{11}{12}$

    $= 10\frac{4}{24} - 3\frac{13}{24} + \dfrac{22}{24} = 7\frac{13}{24} = \dfrac{\mathbf{181}}{\mathbf{24}}$

26.
```
    416.09
  × 0.0011
    41609
   41609
  0.457699
```

27.
$$
\begin{array}{r}
15{,}342.160 \\
-\quad 91.091 \\
\hline
15{,}251.069
\end{array}
$$

28. $\dfrac{419.42}{0.004}$

$$
\begin{array}{r}
104{,}855 \\
4\overline{)419{,}420} \\
\underline{4} \\
019 \\
\underline{16} \\
3\,4 \\
\underline{3\,2} \\
22 \\
\underline{20} \\
20 \\
\underline{20} \\
0
\end{array}
$$

29. $5^2 + \sqrt{64}\left[2^2(2^2 - \sqrt[9]{1})(5^2 - 4^2) + (11)\right]$
$= 25 + 8[4(1)(9) + 11] = 25 + 8[47]$
$= 25 + 376 = \mathbf{401}$

30. $\sqrt[m]{p} + \dfrac{x}{\sqrt{p}} = (\sqrt[4]{16}) + \dfrac{(3)}{\sqrt{(16)}}$

$$= 2 + \frac{3}{4}$$

$$= 2\frac{3}{4}$$

PROBLEM SET 78

1. If $\frac{3}{7}$ had angelic faces, $\frac{4}{7}$ did not have angelic faces.

$$\frac{4}{7} \times 420{,}000 = B$$

$$B = \mathbf{240{,}000 \ cherubs}$$

2. $\dfrac{2}{11} = \dfrac{380}{G}$

$2G = 4180$

$G = \mathbf{2090 \ gargoyles}$

3. Initial speed $= \dfrac{100 \ mi}{5 \ hr} = 20 \ mph$

Doubled speed $= 2 \times 20 \ mph = 40 \ mph$

$\dfrac{1 \ hr}{40 \ mi} \times 200 \ mi = \mathbf{5 \ hr}$

4. $\dfrac{\$24}{48 \ roses} = \dfrac{\$1}{2 \ roses}$

$\dfrac{\$1}{2 \ \text{roses}} \times 6 \ \text{roses} = \mathbf{\$3}$

5. $1.30 \cdot 270 = B$

$\qquad B = \mathbf{351 \ decorations}$

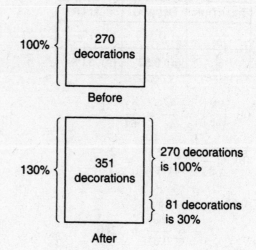

6. $P \cdot 300 = 60$

$\qquad P = \dfrac{60}{300}$

$\qquad P = 0.2$

$\qquad P = \mathbf{20\%}$

7. $P \cdot 8 = 160$

$\qquad P = \dfrac{160}{8}$

$\qquad P = 20$

$\qquad P = \mathbf{2000\%}$

8. $x \le 3$

9. $x > -3$

10. $1.80 \cdot 620 = x$

$\qquad x = \mathbf{1116}$

11. $1.03 \times 10^{10} = \mathbf{10{,}300{,}000{,}000}$

12.
$$2\frac{1}{4} \cdot A = 5\frac{3}{8}$$

$$\frac{9}{4} \cdot A = \frac{43}{8}$$

$$\frac{4}{9} \cdot \frac{9}{4} \cdot A = \frac{43}{8} \cdot \frac{4}{9}$$

$$A = \mathbf{\frac{43}{18}}$$

13.

$V = A_{\text{Base}} \times \text{height}$

$= \left[(20)(8) + \dfrac{(24)(17)}{2}\right] \text{in.}^2 \times 12 \text{ in.}$

$= [160 + 204] \text{ in.}^2 \times 12 \text{ in.}$

$= \mathbf{4368 \text{ in.}^3}$

14. $S.A. = 2A_{\text{Top}} + 2A_{\text{Front}} + 2A_{\text{Side}}$

$= 2(10)(5) \text{ ft}^2 + 2(5)(2) \text{ ft}^2 + 2(10)(2) \text{ ft}^2$

$= 100 \text{ ft}^2 + 20 \text{ ft}^2 + 40 \text{ ft}^2$

$= \mathbf{160 \text{ ft}^2}$

15. $V = \pi r^2 h \approx (3.14)(10)^2(10) \text{ ft}^3$

$\approx \mathbf{3140 \text{ ft}^3}$

$S.A. = 2\pi r^2 + \pi dh$

$\approx 2(3.14)(10^2) \text{ ft}^2 + (3.14)(20)(10) \text{ ft}^2$

$\approx 628 \text{ ft}^2 + 628 \text{ ft}^2$

$\approx \mathbf{1256 \text{ ft}^2}$

16. $10 = 2 \cdot 5$

$12 = 2 \cdot 2 \cdot 3$

$18 = 2 \cdot 3 \cdot 3$

$\text{LCM } (10, 12, 18) = 2 \cdot 2 \cdot 3 \cdot 3 \cdot 5 = \mathbf{180}$

17.

$\dfrac{1\frac{2}{3}}{6\frac{3}{4}} = \dfrac{\frac{3}{4}}{x}$

$\dfrac{\frac{5}{3}}{\frac{27}{4}} = \dfrac{\frac{3}{4}}{x}$

$\dfrac{5}{3}x = \dfrac{81}{16}$

$\dfrac{3}{5} \cdot \dfrac{5}{3}x = \dfrac{81}{16} \cdot \dfrac{3}{5}$

$x = \mathbf{\dfrac{243}{80}}$

18.

$\dfrac{\frac{4}{7}}{\frac{8}{28}} = \dfrac{p}{6}$

$\dfrac{24}{7} = \dfrac{8}{28}p$

$\dfrac{28}{8} \cdot \dfrac{8}{28}p = \dfrac{24}{7} \cdot \dfrac{28}{8}$

$p = \mathbf{12}$

19.

$5\frac{1}{2}x + 2\frac{1}{3} = 4\frac{5}{6}$

$5\frac{1}{2}x + 2\frac{1}{3} - 2\frac{1}{3} = 4\frac{5}{6} - 2\frac{1}{3}$

$\dfrac{11}{2}x = \dfrac{29}{6} - \dfrac{14}{6}$

$\dfrac{2}{11} \cdot \dfrac{11}{2}x = \dfrac{15}{6} \cdot \dfrac{2}{11}$

$x = \mathbf{\dfrac{5}{11}}$

20. There are four negative signs. Four is an even number, so $-\left\{-[-(-3)]\right\} = \mathbf{3}$.

21. There are two negative signs. Two is an even number, so $-\left\{+[+(-15)]\right\} = \mathbf{15}$.

22. There are two negative signs. Two is an even number, so $-[+(-22)] = \mathbf{22}$.

23. $(3.16 \times 10^{-7})(3.3 \times 10^{15}) = 10.428 \times 10^8$

$= \mathbf{1.0428 \times 10^9}$

24. $-8 + 13 - 6 + 9 = (+5) + (-6) + (+9)$

$= (-1) + (9) = \mathbf{8}$

25. $-4 - 6 + -3 + 1 - 6$

$= (-10) + (-3) + (+1) + (-6)$

$= (-13) + (+1) + (-6) = (-12) + (-6) = \mathbf{-18}$

26. $16\frac{9}{20} - 2\frac{1}{2} \cdot 3\frac{2}{5} + \dfrac{9}{10}$

$= \dfrac{329}{20} - \dfrac{5}{2} \cdot \dfrac{17}{5} + \dfrac{9}{10} = \dfrac{329}{20} - \dfrac{85}{10} + \dfrac{9}{10}$

$= \dfrac{329}{20} - \dfrac{170}{20} + \dfrac{18}{20} = \mathbf{\dfrac{177}{20}}$

27. $\dfrac{1}{8}\left(3\frac{1}{2} \cdot \dfrac{1}{2} - \dfrac{4}{5}\right) + \dfrac{9}{40}$

$= \dfrac{1}{8}\left(\dfrac{35}{20} - \dfrac{16}{20}\right) + \dfrac{9}{40} = \dfrac{1}{8}\left(\dfrac{19}{20}\right) + \dfrac{9}{40}$

$= \dfrac{19}{160} + \dfrac{36}{160} = \dfrac{55}{160} = \mathbf{\dfrac{11}{32}}$

28. $3^2 + \sqrt{25}\left[2^2(2^2 - 3)(5^2 - 4^2) - 20\right]$

$= 9 + 5[4(1)(9) - 20] = 9 + 5[16] = 9 + 80$

$= \mathbf{89}$

29. $\left(\dfrac{3}{4}\right)^3 + \left(\dfrac{5}{8}\right)^2 - \left(\dfrac{3}{2}\right)^6 = \dfrac{3^3}{4^3} + \dfrac{5^2}{8^2} - \dfrac{3^6}{2^6}$

$= \dfrac{27}{64} + \dfrac{25}{64} - \dfrac{729}{64} = \mathbf{-\dfrac{677}{64}}$

30. $\sqrt{x} + x^y + xy = (\sqrt{4}) + (4^3) + (4)(3)$

$= 2 + 64 + 12$

$= \mathbf{78}$

PROBLEM SET 79

1. $\dfrac{65 \text{ gal}}{5 \text{ weeks}} \times 52 \text{ weeks} = \mathbf{676 \text{ gal}}$

2. $\dfrac{4}{6} = \dfrac{86}{E}$

$4E = 516$

$E = \mathbf{129 \text{ eggs}}$

3. $\dfrac{11}{9} = \dfrac{121{,}000}{E}$

$11E = 1{,}089{,}000$

$E = \mathbf{99{,}000 \text{ exotic ones}}$

4. $\dfrac{\$11{,}300}{5 \text{ bikes}} \times 12 \text{ bikes} = \mathbf{\$27{,}120}$

5. $P \cdot 5000 = 8000$

$P = \dfrac{8000}{5000}$

$P = 1.6$

$P = \mathbf{160\%}$

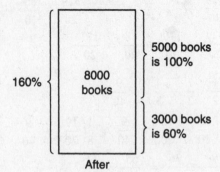

6. $1.12 \cdot 250 = S$

$S = \mathbf{280}$

7. $P \cdot 320 = 1600$

$P = \dfrac{1600}{320}$

$P = 5$

$P = \mathbf{500\%}$

8. $x \le -4$

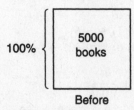

9. $x > 0$

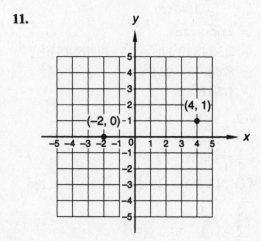

10. $\dfrac{3}{4\frac{1}{3}} = \dfrac{5\frac{1}{2}}{x}$

$3x = \dfrac{11}{2} \cdot \dfrac{13}{3}$

$\dfrac{1}{3} \cdot 3x = \dfrac{143}{6} \cdot \dfrac{1}{3}$

$x = \dfrac{143}{18}$

$x = \mathbf{7\dfrac{17}{18}}$

11.

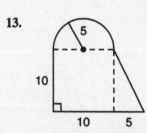

12. $3\dfrac{1}{2} \cdot A = 4200$

$\dfrac{2}{7} \cdot \dfrac{7}{2} \cdot A = 4200 \cdot \dfrac{2}{7}$

$A = \mathbf{1200}$

13.

$V = A_{\text{Base}} \times \text{height}$

$= [A_{\text{Semicircle}} + A_{\text{Square}} + A_{\text{Triangle}}] \times \text{height}$

$= \left[\dfrac{\pi(5)^2}{2} + (10)(10) + \dfrac{(5)(10)}{2} \right](5) \text{ m}^3$

$\approx [(12.5)(3.14) + 100 + 25](5) \text{ m}^3$

$\approx \mathbf{821.25 \text{ m}^3}$

14. $S.A. = 2\pi r^2 + \pi dh$
$\approx 2(3.14)(2^2)\text{ m}^2 + (3.14)(4)(10)\text{ m}^2$
$\approx 25.12\text{ m}^2 + 125.6\text{ m}^2$
$\approx \mathbf{150.72\text{ m}^2}$

15. $\dfrac{2\frac{1}{2}}{5\frac{1}{3}} = \dfrac{8}{x}$

$\dfrac{\frac{5}{2}}{\frac{16}{3}} = \dfrac{8}{x}$

$\dfrac{5}{2}x = \dfrac{128}{3}$

$x = \dfrac{128}{3} \cdot \dfrac{2}{5}$

$x = \mathbf{\dfrac{256}{15}}$

16. $4\frac{1}{8}m + 1\frac{1}{5} = 5\frac{2}{3}$

$4\frac{1}{8}m + 1\frac{1}{5} - 1\frac{1}{5} = 5\frac{2}{3} - 1\frac{1}{5}$

$\dfrac{33}{8}m = \dfrac{85}{15} - \dfrac{18}{15}$

$\dfrac{8}{33} \cdot \dfrac{33}{8}m = \dfrac{67}{15} \cdot \dfrac{8}{33}$

$m = \mathbf{\dfrac{536}{495}}$

17. (a) $-\{+[-(-2)]\} = \mathbf{-2}$ (since there are an odd number of negative signs)

(b) $-\big(+\{-[-(+2)]\}\big) = \mathbf{-2}$ (since there are an odd number of negative signs)

(c) $+[-(-2)] = \mathbf{2}$ (since there are an even number of negative signs)

18. (a) $-\{-[-(-19)]\} = \mathbf{19}$ (since there are four negative signs, an even number)

(b) $+[+(+19)] = \mathbf{19}$ (since there are no negative signs)

(c) $+[-(+19)] = \mathbf{-19}$ (since there is one negative sign, an odd number)

19. $+7 - (-3) + (-2) = 7 + 3 - 2$
$= 10 - 2 = \mathbf{8}$

20. $-3 + (-2) - (-3) = -5 + 3 = \mathbf{-2}$

21. $-(-3) - [-(-4)] - 2 + 7 = 3 - 4 - 2 + 7$
$= -1 - 2 + 7 = -3 + 7 = \mathbf{4}$

22. $\dfrac{78.045}{0.06}$

$\begin{array}{r} 1300.75 \\ 6\overline{)7804.50} \\ \underline{6} \\ 18 \\ \underline{18} \\ 004\ 5 \\ \underline{4\ 2} \\ 30 \\ \underline{30} \\ 0 \end{array}$

23. $\begin{array}{r} 83{,}041.7600 \\ -76.0915 \\ \hline \mathbf{82{,}965.6685} \end{array}$

24. $\dfrac{1}{4}\left(2\frac{1}{3} \cdot \frac{1}{4} + \frac{5}{6}\right) + \frac{7}{9}$

$= \dfrac{1}{4}\left(\frac{7}{12} + \frac{10}{12}\right) + \frac{7}{9} = \dfrac{1}{4}\left(\frac{17}{12}\right) + \frac{7}{9}$

$= \dfrac{17}{48} + \dfrac{7}{9} = \dfrac{51}{144} + \dfrac{112}{144} = \mathbf{\dfrac{163}{144}}$

25. $\left(\dfrac{3}{4}\right)^4 = \dfrac{3^4}{4^4} = \mathbf{\dfrac{81}{256}}$

26. $\sqrt{\dfrac{169}{225}} = \dfrac{\sqrt{169}}{\sqrt{225}} = \mathbf{\dfrac{13}{15}}$

27. $2^3 + \sqrt{25}\,[3^2(2 - 1^{11})(2^2 - \sqrt[5]{1}) + (-9)]$
$= 8 + 5[9(1)(3) - 9] = 8 + 5[18]$
$= 8 + 90 = \mathbf{98}$

28. $(3 \times 10^{12})(15 \times 10^{-2})$
$= (3 \times 15) \times (10^{12} \times 10^{-2})$
$= 45 \times 10^{10} = \mathbf{4.5 \times 10^{11}}$

29. $1{,}000{,}000 \;\cancel{\text{m}} \times \dfrac{100\text{ cm}}{1\;\cancel{\text{m}}}$

$= 100{,}000{,}000\text{ cm} = \mathbf{1.0 \times 10^8\text{ cm}}$

30. $\sqrt[m]{p} + p^x - x^m = \sqrt[3]{27} + 27^1 - 1^3$
$= 3 + 27 - 1$
$= \mathbf{29}$

PROBLEM SET 80

1. $\dfrac{2}{153} = \dfrac{E}{1071}$
$153E = 2142$
$E = \mathbf{14\text{ eggs}}$

2. $\dfrac{13}{5} = \dfrac{3900}{Q}$
$13Q = 19{,}500$
$Q = \mathbf{1500\text{ quiet ones}}$

3. Initial speed = $\dfrac{200 \text{ mi}}{4 \text{ hr}} = 50 \dfrac{\text{mi}}{\text{hr}}$

   Reduced speed = $\left(\dfrac{1}{2} \times 50\right) \dfrac{\text{mi}}{\text{hr}} = 25 \dfrac{\text{mi}}{\text{hr}}$

   $\dfrac{1 \text{ hr}}{25 \text{ mi}} \times 100 \text{ mi} = 4 \text{ hr}$

   Total = 4 hr + 4 hr = **8 hr**

4. If $\dfrac{4}{5}$ got excited, $\dfrac{1}{5}$ did not get excited.

   $\dfrac{1}{5} \times 1650 = B$

   $B = $ **330 spelunkers**

5. $0.12 \cdot 250 = B$

   $B = $ **30 students**

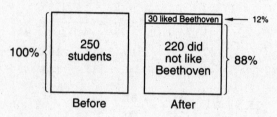

Before   After

6. $P \cdot \$60 = \$84$

   $P = \dfrac{\$84}{\$60}$

   $P = 1.4$

   $P = $ **140%**

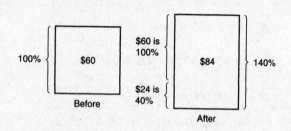

Before   After

7. $1.40 \cdot Z = 10{,}500$

   $Z = \dfrac{10{,}500}{1.4}$

   $Z = $ **7500**

8. Increase = $\$84 - \$60 = \$24$

   $P \cdot 60 = 24$

   $P = \dfrac{24}{60}$

   $P = \dfrac{2}{5}$

   $P = $ **40%**

9. $33 - 22 = 11$ more pairs of wood thrushes

   $P \cdot 22 = 11$

   $P = \dfrac{11}{22}$

   $P = 0.5$

   $P = $ **50%**

10. 35% of 2200 was the increase

    $0.35 \cdot 2200 = B$

    $B = 770$ calories

    $2200 + 770 = $ **2970 calories**

11. $x \geq -2$

    ```
    ◄———————●———————————►
       -3   -2   -1    0    1
    ```

12. $0.28 = \dfrac{28}{100} = \dfrac{7}{25}$

Fraction	Decimal	Percent
$\dfrac{51}{100}$	0.51	51%
(a) $\dfrac{7}{25}$	0.28	(b) **28%**

13. $90{,}000 \text{ mi}^2 \times \dfrac{5280 \text{ ft}}{1 \text{ mi}} \times \dfrac{5280 \text{ ft}}{1 \text{ mi}}$

    $= (90{,}000)(5280)(5280) \text{ ft}^2$

    $= $ **2,509,056,000,000 ft²**

14. $\dfrac{\$1.89}{7 \text{ oz}} = \$0.27$ per oz

    $\dfrac{\$2.99}{11 \text{ oz}} = \$0.271\overline{8}$ per oz

    A thrifty shopper who saves every penny should choose the **7 oz box.**

15. $V = A_{\text{Base}} \times \text{height}$

    $= [A_{\text{Triangle}} + A_{\text{Triangle}}] \times \text{height}$

    $= \left[\dfrac{(5)(9)}{2} + \dfrac{(5)(9)}{2}\right] \text{ft}^2 \times 8 \text{ ft}$

    $= 45 \text{ ft}^2 \times 8 \text{ ft}$

    $= $ **360 ft³**

16. $V = \pi r^2 h$

    $= \pi(1)^2 \text{ in.}^2 \times 1 \text{ in.}$

    $= \pi \text{ in.}^3 \approx$ **3.14 in.³**

    $\text{S.A.} = 2\pi r^2 + \pi d h$

    $= [2\pi(1)^2 + \pi(2)(1)] \text{ in.}^2$

    $= 4\pi \text{ in.}^2$

    $\approx$ **12.56 in.²**

17.
$$\frac{\frac{4}{7}}{1\frac{2}{14}} = \frac{m}{\frac{2}{9}}$$

$$\frac{\frac{4}{7}}{\frac{16}{14}} = \frac{m}{\frac{2}{9}}$$

$$\frac{8}{63} = \frac{16}{14}m$$

$$\frac{14}{16} \cdot \frac{16}{14}m = \frac{8}{63} \cdot \frac{14}{16}$$

$$m = \frac{1}{9}$$

18.
$$11\frac{1}{4}x - 3\frac{1}{2} = \frac{1}{9}$$

$$11\frac{1}{4}x - 3\frac{1}{2} + 3\frac{1}{2} = \frac{1}{9} + 3\frac{1}{2}$$

$$\frac{45}{4}x = \frac{2}{18} + \frac{63}{18}$$

$$\frac{4}{45} \cdot \frac{45}{4}x = \frac{65}{18} \cdot \frac{4}{45}$$

$$x = \frac{26}{81}$$

19. (a) $-----20 = \mathbf{-20}$ (since there are five negative signs, an odd number)

(b) $-\{-[+(-20)]\} = \mathbf{-20}$ (since there are three negative signs, an odd number)

(c) $+-+-(+20) = \mathbf{20}$ (since there are two negative signs, an even number)

20. (a) $-(-3) = \mathbf{3}$ (since there are two negative signs, an even number)

(b) $-\{-[-(+3)]\} = \mathbf{-3}$ (since there are three negative signs, an odd number)

(c) $+[-(+3)] = \mathbf{-3}$ (since there is one negative sign, an odd number)

21. $(4.2 \times 10^4)(3 \times 10^{-2}) = 12.6 \times 10^2$
$= \mathbf{1.26 \times 10^3}$

22.
$$\begin{array}{r} 11,314.0910 \\ - \quad 91.3621 \\ \hline \mathbf{11,222.7289} \end{array}$$

23. $\left(\frac{3}{8}\right)^2 = \frac{3}{8} \cdot \frac{3}{8} = \mathbf{\frac{9}{64}}$

24. $\sqrt[5]{\frac{1}{32}} = \sqrt[5]{\left(\frac{1}{2}\right)^5} = \mathbf{\frac{1}{2}}$

25. $-2 - (-3) - \{-[-(-4)]\} = -2 + 3 + 4$
$= 1 + 4 = \mathbf{5}$

26. $-2 - \{-[+(-6)]\} = -2 - 6 = \mathbf{-8}$

27.
$$\begin{array}{r} 114.09 \\ \times \quad 0.03 \\ \hline \mathbf{3.4227} \end{array}$$

28. $\dfrac{623.89}{0.02}$

$$\begin{array}{r} \mathbf{31,194.5} \\ 2\overline{)62,389.0} \\ \underline{6} \\ 02 \\ \underline{2} \\ 03 \\ \underline{2} \\ 18 \\ \underline{18} \\ 09 \\ \underline{8} \\ 10 \\ \underline{10} \\ 0 \end{array}$$

29. $3^2 + \sqrt{121}\left[4^3(5-3)(3^2-5) + (-400)\right]$
$= 9 + 11[64(2)(4) + (-400)] = 9 + 11[112]$
$= 9 + 1232 = \mathbf{1241}$

30. $x^m + xm + \sqrt{m} = 2^4 + 2(4) + \sqrt{4}$
$= 16 + 8 + 2$
$= \mathbf{26}$

PROBLEM SET 81

1.
$$\frac{4}{9} \cdot A = 440$$

$$\frac{9}{4} \cdot \frac{4}{9} \cdot A = 440 \cdot \frac{9}{4}$$

$$A = \mathbf{990 \text{ sheep}}$$

2.
$$\frac{5}{7} = \frac{450}{R}$$

$$5R = 3150$$

$$R = \mathbf{630 \text{ robbers}}$$

3. $B = 3\frac{1}{2} \times 160$

$$B = \frac{7}{2} \times \frac{160}{1}$$

$$B = \mathbf{560 \text{ oz}}$$

4. $\dfrac{1 \text{ hr}}{20 \text{ km}} \times 480 \text{ km} = \mathbf{24 \text{ hr}}$

5. 60% of 90 is the increase

$0.6 \times 90 = B$

$54 = B$

$90 + 54 = \mathbf{144 \text{ yd}}$

6. $x < 5$

7. $P \cdot 90 = 405$

$\dfrac{P \cdot 90}{90} = \dfrac{405}{90}$

$P = \dfrac{9}{2}$

$P = 4.5$

$P = \mathbf{450\%}$

8. $250\% \cdot A = 80$

$2.5 \cdot A = 80$

$\dfrac{2.5 \cdot A}{2.5} = \dfrac{80}{2.5}$

$A = \mathbf{32}$

9. $\dfrac{7}{5} = 1\dfrac{2}{5} = 1\dfrac{4}{10} = 1.4 = 140\%$

Fraction	Decimal	Percent
$\dfrac{51}{100}$	0.51	51%
$\dfrac{7}{5}$	(a) **1.4**	(b) **140%**

10. $12 \text{ mi}^2 \times \dfrac{5280 \text{ ft}}{1 \text{ mi}} \times \dfrac{5280 \text{ ft}}{1 \text{ mi}}$

$= 12(5280)(5280) \text{ ft}^2 = \mathbf{334{,}540{,}800 \text{ ft}^2}$

11.

$V = A_{\text{Base}} \times \text{height}$

$= \left[A_{\text{Rectangle}} + A_{\text{Triangle}} \right] \times \text{height}$

$= \left[(10)(20) + \dfrac{(20)(20)}{2} \right] \text{ft}^2 \times 2(3) \text{ ft}$

$= 400 \text{ ft}^2 \times 6 \text{ ft}$

$= \mathbf{2400 \text{ ft}^3}$

12. $S.A. = 2A_{\text{End}} + 2A_{\text{Side}} + A_{\text{Bottom}}$

$= \left[2\left(\dfrac{(10)(12)}{2} \right) + 2(22)(13) + (22)(10) \right] \text{ft}^2$

$= [120 + 572 + 220] \text{ ft}^2 = \mathbf{912 \text{ ft}^2}$

13. $\dfrac{15}{17} = \dfrac{A}{51}$

$17 \cdot A = 51 \cdot 15$

$\dfrac{17 \cdot A}{17} = \dfrac{765}{17}$

$A = \mathbf{45 \text{ ants}}$

14. $12 = 2 \cdot 2 \cdot 3$

$18 = 2 \cdot 3 \cdot 3$

$27 = 3 \cdot 3 \cdot 3$

$\text{LCM} (12, 18, 27) = 2 \cdot 2 \cdot 3 \cdot 3 \cdot 3 = \mathbf{108}$

15. $\dfrac{\dfrac{5}{6}}{\dfrac{25}{36}} = \dfrac{p}{12}$

$10 = \dfrac{25}{36}p$

$\dfrac{36}{25} \cdot \dfrac{25}{36}p = 10 \cdot \dfrac{36}{25}$

$p = \mathbf{\dfrac{72}{5}}$

16. $6\dfrac{1}{3}x + 1\dfrac{5}{6} = 7\dfrac{2}{3}$

$\dfrac{19}{3}x + \dfrac{11}{6} = \dfrac{23}{3}$

$\dfrac{19}{3}x + \dfrac{11}{6} - \dfrac{11}{6} = \dfrac{46}{6} - \dfrac{11}{6}$

$\dfrac{19}{3}x = \dfrac{35}{6}$

$\dfrac{3}{19} \cdot \dfrac{19}{3}x = \dfrac{35}{6} \cdot \dfrac{3}{19}$

$x = \mathbf{\dfrac{35}{38}}$

17. (a) $(4.5)(-2) = \mathbf{-9}$

(b) $(-4.5)(-2) = \mathbf{9}$

(c) $(-4.5)(2) = \mathbf{-9}$

18. (a) $\dfrac{-5.5}{5} = -1.1$

(b) $\dfrac{5.5}{-5} = -1.1$

(c) $\dfrac{-5.5}{-5} = 1.1$

19. $2^3 + \sqrt{36}\left[3^2(3^2 - 2^3)(4^2 - 2^4) + 3\right]$
$= 8 + 6[9(1)(0) + 3] = 8 + 6[3] = 8 + 18$
$= \mathbf{26}$

20. $\sqrt[4]{\dfrac{16}{81}} + \left(\dfrac{4}{5}\right)^2 = \dfrac{\sqrt[4]{16}}{\sqrt[4]{81}} + \dfrac{4^2}{5^2} = \dfrac{2}{3} + \dfrac{16}{25}$

$= \dfrac{50}{75} + \dfrac{48}{75} = \dfrac{\mathbf{98}}{\mathbf{75}}$

21. $-(+2) - 2 + -3 - (+6) + 3$
$= (-2) + (-2) + (-3) + (-6) + (+3)$
$= (-4) + (-3) + (-6) + (+3)$
$= (-7) + (-6) + (+3) = (-13) + (+3) = \mathbf{-10}$

22. $-6 + -3 - 6 - [-(-10)] + 1$
$= (-6) + (-3) + (-6) + (-10) + (+1)$
$= (-9) + (-6) + (-10) + (+1)$
$= (-15) + (-10) + (+1) = (-25) + (+1) = \mathbf{-24}$

23. There are three negative signs. Three is an odd number, so $-\left\{-[-(+2)]\right\} = \mathbf{-2}$.

24. There are five negative factors, an odd number, so the product is negative:
$(-1)(-1)(-2)(-1)(-2) = \mathbf{-4}$.

25. $17\dfrac{3}{5} - 1\dfrac{3}{5} \cdot \dfrac{1}{2} + 3\dfrac{1}{5} = 17\dfrac{3}{5} - \dfrac{8}{5} \cdot \dfrac{1}{2} + 3\dfrac{1}{5}$

$= 17\dfrac{3}{5} - \dfrac{4}{5} + 3\dfrac{1}{5} = \mathbf{20}$

26.
$$
\begin{array}{r}
0.1315 \\
\times\ 0.0012 \\
\hline
2630 \\
1315 \\
\hline
\mathbf{0.00015780}
\end{array}
$$

27. $(5 \times 10^{-15}) \times (1 \times 10^8)$
$= (5 \times 1) \times (10^{-15} \times 10^8) = \mathbf{5 \times 10^{-7}}$

28. $(1 \times 10^{-27}) \times (4 \times 10^{-20})$
$= (1 \times 4) \times (10^{-27} \times 10^{-20}) = \mathbf{4 \times 10^{-47}}$

29. $xyz + yz + y^z = \left(\dfrac{1}{3}\right)(6)(2) + (6)(2) + (6^2)$
$= 4 + 12 + 36$
$= \mathbf{52}$

30. $\sqrt[3]{x} + \sqrt{y} + xy = \sqrt[3]{(8)} + \sqrt{(16)} + (8)(16)$
$= 2 + 4 + 128$
$= \mathbf{134}$

PROBLEM SET 82

1. $\dfrac{126}{3} = \dfrac{P}{9}$

$3P = 9 \cdot 126$

$\dfrac{3P}{3} = \dfrac{1134}{3}$

$P = \mathbf{378\ patients}$

2. If $\dfrac{7}{8}$ were chicaned, $\dfrac{1}{8}$ were not chicaned.

$\dfrac{1}{8} \times 3200 = B$

$\mathbf{400\ customers} = B$

3. $A = 1200$

$A = 1200 \times \dfrac{6}{5}$

$A = \mathbf{1440\ were\ prejudiced}$

4. Original rate $= \dfrac{120\ mi}{3\ hr} = \dfrac{40\ mi}{1\ hr}$

New rate $= \dfrac{40\ mi}{1\ hr} - \dfrac{10\ mi}{1\ hr} = \dfrac{30\ mi}{1\ hr}$

$\dfrac{1\ hr}{30\ \cancel{mi}} \times 450\ \cancel{mi} = \mathbf{15\ hr}$

5. $P \cdot 18 = 5.4$

$\dfrac{P \cdot 18}{18} = \dfrac{5.4}{18}$

$P = 0.3$

$P = \mathbf{30\%}$

6. $x \geq -3$

7. $260\% \cdot A = 143$

$2.6 \cdot A = 143$

$\dfrac{2.6A}{2.6} = \dfrac{143}{2.6}$

$A = \mathbf{55}$

8. $B = 180\% \cdot 60$
$B = 1.8 \cdot 60$
$B = \mathbf{108}$

9. $65\% \cdot 300 = B$
$0.65 \cdot 300 = B$
$\mathbf{195 \text{ students}} = B$

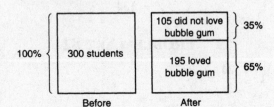

100% { 300 students (Before)

105 did not love bubble gum } 35%
195 loved bubble gum } 65% (After)

10. $P \cdot 80 = 200$
$\dfrac{P \cdot 80}{80} = \dfrac{200}{80}$
$P = \dfrac{5}{2}$
$P = 2.5$
$P = \mathbf{250\%}$

11. $108{,}000 = \mathbf{1.08 \times 10^5}$

12. $D \times 480 = 360$
$D = \dfrac{360}{480} = \dfrac{3}{4} = \mathbf{0.75}$

13.

```
       10
   21      
        10
   31
```

$V = A_{\text{Base}} \times \text{height}$
$\quad = [A_{\text{Rectangle}} + A_{\text{Triangle}}] \times \text{height}$
$\quad = \left[(31)(10) + \dfrac{(21)(10)}{2}\right] \text{cm}^2 \times 100 \text{ cm}$
$\quad = 415 \text{ cm}^2 \times 100 \text{ cm}$
$\quad = \mathbf{41{,}500 \text{ cm}^3}$

14. $P = \left[\dfrac{\pi(20)}{2} + \dfrac{\pi(20)}{2} + 20 + 20\right] \text{ft}$
$\quad \approx \mathbf{102.8 \text{ ft}}$

15. $\dfrac{2\frac{6}{7}}{1\frac{1}{14}} = \dfrac{1}{\frac{2}{y}}$

$\dfrac{\frac{20}{7}}{\frac{15}{14}} = \dfrac{1}{\frac{2}{y}}$

$\dfrac{20}{7}y = \dfrac{15}{28}$

$\dfrac{7}{20} \cdot \dfrac{20}{7}y = \dfrac{15}{28} \cdot \dfrac{7}{20}$

$y = \dfrac{\mathbf{3}}{\mathbf{16}}$

16. $\dfrac{\frac{6}{5}}{\frac{12}{25}} = \dfrac{5}{p}$

$\dfrac{6}{5}p = \dfrac{12}{5}$

$\dfrac{5}{6} \cdot \dfrac{6}{5}p = \dfrac{12}{5} \cdot \dfrac{5}{6}$

$p = \mathbf{2}$

17. $2\dfrac{1}{5}x + 2\dfrac{1}{10} = 6\dfrac{3}{20}$

$2\dfrac{1}{5}x + 2\dfrac{1}{10} - 2\dfrac{1}{10} = 6\dfrac{3}{20} - 2\dfrac{1}{10}$

$\dfrac{11}{5}x = 6\dfrac{3}{20} - 2\dfrac{2}{20}$

$\dfrac{11}{5}x = 4\dfrac{1}{20}$

$\dfrac{5}{11} \cdot \dfrac{11}{5}x = \dfrac{81}{20} \cdot \dfrac{5}{11}$

$x = \dfrac{\mathbf{81}}{\mathbf{44}}$

18. $2^4 + \sqrt{25}\left[2^2(3^2 - 2^3)(2^2 + 1) - 1\right]$
$= 16 + 5[4(1)(5) - 1] = 16 + 5[19]$
$= 16 + 95 = \mathbf{111}$

19. (a) $(-6)(-3) = \mathbf{18}$

(b) $(-6)(3) = \mathbf{-18}$

(c) $(6)(-3) = \mathbf{-18}$

20. (a) $\dfrac{-6}{-2} = 3$

    (b) $\dfrac{6}{-2} = -3$

    (c) $\dfrac{-6}{2} = -3$

21. There are four negative factors, an even number, so the product is positive:
$(-3)(-1)(-1)(-1)(5) = \mathbf{15}.$

22. There are four negative signs. Four is an even number, so $-\left\{-[-(-5)]\right\} = \mathbf{5}.$

23. $-6 + -\left\{+[-(-11)]\right\} + 3 = (-6) + (-11) + (3)$
$= \mathbf{-14}$

24. $-7 - 31 - \left\{-[+(+6)]\right\} = (-7) + (-31) + 6$
$= \mathbf{-32}$

25. $(7 \times 10^{14}) \times (1 \times 10^{-12})$
$= (7 \times 1) \times (10^{14} \times 10^{-12}) = \mathbf{7 \times 10^2}$

26.
```
    0.1319
  × 0.0014
    5276
    1319
  0.00018466
```

27. (a) $x = 5,\ y = 4$

    (b) $x = -3,\ y = 2$

    (c) $x = 4,\ y = -4$

    (d) $x = -4,\ y = -4$

28. $qrq - r = (-5)(2)(-5) - (2)$
$= 50 - 2$
$= \mathbf{48}$

29. $tv + vt = (-4)(3) + (3)(-4)$
$= (-12) + (-12)$
$= \mathbf{-24}$

30. $xy - \dfrac{x}{y} = (-4)(2) - \dfrac{(-4)}{(2)} = -8 + 2 = \mathbf{-6}$

PROBLEM SET 83

1. $\dfrac{7}{3} = \dfrac{1400}{G}$

$7G = 4200$

$G = \mathbf{600}$

2. $\dfrac{5}{6} \times \dfrac{3}{11} \times A = 2202$

$\dfrac{11}{3} \times \dfrac{3}{11} \times A = 2202 \times \dfrac{11}{3}$

$A = \mathbf{8074\ fans}$

3. $\dfrac{77\ \text{mi}}{1\ \text{hr}} \times 5\ \text{hr} = \mathbf{385\ mi}$

4. $2\dfrac{1}{2} \cdot A = 1200$

$\dfrac{2}{5} \cdot \dfrac{5}{2} \cdot A = 1200 \cdot \dfrac{2}{5}$

$A = \mathbf{480\ people}$

5. $20\% \cdot \$1.25 = B$

$0.2 \cdot \$1.25 = B$

$\$0.25 = B$

New price $= \$1.25 + \$0.25 = \mathbf{\$1.50\ per\ gal}$

6. Rate method:

$\dfrac{23\ \text{times}}{8\ \text{hr}}, \dfrac{8\ \text{hr}}{23\ \text{times}}$

$\dfrac{23\ \text{times}}{8\ \text{hr}} \times 24\ \text{hr} = \mathbf{69\ times}$

Proportion method:

$\dfrac{23}{8} = \dfrac{T}{24}$

$8T = 23 \cdot 24$

$T = \mathbf{69\ times}$

7. $x > -2$

8. $240 = 240\% \cdot A$

$240 = 2.4A$

$\dfrac{240}{2.4} = A$

$\mathbf{100} = A$

9. $P \cdot 60 = 105$

$P = \dfrac{105}{60}$

$P = 1.75$

$P = \mathbf{175\%}$

10. $2.45 = 2\dfrac{45}{100} = 2\dfrac{9}{20} = \dfrac{49}{20}$

Fraction	Decimal	Percent
$\dfrac{51}{100}$	0.51	51%
(a) $\dfrac{49}{20}$	2.45	(b) **245%**

11. $9.1 \times 10^{-11} = \mathbf{0.000000000091}$

12.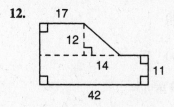

$V = A_{\text{Base}} \times \text{height}$

$= [A_{\text{Rect 1}} + A_{\text{Rect 2}} + A_{\text{Triangle}}] \times \text{height}$

$= \left[42(11) + 17(12) + \dfrac{14(12)}{2} \right] \text{cm}^2 \times 200 \text{ cm}$

$= [462 + 204 + 84](200) \text{ cm}^3$

$= \mathbf{150{,}000 \text{ cm}^3}$

13. $S.A. = 2A_{\text{Top}} + 2A_{\text{Front}} + 2A_{\text{Side}}$

$= 2(5)(10) \text{ m}^2 + 2(5)(3) \text{ m}^2 + 2(10)(3) \text{ m}^2$

$= 100 \text{ m}^2 + 30 \text{ m}^2 + 60 \text{ m}^2$

$= \mathbf{190 \text{ m}^2}$

14. $\dfrac{6\frac{1}{4}}{2\frac{1}{3}} = \dfrac{\frac{5}{12}}{x}$

$\dfrac{\frac{25}{4}}{\frac{7}{3}} = \dfrac{\frac{5}{12}}{x}$

$\dfrac{25}{4}x = \dfrac{35}{36}$

$\dfrac{4}{25} \cdot \dfrac{25}{4}x = \dfrac{35}{36} \cdot \dfrac{4}{25}$

$x = \dfrac{\mathbf{7}}{\mathbf{45}}$

15. $2\dfrac{1}{2}x + 2\dfrac{1}{2} = 6\dfrac{3}{8}$

$2\dfrac{1}{2}x + 2\dfrac{1}{2} - 2\dfrac{1}{2} = 6\dfrac{3}{8} - 2\dfrac{1}{2}$

$\dfrac{5}{2}x = \dfrac{51}{8} - \dfrac{20}{8}$

$\dfrac{2}{5} \cdot \dfrac{5}{2}x = \dfrac{31}{8} \cdot \dfrac{2}{5}$

$x = \dfrac{\mathbf{31}}{\mathbf{20}}$

16. $\dfrac{\frac{2}{21}}{\frac{24}{}} = \dfrac{p}{\frac{7}{12}}$

$\dfrac{7}{6} = \dfrac{21}{24}p$

$\dfrac{24}{21} \cdot \dfrac{21}{24}p = \dfrac{7}{6} \cdot \dfrac{24}{21}$

$p = \dfrac{\mathbf{4}}{\mathbf{3}}$

17. $(2 \times 10^{-14}) \times (3 \times 10^{-13})$

$= (2 \times 3) \times (10^{-14} \times 10^{-13}) = \mathbf{6 \times 10^{-27}}$

18. $2^3 + \sqrt{36}\left[2^3(2^2 - 1)(5^2 - 22) - 1\right]$

$= 8 + 6[8(3)(3) - 1] = 8 + 6[71]$

$= 8 + 426 = \mathbf{434}$

19. (a) $(3)(-2) = \mathbf{-6}$

(b) $(-3)(2) = \mathbf{-6}$

(c) $(-3)(-2) = \mathbf{6}$

20. (a) $\dfrac{12}{-4} = \mathbf{-3}$

(b) $\dfrac{-12}{4} = \mathbf{-3}$

(c) $\dfrac{-12}{-4} = \mathbf{3}$

21. $-2 + 3 - 6 + -2 - 5 = 1 - 6 - 2 - 5$

$= -5 - 2 - 5 = -7 - 5 = \mathbf{-12}$

22. There are seven negative factors, an odd number, so the product is negative:

$(-2)(-1)(-1)(-1)(3)(-1)(-1)(-4) = \mathbf{-24}$.

23. There are six negative signs. Six is an even number, so

$$-\left[-\left(-\left\{-\left[-(-69)\right]\right\}\right)\right] = \textbf{69.}$$

24.
$$
\begin{array}{r}
1.218 \\
\times\ 0.0016 \\
\hline
7308 \\
1218 \\
\hline
\textbf{0.0019488}
\end{array}
$$

25. $16\dfrac{2}{7} - 2\dfrac{1}{2} \cdot 1\dfrac{1}{7} + \dfrac{9}{14}$

$= \dfrac{114}{7} - \dfrac{5}{2} \cdot \dfrac{8}{7} + \dfrac{9}{14} = \dfrac{114}{7} - \dfrac{40}{14} + \dfrac{9}{14}$

$= \dfrac{228}{14} - \dfrac{40}{14} + \dfrac{9}{14} = \dfrac{\textbf{197}}{\textbf{14}}$

26. $\dfrac{1}{3}\left(1\dfrac{1}{4} \cdot \dfrac{7}{8} - \dfrac{1}{4}\right) + \dfrac{17}{32}$

$= \dfrac{1}{3}\left(\dfrac{5}{4} \cdot \dfrac{7}{8} - \dfrac{1}{4}\right) + \dfrac{17}{32}$

$= \dfrac{1}{3}\left(\dfrac{35}{32} - \dfrac{8}{32}\right) + \dfrac{17}{32}$

$= \dfrac{9}{32} + \dfrac{17}{32} = \dfrac{26}{32} = \dfrac{\textbf{13}}{\textbf{16}}$

27. $xy + x^y + y^x + \sqrt{x}$

$= (4)(2) + (4^2) + (2^4) + \sqrt{(4)}$

$= 8 + 16 + 16 + 2$

$= \textbf{42}$

28. $ab - a - b = (10)(-2) - (10) - (-2)$

$\qquad\qquad\quad = -20 - 10 + 2$

$\qquad\qquad\quad = \textbf{-28}$

29. $\dfrac{c}{d} + c - d = \dfrac{(-12)}{(-4)} + (-12) - (-4)$

$\qquad\qquad\quad = 3 + (-12) + 4$

$\qquad\qquad\quad = \textbf{-5}$

30. $\dfrac{xz}{y} + y = \dfrac{(4)(3)}{(-6)} + (-6)$

$\qquad\qquad = -2 + (-6)$

$\qquad\qquad = \textbf{-8}$

PROBLEM SET 84

1. $225\% \times 10{,}000{,}000 = B$

$\quad 2.25 \cdot 10{,}000{,}000 = B$

$\qquad\quad 22{,}500{,}000 = B$

$10{,}000{,}000 + 22{,}500{,}000$

$= \textbf{32,500,000 school-age children}$

2. $\qquad \dfrac{9}{5} = \dfrac{L}{25{,}000}$

$225{,}000 = 5L$

$\textbf{45,000} = L$

3. If $\dfrac{2}{3}$ were peripatetic, $\dfrac{1}{3}$ were not peripatetic.

$\dfrac{1}{3} \cdot A = 49$

$\dfrac{3}{1} \cdot \dfrac{1}{3} \cdot A = 49 \cdot \dfrac{3}{1}$

$\qquad\quad A = \textbf{147 philosophers}$

4. Original speed: $\dfrac{20 \text{ mi}}{1 \text{ hr}}$

Distance covered in first 2 hr:

$\dfrac{20 \text{ mi}}{1 \text{ hr}} \times 2 \text{ hr} = 40 \text{ mi}$

New speed: $\dfrac{(4 \times 20) \text{ mi}}{1 \text{ hr}} = \dfrac{80 \text{ mi}}{1 \text{ hr}}$

Time to cover remaining distance:

$\dfrac{1 \text{ hr}}{80 \text{ mi}} \times 560 \text{ mi} = 7 \text{ hr}$

Total time $= 2 \text{ hr} + 7 \text{ hr} = \textbf{9 hr}$

5. $x \geq -7$

6. $\dfrac{17}{3} = \dfrac{F}{24}$

$3F = 17 \cdot 24$

$F = \dfrac{408}{3}$

$F = \textbf{136 fish}$

7. $\quad 90 = 2.25 \cdot A$

$\dfrac{90}{2.25} = A$

$\quad \textbf{40} = A$

8.

$$6\frac{1}{4} \cdot A = 8\frac{4}{7}$$

$$\frac{25}{4} \cdot A = \frac{60}{7}$$

$$\frac{4}{25} \cdot \frac{25}{4} \cdot A = \frac{60}{7} \cdot \frac{4}{25}$$

$$A = \frac{48}{35}$$

9.

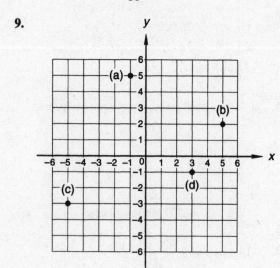

10. $(2 \times 10^{-14}) \times (2 \times 10^{-6})$
$= (2 \times 2) \times (10^{-14} \times 10^{-6}) = \mathbf{4 \times 10^{-20}}$

11.

$V = A_{\text{Base}} \times \text{height}$

$= [A_{\text{Semicircle}} + A_{\text{Square}} + A_{\text{Triangle}}] \times \text{height}$

$= \left[\dfrac{\pi(1)^2}{2} + (2)(2) + \dfrac{(1)(2)}{2}\right] \text{ft}^2 \times 4 \text{ ft}$

$= \left[\dfrac{\pi}{2} + 4 + 1\right] \text{ft}^2 \times 4 \text{ ft}$

$\approx \mathbf{26.28 \text{ ft}^3}$

12. $V = \pi r^2 h$

$= \pi(2)^2 \text{ cm}^2 \times 1 \text{ cm}$

$= 4\pi \text{ cm}^3$

$\approx \mathbf{12.56 \text{ cm}^3}$

$S.A. = 2A_{\text{Base}} + \text{L.A.}$

$= 2\pi r^2 + \pi dh$

$= 2\left[\pi(2)^2\right] \text{ cm}^2 + \pi(4)(1) \text{ cm}^2$

$= 8\pi \text{ cm}^2 + 4\pi \text{ cm}^2$

$\approx \mathbf{37.68 \text{ cm}^2}$

13.

$$6a + 15 = 33$$
$$6a + 15 - 15 = 33 - 15$$
$$6a = 18$$
$$a = \mathbf{3}$$

14.

$$7\frac{1}{2}m + 2\frac{1}{3} = 9\frac{5}{6}$$

$$7\frac{1}{2}m + 2\frac{1}{3} - 2\frac{1}{3} = 9\frac{5}{6} - 2\frac{1}{3}$$

$$7\frac{1}{2}m = 7\frac{1}{2}$$

$$m = \mathbf{1}$$

15.

$$\frac{3\frac{1}{2}}{2\frac{1}{5}} = \frac{\frac{1}{4}}{p}$$

$$\frac{\frac{7}{2}}{\frac{11}{5}} = \frac{\frac{1}{4}}{p}$$

$$\frac{7}{2}p = \frac{11}{20}$$

$$\frac{2}{7} \cdot \frac{7}{2}p = \frac{11}{20} \cdot \frac{2}{7}$$

$$p = \mathbf{\frac{11}{70}}$$

16.

$$\frac{\frac{2}{3}}{\frac{1}{7}} = \frac{s}{21}$$

$$14 = \frac{1}{7}s$$

$$\frac{7}{1} \cdot \frac{1}{7}s = 14 \cdot \frac{7}{1}$$

$$s = \mathbf{98}$$

17.

$$\begin{array}{r} -5x - 3 = 12 \\ \underline{+3 \quad +3} \\ -5x \quad = 15 \end{array}$$

$$\frac{-5x}{-5} = \frac{15}{-5}$$

$$x = \mathbf{-3}$$

18.

$$\begin{array}{r} -2x + \dfrac{1}{2} = 8\dfrac{1}{2} \\ \underline{-\dfrac{1}{2} \quad -\dfrac{1}{2}} \\ -2x \quad = 8 \end{array}$$

$$\frac{-2x}{-2} = \frac{8}{-2}$$

$$x = \mathbf{-4}$$

19. For the method shown here, the **Subtraction Property of Equality** and the **Division Property of Equality** were used. (Answers may vary. See student work.)

20. $3^3 + \sqrt{144}\left[4^2(3^2 - 8)(5^2 - \sqrt{225}) + (-110)\right]$
$= 27 + 12[16(1)(10) - 110] = 27 + 12[50]$
$= 27 + 600 = \mathbf{627}$

21. (a) $(4)(-5) = \mathbf{-20}$

(b) $(-4)(5) = \mathbf{-20}$

(c) $(-4)(-5) = \mathbf{20}$

22. $-(+9) - 12 + 16 - -(-1)$
$= (-9) + (-12) + 16 + (-1)$
$= (-21) + 16 + (-1) = (-5) + (-1) = \mathbf{-6}$

23. There are five negative signs. Five is an odd number, so
$-\left(-\left\{-\left[-\left(-2.54\right)\right]\right\}\right) = \mathbf{-2.54}$.

24. $\dfrac{17.658}{0.009}$

$$
\begin{array}{r}
1962 \\
9\overline{)17{,}658} \\
\underline{9} \\
8\,6 \\
\underline{8\,1} \\
5\,5 \\
\underline{5\,4} \\
1\,8 \\
\underline{1\,8} \\
0
\end{array}
$$

25. $11\dfrac{1}{4} - 3\dfrac{2}{3} \cdot 1\dfrac{1}{5} + \sqrt{\dfrac{4}{121}}$

$= \dfrac{45}{4} - \dfrac{11}{3} \cdot \dfrac{6}{5} + \dfrac{2}{11} = \dfrac{45}{4} - \dfrac{22}{5} + \dfrac{2}{11}$

$= \dfrac{2475}{220} - \dfrac{968}{220} + \dfrac{40}{220} = \dfrac{\mathbf{1547}}{\mathbf{220}}$

26. $\dfrac{1}{4}\left(\dfrac{1}{2} \cdot \dfrac{1}{8} - \dfrac{1}{3}\right) + \dfrac{11}{24}$

$= \dfrac{1}{4}\left(\dfrac{1}{16} - \dfrac{1}{3}\right) + \dfrac{11}{24}$

$= \dfrac{1}{4}\left(\dfrac{3}{48} - \dfrac{16}{48}\right) + \dfrac{11}{24} = \dfrac{1}{4}\left(-\dfrac{13}{48}\right) + \dfrac{11}{24}$

$= -\dfrac{13}{192} + \dfrac{88}{192} = \dfrac{75}{192} = \dfrac{\mathbf{25}}{\mathbf{64}}$

27. Mean $= \dfrac{\dfrac{1}{4} + \dfrac{3}{20} + \dfrac{7}{25} + 1\dfrac{1}{2}}{4}$

$= \dfrac{\dfrac{25}{100} + \dfrac{15}{100} + \dfrac{28}{100} + \dfrac{150}{100}}{4}$

$= \dfrac{\dfrac{218}{100}}{4} = \dfrac{109}{50} \cdot \dfrac{1}{4} = \dfrac{\mathbf{109}}{\mathbf{200}}$

28. $p^x + \sqrt[m]{x} = (2^8) + (\sqrt[3]{8}) = 256 + 2 = \mathbf{258}$

29. $ab - \dfrac{b}{ab} = (-2)(5) - \dfrac{(5)}{(-2)(5)}$

$= -10 + \dfrac{1}{2}$

$= \mathbf{-9\dfrac{1}{2}}$

30. $x - xy + \dfrac{y}{x} = (-2) - (-2)(6) + \dfrac{(6)}{(-2)}$

$= -2 + 12 - 3$

$= \mathbf{7}$

PROBLEM SET 85

1. $1.375 \times .368 = B$
$\mathbf{.506} = B$

2. Rate method:
$\dfrac{102 \text{ hits}}{3 \text{ hr}} = \dfrac{34 \text{ hits}}{1 \text{ hr}}, \dfrac{1 \text{ hr}}{34 \text{ hits}}$

$\dfrac{1 \text{ hr}}{34 \text{ hits}} \times 612 \text{ hits} = \mathbf{18 \text{ hr}}$

Proportion method:
$\dfrac{102}{3} = \dfrac{612}{T}$

$102T = 612 \cdot 3$

$\dfrac{102T}{102} = \dfrac{1836}{102}$

$T = \mathbf{18 \text{ hr}}$

3. Original rates: $\dfrac{40 \text{ mi}}{1 \text{ hr}}, \dfrac{1 \text{ hr}}{40 \text{ mi}}$

Distance traveled: $\dfrac{40 \text{ mi}}{1 \text{ hr}} \times 4 \text{ hr} = 160 \text{ mi}$

New rates: $\dfrac{60 \text{ mi}}{1 \text{ hr}}, \dfrac{1 \text{ hr}}{60 \text{ mi}}$

Distance traveled: $\dfrac{60 \text{ mi}}{1 \text{ hr}} \times 3 \text{ hr} = 180 \text{ mi}$

Total distance traveled $= 160 \text{ mi} + 180 \text{ mi}$
$= \mathbf{340 \text{ mi}}$

4. $\frac{1}{35} \cdot A = 14$

$A = 14 \cdot \frac{35}{1}$

$A = \mathbf{490 \ oz}$

5. 5% of $21 is the sales tax.

$0.05 \cdot \$21 = t$

$\mathbf{\$1.05} = t$

6. $x \le 2$

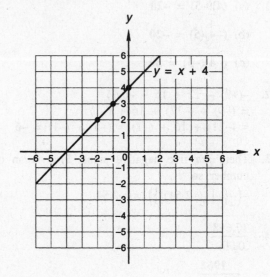

13. $y = x + 4$

x	y
−2	2
−1	3
0	4

7. $B = 160\% \times 120$

$B = 1.6 \times 120$

$B = \mathbf{192}$

8. $120\% \cdot 220 = B$

$1.2 \cdot 220 = B$

$264 = B$

$220 + 264 = \mathbf{484}$

9. $P \cdot 50 = 125$

$P = \frac{125}{50}$

$P = 2.5$

$P = \mathbf{250\%}$

10. $D \times 800 = 280$

$D = \frac{280}{800} = \frac{7}{20} = \mathbf{0.35}$

11.

20, 10, 10, 10, 10

$V = A_{\text{Base}} \times \text{height}$

$= \left[A_{\text{Rectangle}} + A_{\text{Square}} + A_{\text{Triangle}} \right] \times \text{height}$

$= \left[10(20) + 10(10) + \frac{10(10)}{2} \right] \text{cm}^2 \times 150 \ \text{cm}$

$= [200 + 100 + 50](150) \ \text{cm}^3$

$= \mathbf{52{,}500 \ cm^3}$

12. $P = \left[\frac{\pi(24)}{2} + \frac{\pi(12)}{2} + 12 \right] \text{yd}$

$\approx \mathbf{68.52 \ yd}$

14. $y = 2x - 1$

x	y
−2	−5
0	−1
1	1

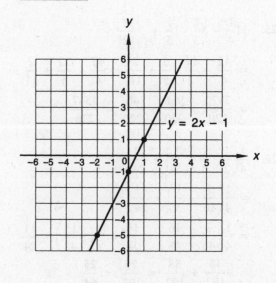

15.
$$\frac{7\frac{1}{4}}{2\frac{1}{2}} = \frac{\frac{1}{6}}{x}$$

$$\frac{\frac{29}{4}}{\frac{5}{2}} = \frac{\frac{1}{6}}{x}$$

$$\frac{29}{4}x = \frac{5}{12}$$

$$\frac{4}{29} \cdot \frac{29}{4}x = \frac{5}{12} \cdot \frac{4}{29}$$

$$x = \frac{5}{87}$$

16.
$$\frac{\frac{3}{4}}{\frac{9}{16}} = \frac{z}{6}$$

$$\frac{9}{2} = \frac{9}{16}z$$

$$\frac{16}{9} \cdot \frac{9}{16}z = \frac{9}{2} \cdot \frac{16}{9}$$

$$z = 8$$

17.
$$3\frac{1}{5}x + 2\frac{1}{3} = 5\frac{2}{5}$$

$$3\frac{1}{5}x + 2\frac{1}{3} - 2\frac{1}{3} = 5\frac{2}{5} - 2\frac{1}{3}$$

$$\frac{16}{5}x = \frac{81}{15} - \frac{35}{15}$$

$$\frac{5}{16} \cdot \frac{16}{5}x = \frac{46}{15} \cdot \frac{5}{16}$$

$$x = \frac{23}{24}$$

18.
$$-5x - 18 = 32$$
$$\underline{+18 \quad +18}$$
$$-5x = 50$$

$$\frac{-5x}{-5} = \frac{50}{-5}$$

$$x = -10$$

19.
$$5 - 2x = -7$$
$$\underline{-5 -5}$$
$$-2x = -12$$

$$\frac{-2x}{-2} = \frac{-12}{-2}$$

$$x = 6$$

20. For the method shown here, the **Addition Property of Equality** and the **Division Property of Equality** were used. (Answers may vary. See student work.)

21. $3^2 + 2^3\left[\sqrt{36}(2^2 - 3)(2^2 + 1) - 20\right]$
$= 9 + 8[6(1)(5) - 20] = 9 + 8[10]$
$= 9 + 80 = \mathbf{89}$

22. (a) $\dfrac{24}{-6} = \mathbf{-4}$

(b) $\dfrac{-24}{6} = \mathbf{-4}$

(c) $\dfrac{-24}{-6} = \mathbf{4}$

23. There are five negative signs. Five is an odd number, so
$$-\left(-\left\{-\left[-(-9)\right]\right\}\right) = \mathbf{-9}.$$

24. $-5 - 4 - 2 + 6 + -3 = -9 - 2 + 6 - 3$
$= -11 + 3 = \mathbf{-8}$

25. $-2 - 2 + -4 - 2 = -4 - 4 - 2 = -8 - 2$
$= \mathbf{-10}$

26. $3\frac{3}{7} - 1\frac{1}{7} \cdot 3\frac{1}{3} + \frac{8}{21}$

$= \frac{24}{7} - \frac{8}{7} \cdot \frac{10}{3} + \frac{8}{21} = \frac{24}{7} - \frac{80}{21} + \frac{8}{21}$

$= \frac{72}{21} - \frac{80}{21} + \frac{8}{21} = \mathbf{0}$

27. $(3 \times 10^{26}) \times (2 \times 10^{-9})$
$(3 \times 2) \times (10^{26} \times 10^{-9}) = \mathbf{6 \times 10^{17}}$

28. $tv + t^v + v^t + v^2$
$= (3)(3) + (3^3) + (3^3) + (3)^2$
$= 9 + 27 + 27 + 9$
$= \mathbf{72}$

29. $s - sts = (-3) - (-3)(5)(-3)$
$= -3 - 45$
$= \mathbf{-48}$

30. $qr + r = (-0.5)(-6) + (-6) = 3 - 6 = \mathbf{-3}$

PROBLEM SET 86

1. Rates for last 240 mi: $\dfrac{60 \text{ mi}}{1 \text{ hr}}, \dfrac{1 \text{ hr}}{60 \text{ mi}}$

   Time for last 240 mi: $\dfrac{1 \text{ hr}}{60 \text{ mi}} \times 240 \text{ mi} = 4 \text{ hr}$

   Time for first 240 mi: $10 \text{ hr} - 4 \text{ hr} = 6 \text{ hr}$

   Rate for first 240 mi: $\dfrac{240 \text{ mi}}{6 \text{ hr}} = \dfrac{40 \text{ mi}}{1 \text{ hr}}$
   $= \textbf{40 mph}$

2. $\dfrac{40}{\$20,000} = \dfrac{12}{D}$

   $40D = 12 \cdot \$20,000$

   $\dfrac{40D}{40} = \dfrac{\$240,000}{40}$

   $D = \textbf{\$6000}$

3. $\dfrac{2}{17} = \dfrac{B}{68,000}$

   $136,000 = 17B$

   $B = \textbf{8000 screaming fans}$

4. $\dfrac{1}{7} \cdot A = 1400$

   $\dfrac{7}{1} \cdot \dfrac{1}{7} \cdot A = 1400 \cdot \dfrac{7}{1}$

   $A = 9800$

   $9800 - 1400 = \textbf{8400 stockbrokers}$

5. (a) Discount $= 0.2 \cdot \$79 = \textbf{\$15.80}$

   (b) New price $= \$79 - \$15.80 = \textbf{\$63.20}$

6. (a) $-\dfrac{1}{8}N$

   (b) $7(x + 1)$

   (c) $\dfrac{3}{10}N + 1$

   (d) $x - 4$

7. $x > -1$

   ![number line from -2 to 2 with open circle at -1 and arrow to the right]

8. $B = 140\% \cdot 140$
   $B = 1.4 \cdot 140$
   $B = \textbf{196}$

9. $160\% \cdot 230 = B$
   $1.6 \cdot 230 = B$
   $368 = B$

   $230 + 368 = \textbf{598}$

10. $P \cdot 70 = 98$

    $P = \dfrac{98}{70}$

    $P = 1.4$

    $P = \textbf{140\%}$

11. $\dfrac{4}{5} = \dfrac{8}{10} = 0.8$

FRACTION	DECIMAL	PERCENT
$\dfrac{51}{100}$	0.51	51%
$\dfrac{4}{5}$	(a) **0.8**	(b) **80%**

12. $\dfrac{6}{10} \cdot A = 750$

    $\dfrac{10}{6} \cdot \dfrac{6}{10} \cdot A = 750 \cdot \dfrac{10}{6}$

    $A = \textbf{1250}$

13.

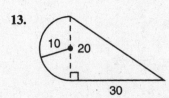

$V = A_{\text{Base}} \times \text{height}$
$= [A_{\text{Semicircle}} + A_{\text{Triangle}}] \times \text{height}$
$= \left[\dfrac{\pi(10)^2}{2} + \dfrac{(30)(20)}{2} \right] \text{in.}^2 \times 3 \text{ in.}$
$= [50\pi + 300] \text{ in.}^2 \times 3 \text{ in.}$
$\approx \textbf{1371 in.}^3$

14. $4 \text{ mi} \times \dfrac{5280 \text{ ft}}{1 \text{ mi}} \times \dfrac{12 \text{ in.}}{1 \text{ ft}} = 4(5280)(12) \text{ in.}$
    $= \textbf{253,440 in.}$

15. $\dfrac{1357}{12} = 113\dfrac{1}{12}$

$$
\begin{array}{r}
113 \\
12\overline{)1357} \\
\underline{12} \\
15 \\
\underline{12} \\
37 \\
\underline{36} \\
1
\end{array}
$$

16. $y = x + 4$

x	y
–4	0
0	4
2	6

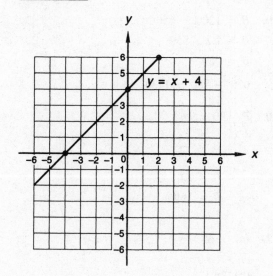

$y = x + 4$

17.

$$\dfrac{3\frac{3}{4}}{6\frac{3}{8}} = \dfrac{\frac{1}{4}}{x}$$

$$\dfrac{\frac{15}{4}}{\frac{51}{8}} = \dfrac{\frac{1}{4}}{x}$$

$$\dfrac{15}{4}x = \dfrac{51}{32}$$

$$\dfrac{4}{15} \cdot \dfrac{15}{4}x = \dfrac{\overset{17}{\cancel{51}}}{\underset{8}{\cancel{32}}} \cdot \dfrac{\cancel{4}}{\underset{5}{\cancel{15}}}$$

$$x = \dfrac{17}{40}$$

18. $$2\frac{1}{3}x + 3\frac{1}{2} = 6\frac{1}{6}$$

$$2\frac{1}{3}x + 3\frac{1}{2} - 3\frac{1}{2} = 6\frac{1}{6} - 3\frac{1}{2}$$

$$\frac{7}{3}x = \frac{37}{6} - \frac{21}{6}$$

$$\frac{3}{7} \cdot \frac{7}{3}x = \frac{16}{6} \cdot \frac{3}{7}$$

$$x = \frac{8}{7}$$

19.
$$-7x + \frac{1}{4} = \frac{3}{2}$$
$$ \; \underline{-\frac{1}{4} \quad -\frac{1}{4}}$$
$$-7x = \frac{5}{4}$$

$$\left(-\frac{1}{7}\right)(-7x) = \left(\frac{5}{4}\right)\left(-\frac{1}{7}\right)$$

$$x = -\frac{5}{28}$$

20.
$$11x - 3 = -58$$
$$\underline{ +3 \qquad +3}$$
$$11x = -55$$

$$\frac{11x}{11} = \frac{-55}{11}$$

$$x = -5$$

21. For the method shown here, the **Addition Property of Equality** and the **Division Property of Equality** were used. (Answers may vary. See student work.)

22. $(8 \times 10^{10}) \times (1 \times 10^{-10})$
$= (8 \times 1) \times (10^{10} \times 10^{-10}) = \mathbf{8}$

23. $3^2 + 2^2\left[2^3(\sqrt{36} - 2^2)(1 + \sqrt{16}) - \sqrt[3]{\dfrac{27}{125}}\right]$

$= 9 + 4\left[8(2)(5) - \dfrac{3}{5}\right] = 9 + 4\left[\dfrac{397}{5}\right]$

$= \dfrac{45}{5} + \dfrac{1588}{5} = \dfrac{\mathbf{1633}}{\mathbf{5}}$

24. (a) $3(-6) = \mathbf{-18}$

(b) $(-2)(-5) = \mathbf{10}$

(c) $(-1)(+7) = \mathbf{-7}$

25. There are four negative signs, an even number, so the product is positive.
$$(-1)(-2)(-3)(-4)(+5) = \mathbf{120}$$

26. $-6 - 3 + (-2) - (-3) = -9 - 2 + 3$
$= -11 + 3 = \mathbf{-8}$

27. There are five negative signs, an odd number, so the product is **negative.**

28. $-x - xy = -(-2) - (-2)(4) = 2 + 8 = \mathbf{10}$

29. $ay - a - y = (-7)(-3) - (-7) - (-3)$
$$= 21 + 7 + 3$$
$$= 28 + 3$$
$$= \textbf{31}$$

30. $x^y + y^x + 2xy = (4^3) + (3^4) + 2(4)(3)$
$$= 64 + 81 + 24$$
$$= \textbf{169}$$

PROBLEM SET 87

1. Rates for last 220 mi: $\dfrac{55 \text{ mi}}{1 \text{ hr}}, \dfrac{1 \text{ hr}}{55 \text{ mi}}$

Time for last 220 mi: $\dfrac{1 \text{ hr}}{55 \text{ mi}} \times 220 \text{ mi} = 4 \text{ hr}$

Time for first 210 mi: $7 \text{ hr} - 4 \text{ hr} = 3 \text{ hr}$

Rate for first 210 mi: $\dfrac{210 \text{ mi}}{3 \text{ hr}} = \dfrac{70 \text{ mi}}{1 \text{ hr}}$
$$= \textbf{70 mph}$$

2. $\dfrac{\$15,000}{50 \text{ dresses}} \times 19 \text{ dresses} = \textbf{\$5700}$

3. $\dfrac{2}{120} = \dfrac{M}{228,000}$
$456,000 = 120M$
$M = \textbf{3800 masterpieces}$

4. $\dfrac{1}{20} \times T = 1100$

$\dfrac{20}{1} \cdot \dfrac{1}{20} \cdot T = 1100 \cdot \dfrac{20}{1}$

$T = \textbf{22,000 fans}$

5. Rate method:

$\dfrac{7 \text{ hr}}{25 \text{ pools}} \times 275 \text{ pools} = \textbf{77 hr}$

Proportion method:

$\dfrac{7}{25} = \dfrac{H}{275}$
$1925 = 25H$
$H = \textbf{77 hr}$

6. (a) Tariff $= 0.30 \cdot \$9000 = \textbf{\$2700}$

(b) New cost $= \$9000 + \$2700 = \textbf{\$11,700}$

7. $x \le 1$

8. $B = 260\% \cdot 90$
$B = 2.6 \cdot 90$
$B = \textbf{234}$

9. $B = 120\% \cdot 650$
$B = 1.2 \cdot 650$
$B = 780$
$650 + 780 = \textbf{1430}$

10. $P \cdot 90 = 144$
$P = \dfrac{144}{90}$
$P = 1.6$
$P = \textbf{160\%}$

11. (a) $\dfrac{1}{2}N$

(b) $\textbf{2x}$

(c) $\textbf{(x)(-x)}$

(d) $\dfrac{2}{5}\sqrt{x}$

12. $20,000 \times 0.000004 = (2 \times 10^4)(4 \times 10^{-6})$
$= (2 \times 4) \times (10^4 \times 10^{-6}) = \textbf{8} \times \textbf{10}^{-2}$

13. $0.03 \times A = 9.3$
$A = \dfrac{9.3}{0.03}$
$A = \textbf{310}$

14. $10,000,000 \text{ in.}^2 \times \dfrac{1 \text{ ft}}{12 \text{ in.}} \times \dfrac{1 \text{ ft}}{12 \text{ in.}}$
$= \dfrac{(10,000,000)}{(12)(12)} \text{ ft}^2 = \textbf{69,444} \dfrac{4}{9} \text{ ft}^2$

15. $V = A_{\text{Base}} \times \text{height}$
$= [A_{\text{Rectangle}} + 2A_{\text{Triangle}}] \times \text{height}$
$= \left[(6)(2) + 2\left(\dfrac{(1)(2)}{2} \right) \right] \text{m}^2 \times 2 \text{ m}$
$= [12 + 2] \text{ m}^2 \times 2 \text{ m}$
$= \textbf{28 m}^3$

16. $V = A_{\text{Base}} \times \text{height} = \pi r^2 h$
$$= \pi(1.5)^2 \text{ yd}^2 \times 3 \text{ yd}$$
$$= 2.25\pi \text{ yd}^2 \times 3 \text{ yd}$$
$$\approx \mathbf{21.195 \text{ yd}^3}$$

17. $y = -3x + 2$

x	y
-1	5
0	2
1	-1

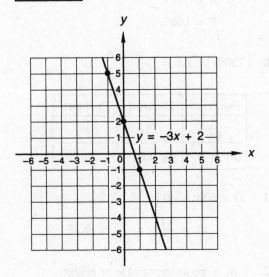

$y = -3x + 2$

18. $\dfrac{2}{3} \cdot \dfrac{1}{5} \cdot \dfrac{3}{2} \cdot 5 = \left(\dfrac{2}{3} \cdot \dfrac{3}{2}\right) \cdot \left(\dfrac{1}{5} \cdot 5\right)$
$$= 1 \cdot 1 = \mathbf{1}$$

19. $-9x + 11 = -92$
$$\underline{ -11 \quad -11}$$
$$-9x \quad = -103$$
$$\dfrac{-9x}{-9} = \dfrac{-103}{-9}$$
$$x = \dfrac{\mathbf{103}}{\mathbf{9}}$$

20. $-\dfrac{1}{4}x - 7 = -\dfrac{7}{2}$
$$\underline{\phantom{-\dfrac{1}{4}x} +7 \quad +7}$$
$$-\dfrac{1}{4}x \quad = \dfrac{7}{2}$$
$$\left(-\dfrac{4}{1}\right)\left(-\dfrac{1}{4}x\right) = \left(\dfrac{7}{2}\right)\left(-\dfrac{4}{1}\right)$$
$$x = \mathbf{-14}$$

21. $\dfrac{4\frac{1}{5}}{3\frac{1}{2}} = \dfrac{1\frac{1}{8}}{x}$
$$\dfrac{\frac{21}{5}}{\frac{7}{2}} = \dfrac{\frac{9}{8}}{x}$$
$$\dfrac{21}{5}x = \dfrac{63}{16}$$
$$\dfrac{5}{21} \cdot \dfrac{21}{5}x = \dfrac{63}{16} \cdot \dfrac{5}{21}$$
$$x = \dfrac{\mathbf{15}}{\mathbf{16}}$$

22. $\dfrac{\frac{6}{7}}{\frac{54}{42}} = \dfrac{\frac{2}{3}}{x}$
$$\dfrac{6}{7}x = \dfrac{18}{21}$$
$$\dfrac{7}{6} \cdot \dfrac{6}{7}x = \dfrac{18}{21} \cdot \dfrac{7}{6}$$
$$x = \mathbf{1}$$

23. $8\dfrac{3}{14}x + 1\dfrac{1}{5} = 1\dfrac{1}{30}$
$$8\dfrac{3}{14}x + 1\dfrac{1}{5} - 1\dfrac{1}{5} = 1\dfrac{1}{30} - 1\dfrac{1}{5}$$
$$\dfrac{115}{14}x = \dfrac{31}{30} - \dfrac{36}{30}$$
$$\dfrac{14}{115} \cdot \dfrac{115}{14}x = -\dfrac{5}{30} \cdot \dfrac{14}{115}$$
$$x = -\dfrac{\mathbf{7}}{\mathbf{345}}$$

24. $2^3 + 2^2[2^1(\sqrt{64} - 7)(1 + 3^2) + (-3)]$
$$= 8 + 4[2(1)(10) - 3] = 8 + 4[17]$$
$$= 8 + 68 = \mathbf{76}$$

25. (a) $\dfrac{27}{-9} = \mathbf{-3}$

(b) $\dfrac{-27}{9} = \mathbf{-3}$

(c) $\dfrac{-27}{-9} = \mathbf{3}$

26. There are five negative signs, so
$$-\left(-\left\{-[-(-4)]\right\}\right) = \mathbf{-4}.$$

27. $x + xy = (-1) + (-1)(-9) = -1 + 9 = \mathbf{8}$

28. $-m - mx = -(-2) - (-2)(-3) = 2 - 6 = \mathbf{-4}$

29. $ap - p + a = (-4)(-3) - (-3) + (-4)$

$$= 12 + 3 - 4$$
$$= 15 - 4$$
$$= \mathbf{11}$$

30. $0.6 \, \cancel{cm} \times \dfrac{1 \, m}{100 \, \cancel{cm}} = \mathbf{0.006 \; m}$

PROBLEM SET 88

1. $F \times 16{,}000 = 2000$

$$F = \frac{2000}{16{,}000} = \mathbf{\frac{1}{8}}$$

2. $\dfrac{2}{19} = \dfrac{D}{380}$

$760 = 19D$

$D = \mathbf{40 \; components}$

3. $1\frac{1}{4} \times 800 = B$

$\dfrac{5}{4} \times 800 = B$

$B = \mathbf{1000 \; delegates}$

4. Rates for first 50 mi: $\dfrac{25 \, mi}{1 \, hr}, \dfrac{1 \, hr}{25 \, mi}$

Time for first 50 mi: $\dfrac{1 \, hr}{25 \, \cancel{mi}} \times 50 \, \cancel{mi} = 2 \, hr$

Rates for last 400 mi: $\dfrac{50 \, mi}{1 \, hr}, \dfrac{1 \, hr}{50 \, mi}$

Time for last 400 mi: $\dfrac{1 \, hr}{50 \, \cancel{mi}} \times 400 \, \cancel{mi} = 8 \, hr$

Total time for trip: $2 \, hr + 8 \, hr = \mathbf{10 \; hr}$

5. Overall avg $= \dfrac{4(19) + 9(22)}{4 + 9} = \dfrac{76 + 198}{13}$

$$= \dfrac{274}{13} \approx \mathbf{21.08}$$

6. (a) $\mathbf{2N}$

(b) $\mathbf{2N + 2}$

(c) $\mathbf{\dfrac{3}{2}x}$

(d) $\mathbf{\dfrac{3}{2}x + 6}$

7. $x \geq -2$

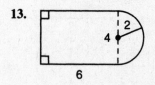

8. $B = 185\% \cdot 160$

$B = 1.85 \cdot 160$

$B = 296$

$160 + 296 = \mathbf{456}$

9. $P \cdot 60 = 96$

$$P = \frac{96}{60}$$

$$P = 1.6$$

$$P = \mathbf{160\%}$$

10. $250\% = 2.5 = 2\frac{1}{2} = \dfrac{5}{2}$

FRACTION	DECIMAL	PERCENT
$\frac{51}{100}$	0.51	51%
(a) $\mathbf{\frac{5}{2}}$	(b) $\mathbf{2.5}$	250%

11. $D \times 620 = 217$

$$D = \frac{217}{620} = \frac{7}{20} = \mathbf{0.35}$$

12. $L.A.$ = perimeter of base $\times$ height

$$= \left[6 + 4 + 6 + \frac{\pi(4)}{2} \right] cm \times 7 \, cm$$
$$= (16 + 2\pi)(7) \; cm^2$$
$$\approx \mathbf{155.96 \; cm^2}$$

13.

$S.A. = L.A. + 2A_{Base}$

$\approx (155.96 \, cm^2) + 2[A_{Rectangle} + A_{Semicircle}]$

$\approx 155.96 \; cm^2 + 2\left[6(4) + \dfrac{(3.14)2^2}{2} \right] cm^2$

$\approx 155.96 \; cm^2 + 60.56 \; cm^2$

$\approx \mathbf{216.52 \; cm^2}$

14.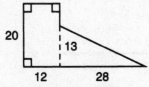

$V = A_{Base} \times \text{height}$

$= [A_{Rectangle} + A_{Triangle}] \times \text{height}$

$= \left[(12)(20) + \dfrac{(28)(13)}{2}\right]\left(\dfrac{10}{100}\right) \text{m}^3$

$= [240 + 182](0.1) \text{ m}^3$

$= \mathbf{42.2 \text{ m}^3}$

15. $(3 \times 10^{-11}) \times (2 \times 10^{15})$

$= (3 \times 2)(10^{-11} \times 10^{15}) = \mathbf{6 \times 10^4}$

16. $1385.31 \text{ km} \times \dfrac{1000 \text{ m}}{1 \text{ km}} \times \dfrac{100 \text{ cm}}{1 \text{ m}}$

$= (1385.31)(1000)(100) \text{ cm} = \mathbf{138{,}531{,}000 \text{ cm}}$

17. $y = -3x - 1$

x	y
-2	5
0	-1
1	-4

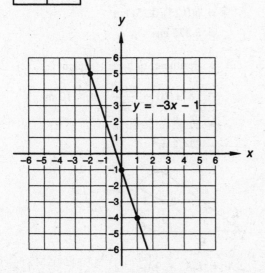

$y = -3x - 1$

18. $88 + 75 + 12 + 25 = (88 + 12) + (75 + 25)$

$= 100 + 100 = \mathbf{200}$

19. $-8x + 4 = 28$

$ \underline{-4 \quad\; -4}$

$-8x = 24$

$\dfrac{-8x}{-8} = \dfrac{24}{-8}$

$x = \mathbf{-3}$

20. $-2\dfrac{1}{4}x + \dfrac{1}{2} = 3\dfrac{3}{4}$

$-2\dfrac{1}{4}x + \dfrac{1}{2} - \dfrac{1}{2} = 3\dfrac{3}{4} - \dfrac{1}{2}$

$-\dfrac{9}{4}x = \dfrac{15}{4} - \dfrac{2}{4}$

$\left(-\dfrac{4}{9}\right)\left(-\dfrac{9}{4}\right)x = \dfrac{13}{4}\left(-\dfrac{4}{9}\right)$

$x = \mathbf{-\dfrac{13}{9}}$

21. $\dfrac{2\frac{2}{3}}{\frac{3}{4}} = \dfrac{\frac{4}{5}}{x}$

$\dfrac{8}{3}x = \dfrac{3}{5}$

$\dfrac{3}{8} \cdot \dfrac{8}{3}x = \dfrac{3}{5} \cdot \dfrac{3}{8}$

$x = \mathbf{\dfrac{9}{40}}$

22. For the method shown here, the **Subtraction Property of Equality** and the **Division Property of Equality** were used. (Answers may vary. See student work.)

23. $2^3 + 2^2\left[2(\sqrt{49} - \sqrt{36})(1 - \sqrt{9}) - \sqrt[3]{\dfrac{8}{64}}\right]$

$= 8 + 4\left[2(1)(-2) - \dfrac{2}{4}\right] = 8 + 4\left[\dfrac{-18}{4}\right]$

$= 8 - 18 = \mathbf{-10}$

24. $1\dfrac{3}{4} \cdot 2\dfrac{1}{3} - \dfrac{5}{12} = \dfrac{7}{4} \cdot \dfrac{7}{3} - \dfrac{5}{12} = \dfrac{49}{12} - \dfrac{5}{12}$

$= \dfrac{44}{12} = \mathbf{\dfrac{11}{3}}$

25. $\dfrac{1}{2}\left(1\dfrac{2}{3} \cdot 1\dfrac{4}{5} - \dfrac{1}{2}\right) + \dfrac{5}{6}$

$= \dfrac{1}{2}\left(\dfrac{5}{3} \cdot \dfrac{9}{5} - \dfrac{1}{2}\right) + \dfrac{5}{6}$

$= \dfrac{1}{2}\left(3 - \dfrac{1}{2}\right) + \dfrac{5}{6} = \dfrac{5}{4} + \dfrac{5}{6}$

$= \dfrac{15}{12} + \dfrac{10}{12} = \mathbf{\dfrac{25}{12}}$

26. (a) $(4)(-2) = -8$

(b) $\dfrac{16}{-2} = -8$

(c) $(-6)(-2) = 12$

27. There are four negative signs, so
$-\left(-\left\{-[-(5)]\right\}\right) = 5.$

28. There are six negative signs, an even number, so the product is **positive.**

29. $xyz + xy = (-1)(-2)(-3) + (-1)(-2)$
$= -6 + 2$
$= -4$

30. $xy + y^x + x^x = (2)(3) + (3^2) + (2^2)$
$= 6 + 9 + 4$
$= 19$

PROBLEM SET 89

1. $\dfrac{\$120}{40} = \dfrac{D}{800}$
$40D = \$96,000$
$D = \textbf{\$2400}$

2. $\dfrac{4}{5} = \dfrac{26,000}{S}$
$4S = 130,000$
$S = \textbf{32,500 snapdragons}$

3. $7\dfrac{4}{5} \cdot A = 109,200$
$\dfrac{5}{39} \cdot \dfrac{39}{5} \cdot A = 109,200 \cdot \dfrac{5}{39}$
$A = \textbf{14,000}$

4. $\dfrac{680 \text{ mi}}{1 \text{ hr}} \times 10 \text{ hr} = \textbf{6800 mi}$

5. (a) $2N - 16$

(b) $-3(N + 5)$

(c) $-N - 4$

(d) $5(2N + 6)$

6. $x \not\geq -1$
$x < -1$

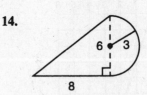

7. $x \not\leq 4$
$x > 4$

8. $B = 190\% \cdot 340$
$B = 1.9 \cdot 340$
$B = \textbf{646}$

9. $B = 160\% \cdot 200$
$B = 1.6 \cdot 200$
$B = 320$
$200 + 320 = \textbf{520}$

10. $P \cdot 80 = 128$
$P = \dfrac{128}{80}$
$P = 1.6$
$P = \textbf{160\%}$

11. $\dfrac{1}{6} + \dfrac{3}{7} + \dfrac{5}{6} + 2\dfrac{4}{7} = \left(\dfrac{1}{6} + \dfrac{5}{6}\right) + \left(\dfrac{3}{7} + 2\dfrac{4}{7}\right)$
$= 1 + 3 = \textbf{4}$

12. $S.A. = 6A_{\text{Side}}$
$= 6(0.25)(0.25) \text{ cm}^2$
$= \textbf{0.375 cm}^2$

13. $L.A. = \text{Perimeter of base} \times \text{height}$
$= \left[8 + 10 + \dfrac{\pi(6)}{2}\right] \text{m} \times 10 \text{ m}$
$\approx 27.42 \text{ m} \times 10 \text{ m}$
$\approx \textbf{274.2 m}^2$

14.

$S.A. = L.A. + 2A_{\text{Base}}$
$\approx 274.2 \text{ m}^2 + 2[A_{\text{Triangle}} + A_{\text{Semicircle}}]$
$\approx 274.2 \text{ m}^2 + 2\left[\dfrac{8(6)}{2} + \dfrac{\pi(9)}{2}\right] \text{m}^2$
$\approx 274.2 \text{ m}^2 + 2[38.13] \text{ m}^2$
$\approx \textbf{350.46 m}^2$

15.

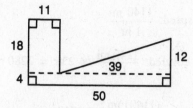

$V = A_{\text{Base}} \times$ height

$= [A_{\text{Rect 1}} + A_{\text{Rect 2}} + A_{\text{Triangle}}] \times$ height

$= \left[(11)(18) + (50)(4) + \dfrac{(39)(12)}{2}\right](2) \text{ yd}^3$

$= [198 + 200 + 234](2) \text{ yd}^3$

$= \mathbf{1264 \text{ yd}^3}$

16. $612 \text{ in.}^2 \times \dfrac{1 \text{ ft}}{12 \text{ in.}} \times \dfrac{1 \text{ ft}}{12 \text{ in.}} \times \dfrac{1 \text{ yd}}{3 \text{ ft}} \times \dfrac{1 \text{ yd}}{3 \text{ ft}}$

$= \dfrac{612}{(12)(12)(3)(3)} \text{ yd}^2 \approx \mathbf{0.472 \text{ yd}^2}$

17. $20\% \cdot 30 = 0.2 \cdot 30 = 6$ middle-eaters

$\dfrac{1}{3} \cdot 30 = 10$ dunkers

$10 - 6 = \mathbf{4 \text{ more dunkers}}$

18. $y = -x - 2$

x	y
-2	0
0	-2
2	-4

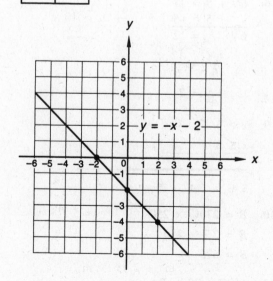

19.
$$\begin{array}{r} -6x - 6 = -3 \\ +6 \quad +6 \\ \hline -6x \quad\;\; = 3 \end{array}$$

$\dfrac{-6x}{-6} = \dfrac{3}{-6}$

$x = \mathbf{-\dfrac{1}{2}}$

20.
$$\begin{array}{r} -7x - 3 = 12 \\ +3 \quad +3 \\ \hline -7x \quad\;\; = 15 \end{array}$$

$\dfrac{-7x}{-7} = \dfrac{15}{-7}$

$x = \mathbf{-\dfrac{15}{7}}$

21. $\dfrac{3\frac{1}{3}}{6\frac{1}{2}} = \dfrac{\frac{4}{13}}{x}$

$\dfrac{\frac{10}{3}}{\frac{13}{2}} = \dfrac{\frac{4}{13}}{x}$

$\dfrac{10}{3}x = 2$

$\dfrac{3}{10} \cdot \dfrac{10}{3}x = 2 \cdot \dfrac{3}{10}$

$x = \mathbf{\dfrac{3}{5}}$

22. $-3\dfrac{1}{3}x + \dfrac{3}{4} = 2\dfrac{5}{12}$

$-3\dfrac{1}{3}x + \dfrac{3}{4} - \dfrac{3}{4} = 2\dfrac{5}{12} - \dfrac{3}{4}$

$-\dfrac{10}{3}x = \dfrac{29}{12} - \dfrac{9}{12}$

$\left(-\dfrac{3}{10}\right)\left(-\dfrac{10}{3}\right)x = \dfrac{20}{12}\left(-\dfrac{3}{10}\right)$

$x = \mathbf{-\dfrac{1}{2}}$

23. For the method shown here, the **Addition Property of Equality** and the **Division Property of Equality** were used. (Answers may vary. See student work.)

24. $2^2 + 2^3[3(\sqrt{49} - \sqrt{36})(\sqrt{16} + \sqrt{4}) - \sqrt{9}]$
$\quad + \sqrt{81}$

$= 4 + 8[3(1)(6) - 3] + 9 = 4 + 8[15] + 9$

$= 4 + 120 + 9 = \mathbf{133}$

25. $2\frac{2}{3} \cdot 3\frac{1}{4} - \frac{5}{12} = \frac{8}{3} \cdot \frac{13}{4} - \frac{5}{12}$

$= \frac{104}{12} - \frac{5}{12} = \frac{99}{12} = \frac{33}{4}$

26. (a) $(3)(-2) = -6$

(b) $\frac{-6}{-3} = 2$

(c) $(-1)(-1) = 1$

27. There are four negative signs, so

$-\left(-\left\{-\left[-(+6)\right]\right\}\right) = 6.$

28. $-4 - (-3) + \left[-(-2)\right] - (+5) - 1$

$= -4 + 3 + 2 - 5 - 1 = -4 - 1 = -5$

29. $yz - xz = (1)(2) - (-1)(2) = 2 + 2 = 4$

30. $y^y + x^x + y^x + x^y$

$= (3^3) + (2^2) + (3^2) + (2^3)$

$= 27 + 4 + 9 + 8$

$= 31 + 9 + 8$

$= 40 + 8$

$= 48$

PROBLEM SET 90

1. Rate method:

$\frac{\$1110}{20 \text{ tickets}} \times 150 \text{ tickets} = \$8325$

Proportion method:

$\frac{\$1110}{20} = \frac{D}{150}$

$\$1110(150) = 20D$

$D = \frac{\$1110(150)}{20}$

$D = \$8325$

2. $\frac{6}{1} = \frac{24{,}600}{S}$

$6S = 24{,}600$

$S = 4100$ *Sequoia semivirons*

3. $260\% \cdot \$240{,}000 = B$

$2.6 \cdot \$240{,}000 = B$

$\$624{,}000 = B$

$\$240{,}000 + \$624{,}000 = \$864{,}000$

4. Original speed: $\frac{1140 \text{ mi}}{1 \text{ hr}}$

Distance covered: $\frac{1140 \text{ mi}}{1 \text{ hr}} \times 2 \text{ hr} = 2280 \text{ mi}$

New speed: $\frac{\frac{3}{2}(1140) \text{ mi}}{1 \text{ hr}} = \frac{1710 \text{ mi}}{1 \text{ hr}}$

Distance covered: $\frac{1710 \text{ mi}}{1 \text{ hr}} \times 6 \text{ hr} = 10{,}260 \text{ mi}$

Total distance covered: $2280 \text{ mi} + 10{,}260 \text{ mi}$
$= 12{,}540 \text{ mi}$

5. (a) $3N - 11$

(b) $-5(N - 2)$

(c) $-3(N - 4)$

6. $2N = 11$

$\frac{2N}{2} = \frac{11}{2}$

$N = \frac{11}{2}$

7. $3N - 1 = 14$

$\underline{+1 \quad +1}$

$3N \quad = 15$

$\frac{3N}{3} = \frac{15}{3}$

$N = 5$

8. $6N + 5 = 17$

$\underline{-5 \quad -5}$

$6N \quad = 12$

$\frac{6N}{6} = \frac{12}{6}$

$N = 2$

9. $x \not> 3$

$x \le 3$

10. $B = 270\% \times 250$

$B = 2.7 \times 250$

$B = 675$

11. $150\% \cdot 78 = B$

$1.5 \cdot 78 = B$

$117 = B$

$78 + 117 = 195$

12. $P \cdot 90 = 360$

$P = \dfrac{360}{90}$

$P = 4$

$P = \mathbf{400\%}$

13.

$V = A_{\text{Base}} \times \text{height}$

$= [A_{\text{Semicircle}} + A_{\text{Square}} + A_{\text{Triangle}}] \times \text{height}$

$= \left[\dfrac{\pi(5)^2}{2} + (10)(10) + \dfrac{(5)(10)}{2} \right] \text{ft}^2 \times 4 \text{ ft}$

$= [12.5\pi + 100 + 25](4) \text{ ft}^3$

$\approx [164.25](4) \text{ ft}^3$

$\approx \mathbf{657 \text{ ft}^3}$

14. $L.A. = \text{Perimeter} \times \text{height}$

$= \left[15 + 10 + \dfrac{\pi(10)}{2} + 11.18 \right] \text{ft} \times 4 \text{ ft}$

$= [36.18 + 5\pi](4) \text{ ft}^2$

$\approx \mathbf{207.52 \text{ ft}^2}$

15. $S.A. = 2A_{\text{Base}} + L.A.$

$\approx 2[164.25] \text{ ft}^2 + 207.52 \text{ ft}^2$

$\approx [328.5 + 207.52] \text{ ft}^2$

$\approx \mathbf{536.02 \text{ ft}^2}$

16. $(1 \times 10^{20})(5 \times 10^{24})$

$= (1 \times 5)(10^{20} \times 10^{24}) = \mathbf{5 \times 10^{44}}$

17. $0.75 \, \text{m}^2 \times \dfrac{100 \text{ cm}}{1 \text{ m}} \times \dfrac{100 \text{ cm}}{1 \text{ m}}$

$= (0.75)(100)(100) \text{ cm}^2 = \mathbf{7500 \text{ cm}^2}$

18. $0.7 \times A = 9.1$

$A = \dfrac{9.1}{0.7} = \mathbf{13}$

19. $\dfrac{11}{2} \cdot \dfrac{3}{7} \cdot 4 \cdot \dfrac{2}{11} \cdot \dfrac{7}{4}$

$= \left(\dfrac{11}{2} \cdot \dfrac{2}{11} \right) \cdot \dfrac{3}{7} \cdot \left(4 \cdot \dfrac{7}{4} \right)$

$= 1 \cdot \dfrac{3}{7} \cdot 7 = \mathbf{3}$

20. $-5x - 4 = 16$

$ \dfrac{+4 \quad +4}{}$

$-5x = 20$

$\dfrac{-5x}{-5} = \dfrac{20}{-5}$

$x = \mathbf{-4}$

21. $-11x + 9 = 42$

$ \dfrac{-9 \quad -9}{}$

$-11x = 33$

$\dfrac{-11x}{-11} = \dfrac{33}{-11}$

$x = \mathbf{-3}$

22. $\dfrac{5\frac{1}{2}}{\frac{2}{3}} = \dfrac{3}{x}$

$\dfrac{\frac{11}{2}}{\frac{2}{3}} = \dfrac{3}{x}$

$\dfrac{11}{2}x = 2$

$\dfrac{2}{11} \cdot \dfrac{11}{2}x = 2 \cdot \dfrac{2}{11}$

$x = \dfrac{\mathbf{4}}{\mathbf{11}}$

23. $-1\dfrac{1}{2}x + 2\dfrac{3}{5} = 4\dfrac{3}{10}$

$-1\dfrac{1}{2}x + 2\dfrac{3}{5} - 2\dfrac{3}{5} = 4\dfrac{3}{10} - 2\dfrac{3}{5}$

$-\dfrac{3}{2}x = \dfrac{43}{10} - \dfrac{26}{10}$

$\left(-\dfrac{2}{3} \right)\left(-\dfrac{3}{2} \right)x = \dfrac{17}{10}\left(-\dfrac{2}{3} \right)$

$x = -\dfrac{\mathbf{17}}{\mathbf{15}}$

24. $5^2 + 4^2[2(\sqrt{121} - \sqrt{81})]$

$= 25 + 16[2(11 - 9)] = 25 + 16[4] = \mathbf{89}$

25. $2\dfrac{1}{5} \cdot 1\dfrac{1}{2} - \dfrac{7}{10} = \dfrac{11}{5} \cdot \dfrac{3}{2} - \dfrac{7}{10}$

$= \dfrac{33}{10} - \dfrac{7}{10} = \dfrac{26}{10} = \dfrac{\mathbf{13}}{\mathbf{5}}$

26. (a) $(-5)(-3) = $ **15**

(b) $\dfrac{14}{-2} = $ **−7**

(c) $\dfrac{-8}{-4} = $ **2**

27. $2 - 5(2) - 4(-2) - 5(-3) = 2 - 10 + 8 + 15$
$= -8 + 8 + 15 = $ **15**

28. $km + m(-k) = (3)(9) + (9)(-3)$
$\qquad\qquad\qquad = 27 - 27$
$\qquad\qquad\qquad = $ **0**

29. $-(km) - k(m) = -(5)(3) - (5)(3)$
$\qquad\qquad\qquad\quad = -15 - 15$
$\qquad\qquad\qquad\quad = $ **−30**

30. $a^b + bax + \sqrt[x]{a} = (8^2) + (2)(8)(3) + (\sqrt[3]{8})$
$\qquad\qquad\qquad\qquad = 64 + 48 + 2$
$\qquad\qquad\qquad\qquad = $ **114**

PROBLEM SET 91

1. $\qquad \dfrac{3}{8} \cdot A = 21{,}000$

$\dfrac{8}{3} \cdot \dfrac{3}{8} \cdot A = 21{,}000 \cdot \dfrac{8}{3}$

$\qquad\quad A = $ **56,000 seats**

2. $\dfrac{2}{230} = \dfrac{5}{M}$

$2M = 1150$

$M = $ **575 megabytes**

3. $\dfrac{40}{\$28{,}000} = \dfrac{72}{D}$

$40D = 72 \cdot \$28{,}000$

$D = 72 \cdot \$700$

$D = $ **\$50,400**

4. Original speed $= \dfrac{480 \text{ mi}}{12 \text{ hr}} = \dfrac{40 \text{ mi}}{1 \text{ hr}}$

New speed $= \dfrac{(40 + 20) \text{ mi}}{1 \text{ hr}} = \dfrac{60 \text{ mi}}{1 \text{ hr}}$

Time $= \dfrac{1 \text{ hr}}{60 \text{ mi}} \times 300 \text{ mi} = $ **5 hr**

5. $\quad 4[3N + (-6)] = 12$

$\dfrac{1}{4} \cdot 4[3N - 6] = 12 \cdot \dfrac{1}{4}$

$\qquad\quad 3N - 6 = 3$
$\qquad\qquad \underline{\quad +6 \quad +6}$
$\qquad\quad 3N \quad\;\; = 9$

$\qquad\quad \dfrac{3N}{3} = \dfrac{9}{3}$

$\qquad\quad N = $ **3**

6. $\quad 25 - 3x = 7$
$\quad \underline{-25 \qquad -25}$
$\qquad\quad -3x = -18$

$\qquad\quad \dfrac{-3x}{-3} = \dfrac{-18}{-3}$

$\qquad\quad x = $ **6**

7. $\quad -N - 25 = 0$
$\quad \underline{+N \qquad\;\; +N}$
$\qquad\; \mathbf{-25 = N}$

8. $x \not< -9$
$x \geq -9$

9. $x \not\geq -4$
$x < -4$

10. $170\% \cdot 250 = B$
$\quad 1.7 \cdot 250 = B$
$\qquad\qquad 425 = B$

$250 + 425 = $ **675**

11. $P \cdot 90 = 72$

$P = \dfrac{72}{90}$

$P = \dfrac{8}{10}$

$P = $ **80%**

12. $(300{,}000{,}000)(0.00000004)$
$= (3 \times 10^8)(4 \times 10^{-8}) = $ **12**

13. $y = \dfrac{2}{3}x + 6$

x	y
-3	4
0	6
3	8

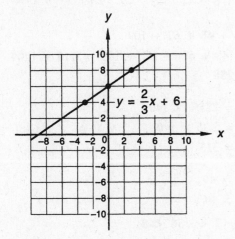

14.

$V = A_{\text{Base}} \times \text{height}$

$= [A_{\text{Triangle}} + A_{\text{Rect 1}} + A_{\text{Rect 2}}] \times \text{height}$

$= \left[\dfrac{36(25)}{2} + 5(50) + 14(20)\right] \text{cm}^2$
$\times 100 \text{ cm}$

$= [450 + 250 + 280](100) \text{ cm}^3$

$= \textbf{98,000 cm}^3$

15. $L.A. = \text{Perimeter} \times \text{height}$

$= [50 + 30 + 43.83 + 20 + 14 + 25] \text{ cm}$

$\times 100 \text{ cm}$

$= [182.83](100) \text{ cm}^2$

$= \textbf{18,283 cm}^2$

16. $2 \text{ mi} \times \dfrac{5280 \text{ ft}}{1 \text{ mi}} \times \dfrac{12 \text{ in.}}{1 \text{ ft}}$

$= 2(5280)(12) \text{ in.} = \textbf{126,720 in.}$

17.
$$-2\dfrac{1}{3}x - \dfrac{3}{4} = 3\dfrac{7}{12}$$

$$-2\dfrac{1}{3}x - \dfrac{3}{4} + \dfrac{3}{4} = 3\dfrac{7}{12} + \dfrac{3}{4}$$

$$-\dfrac{7}{3}x = \dfrac{43}{12} + \dfrac{9}{12}$$

$$\left(-\dfrac{3}{7}\right)\left(-\dfrac{7}{3}\right)x = \dfrac{52}{12}\left(-\dfrac{3}{7}\right)$$

$$x = -\dfrac{13}{7}$$

18.
$$\begin{array}{r} -7x - 7 = -2 \\ +7 \quad +7 \\ \hline -7x \quad = 5 \end{array}$$

$$\dfrac{-7x}{-7} = \dfrac{5}{-7}$$

$$x = -\dfrac{5}{7}$$

19. For the method shown here, the **Addition Property of Equality** and the **Division Property of Equality** were used. (Answers may vary. See student work.)

20. $\dfrac{2\frac{1}{4}}{1\frac{5}{6}} = \dfrac{1\frac{1}{2}}{x}$

$\dfrac{\frac{9}{4}}{\frac{11}{6}} = \dfrac{\frac{3}{2}}{x}$

$\dfrac{9}{4}x = \dfrac{11}{4}$

$x = \dfrac{11}{4} \cdot \dfrac{4}{9} = \dfrac{\textbf{11}}{\textbf{9}}$

21. $2^3 + 2^2\left[3(\sqrt{64} - \sqrt{49})(\sqrt[3]{8} + 1) - 1\right]$
$= 8 + 4[3(8 - 7)(2 + 1) - 1]$
$= 8 + 4[3(1)(3) - 1] = 8 + 4[8] = 8 + 32$
$= \textbf{40}$

22. (a) $2(-3) = \textbf{-6}$

(b) $\dfrac{-12}{-3} = \textbf{4}$

(c) $(-1)(-2)(3) = (2)(3) = \textbf{6}$

23. $-2(-3) + 3[(-2)(-4) - (-2)] = 6 + 3[8 + 2]$
$= 6 + 3[10] = \textbf{36}$

24. $\frac{1}{2}[(-2)(3) - (-2)(4)] - 3(-2) - 1$

$= \frac{1}{2}[-6 - 8] + 6 - 1 = -7 + 6 - 1 = \mathbf{-2}$

25. $(7 - 2 \cdot 2) - 3[(11 - 2) + 3]$
$= (7 - 4) - 3[9 + 3] = 3 - 3[12]$
$= 3 - 36 = \mathbf{-33}$

26. $\frac{1}{2}\left(3\frac{2}{5} \cdot \frac{3}{4} - 1\frac{1}{4}\right) = \frac{1}{2}\left(\frac{17}{5} \cdot \frac{3}{4} - \frac{5}{4}\right)$

$= \frac{1}{2}\left(\frac{51}{20} - \frac{25}{20}\right) = \frac{1}{2}\left(\frac{26}{20}\right) = \mathbf{\frac{13}{20}}$

27. $3\frac{1}{2} \times 2\frac{1}{3} \div \frac{5}{6} \times \frac{1}{3} = \frac{7}{2} \times \frac{7}{3} \times \frac{6}{5} \times \frac{1}{3}$

$= \mathbf{\frac{49}{15}}$

28. $-xy + y = -(-2)(-3) + (-3) = -6 - 3 = \mathbf{-9}$

29. $xy + yz + xz$
$= (-1)(-2) + (-2)(-3) + (-1)(-3)$
$= 2 + 6 + 3$
$= 8 + 3$
$= \mathbf{11}$

30. $6.8 + 4.91 + 5.2 + 1.39$
$= (6.8 + 5.2) + (4.91 + 1.39)$
$= 12 + 6.3 = \mathbf{18.3}$

PROBLEM SET 92

1. $2N + 42 = 126$
$\quad\quad 2N = 84$
$\quad\quad\quad N = \mathbf{42}$

2. $3N - 20 = 223$
$\quad\quad 3N = 243$
$\quad\quad\quad N = \mathbf{81}$

3. $33N + 4 = -260$
$\quad\quad 33N = -264$
$\quad\quad\quad N = \mathbf{-8}$

4. Original speed $= \frac{4 \text{ mi}}{1 \text{ hr}}$

New speed $= \frac{(4 \times 4) \text{ mi}}{1 \text{ hr}} = \frac{16 \text{ mi}}{1 \text{ hr}}$

Time $= \frac{1 \text{ hr}}{16 \text{ mi}} \times 80 \text{ mi} = \mathbf{5 \text{ hr}}$

5. $x \nleq 3$
$x > 3$

$$\begin{array}{c} \longleftrightarrow \\ 2 \quad 3 \quad 4 \quad 5 \quad 6 \end{array}$$

6. $x \ngtr -2$
$x \leq -2$

$$\begin{array}{c} \longleftrightarrow \\ -5 \quad -4 \quad -3 \quad -2 \quad -1 \end{array}$$

7. $11 + 48 + 62 + 109$
$= (48 + 62) + (11 + 109) = 110 + 120$
$= \mathbf{230}$

8. $B = 225\% \times 140$
$B = 2.25 \times 140$
$B = \mathbf{315}$

9. $7^2 = 49$
$8^2 = 64$
So, $\mathbf{7 < \sqrt{56} < 8}$

10. $10^2 = 100$
$11^2 = 121$
So, $\mathbf{10 < \sqrt{108} < 11}$

11. $P \cdot 20 = 5$

$P = \frac{5}{20} = \frac{25}{100} = \mathbf{25\%}$

Old membership $= 20$
New membership $= 20 + 5 = 25$
P of Old $=$ New

$P \cdot 20 = 25$

$P = \frac{25}{20}$

$P = \frac{125}{100}$

$P = \mathbf{125\%}$

P of New $=$ Old

$P \cdot 25 = 20$

$P = \frac{20}{25}$

$P = \frac{80}{100}$

$P = \mathbf{80\%}$

12. Joe's Mack brand: $\dfrac{\$1.20}{8 \text{ oz}} = \dfrac{\$0.15}{1 \text{ oz}}$

El Primo Deluxe: $\dfrac{\$1.52}{9.5 \text{ oz}} = \dfrac{\$0.16}{1 \text{ oz}}$

Joe's Mack brand is the better buy.

13. $y = 2x + 2$

x	y
-1	0
0	2
1	4

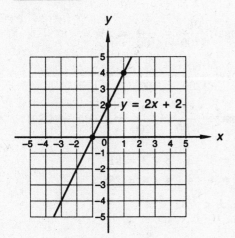

14.

$V = A_{\text{Base}} \times \text{height}$

$= [A_{\text{Triangle}} + A_{\text{Rectangle}} + A_{\text{Semicircle}}]$
$\quad \times \text{height}$

$= \left[\dfrac{21(20)}{2} + 27(20) + \dfrac{\pi(10)^2}{2}\right]\left(\dfrac{250}{100}\right) \text{m}^3$

$= [210 + 540 + 50\pi](2.5) \text{ m}^3$

$\approx \textbf{2267.5 m}^3$

15. $V = A_{\text{Base}} \times \text{height}$

$= (12 \text{ in.})(10 \text{ in.}) \times (3 \text{ in.})$

$= \textbf{360 in.}^3$

16. $L.A. = \text{Perimeter of base} \times \text{height}$

$= [2(12) + 2(10)] \text{ in.} \times 3 \text{ in.}$

$= 44 \text{ in.} \times 3 \text{ in.}$

$= \textbf{132 in.}^2$

17.
$$-2x + 6 = 1$$
$$\underline{\quad\; -6 \quad -6}$$
$$-2x \quad\;\; = -5$$
$$\dfrac{-2x}{-2} = \dfrac{-5}{-2}$$
$$x = \dfrac{5}{2}$$

18.
$$-3\tfrac{1}{3}x - \dfrac{1}{4} = 6\tfrac{1}{2}$$
$$-3\tfrac{1}{3}x - \dfrac{1}{4} + \dfrac{1}{4} = 6\tfrac{1}{2} + \dfrac{1}{4}$$
$$-\dfrac{10}{3}x = \dfrac{26}{4} + \dfrac{1}{4}$$
$$\left(-\dfrac{3}{10}\right)\left(-\dfrac{10}{3}\right)x = \dfrac{27}{4}\left(-\dfrac{3}{10}\right)$$
$$x = -\dfrac{81}{40}$$

19.
$$\dfrac{3\tfrac{1}{6}}{2\tfrac{1}{3}} = \dfrac{\tfrac{1}{2}}{x}$$
$$\dfrac{\tfrac{19}{6}}{\tfrac{7}{3}} = \dfrac{\tfrac{1}{2}}{x}$$
$$\dfrac{19}{6}x = \dfrac{7}{6}$$
$$x = \dfrac{7}{6} \cdot \dfrac{6}{19} = \dfrac{7}{19}$$

20. For the method shown here, the **Subtraction Property of Equality** and the **Division Property of Equality** were used. (Answers may vary. See student work.)

21. $2^2 + 2[\sqrt{9}(2^3 - 2 \cdot 3)(3^2 - 4 \cdot 2)]$
$= 4 + 2[3(8 - 6)(9 - 8)]$
$= 4 + 2[3(2)(1)] = 4 + 2[6] = 4 + 12 = \textbf{16}$

22. (a) $3(-2) = \textbf{-6}$

(b) $\dfrac{-18}{3} = \textbf{-6}$

(c) $(-2)(-2)(5) = 4(5) = \textbf{20}$

23. $-4[-(-3) - (-4)] + 6 - (-5)$
$= -4[3 + 4] + 6 + 5$
$= -4[7] + 6 + 5 = -28 + 6 + 5$
$= -22 + 5 = \textbf{-17}$

24. $-3(-4 - 1) + 2 - 3(6 - 4)$
$= -3(-5) + 2 - 3(2) = 15 + 2 - 6$
$= 17 - 6 = \mathbf{11}$

25. $-[-(-2)] - (-1) - 5 + \left(\sqrt[3]{\dfrac{64}{27}}\right)$

$= -2 + 1 - 5 + \dfrac{4}{3} = -1 - 5 + \dfrac{4}{3}$

$= -6 + \dfrac{4}{3} = -\dfrac{18}{3} + \dfrac{4}{3} = -\dfrac{\mathbf{14}}{\mathbf{3}}$

26. $\dfrac{1}{3}\left(2\dfrac{1}{4} \cdot 1\dfrac{1}{5} - \dfrac{17}{20}\right) = \dfrac{1}{3}\left(\dfrac{9}{4} \cdot \dfrac{6}{5} - \dfrac{17}{20}\right)$

$= \dfrac{1}{3}\left(\dfrac{54}{20} - \dfrac{17}{20}\right) = \dfrac{1}{3}\left(\dfrac{37}{20}\right) = \dfrac{\mathbf{37}}{\mathbf{60}}$

27. $4\dfrac{1}{3} \div 1\dfrac{2}{3} \times 2\dfrac{3}{4} \div \dfrac{1}{4}$

$= \dfrac{13}{3} \times \dfrac{3}{5} \times \dfrac{11}{4} \times \dfrac{4}{1} = \dfrac{\mathbf{143}}{\mathbf{5}}$

28. $xy + yz = (-1)(-2) + (-2)(-3) = 2 + 6 = \mathbf{8}$

29. $zy - zx = (-1)(-3) - (-1)(-2) = 3 - 2 = \mathbf{1}$

30. $x^2 + y^2 + 2xy = (2)^2 + (3)^2 + 2(2)(3)$
$= 4 + 9 + 12$
$= 13 + 12$
$= \mathbf{25}$

PROBLEM SET 93

1. $15N + 4 = 49$
$15N = 45$
$N = \mathbf{3}$

2. $10N - 15 = -105$
$10N = -90$
$N = \mathbf{-9}$

3. $14N + 8 = 50$
$14N = 42$
$N = \mathbf{3}$

4. $\dfrac{2\dfrac{1}{2}}{3} = \dfrac{S}{300}$
$750 = 3S$
$\mathbf{250} = S$

5. $x > 1$

6. $\dfrac{2}{3} \cdot 50 \cdot 48 \cdot \dfrac{2}{5} = \dfrac{2}{3} \cdot 48 \cdot 50 \cdot \dfrac{2}{5}$

$= \left(\dfrac{2}{3} \cdot 48\right) \cdot \left(50 \cdot \dfrac{2}{5}\right) = 32 \cdot 20 = \mathbf{640}$

7. $B = 230\% \cdot 350$
$B = 2.3 \cdot 350$
$B = \mathbf{805}$

8. $130\% \cdot 200 = B$
$1.3 \cdot 200 = B$
$260 = B$
$200 + 260 = \mathbf{460}$

9. rangy: $\dfrac{36}{1200} = \dfrac{3}{100} = \mathbf{3\%}$

lanky: $\dfrac{60}{1200} = \dfrac{5}{100} = \mathbf{5\%}$

gamey: $\dfrac{54}{1200} = \dfrac{9}{200} = 0.045 = \mathbf{4.5\%}$

10. $3^4 = 3 \cdot 3 \cdot 3 \cdot 3 = 81$
$4^4 = 4 \cdot 4 \cdot 4 \cdot 4 = 256$
So, $3 < \sqrt[4]{82} < 4$

11. $0.36 = \dfrac{36}{100} = \dfrac{9}{25}$

FRACTION	DECIMAL	PERCENT
$\dfrac{51}{100}$	0.51	51%
(a) $\dfrac{9}{25}$	0.36	(b) **36%**

12. $S.A. = 2\pi r^2 + \pi dh$
$= 2\pi(6)^2 \text{ in.}^2 + \pi(12)(12) \text{ in.}^2$
$= 72\pi \text{ in.}^2 + 144\pi \text{ in.}^2$
$\approx \mathbf{678.24 \text{ in.}^2}$

13.

$V = A_{\text{Base}} \times \text{height}$
$= [A_{\text{Triangle}} + A_{\text{Rectangle}}] \times \text{height}$
$= \left[\dfrac{(32)(16)}{2} + (53)(5)\right] \text{yd}^2 \times 2 \text{ yd}$
$= [256 + 265](2) \text{ yd}^3$
$= \mathbf{1042 \text{ yd}^3}$

14. $L.A.$ = Perimeter × height
$$= [53 + 21 + 35.78 + 21 + 5] \text{ yd} \times 2 \text{ yd}$$
$$= [135.78](2) \text{ yd}^2$$
$$= \textbf{271.56 yd}^2$$

15. $1234.8792 \; \cancel{m} \times \dfrac{1 \text{ km}}{1000 \; \cancel{m}} = \textbf{1.2348792 km}$

16.
$$-3x - 10 = -1$$
$$\underline{+10 \quad +10}$$
$$-3x \quad\quad = 9$$
$$\frac{-3x}{-3} = \frac{9}{-3}$$
$$x = \textbf{-3}$$

17.
$$\frac{3\frac{1}{3}}{2\frac{1}{4}} = \frac{1\frac{1}{2}}{x}$$
$$\frac{\frac{10}{3}}{\frac{9}{4}} = \frac{\frac{3}{2}}{x}$$
$$\frac{10}{3}x = \frac{27}{8}$$
$$\frac{3}{10} \cdot \frac{10}{3}x = \frac{27}{8} \cdot \frac{3}{10}$$
$$x = \textbf{\frac{81}{80}}$$

18.
$$-3\frac{1}{4}x - \frac{3}{8} = 4\frac{1}{2}$$
$$-3\frac{1}{4}x - \frac{3}{8} + \frac{3}{8} = 4\frac{1}{2} + \frac{3}{8}$$
$$-\frac{13}{4}x = \frac{36}{8} + \frac{3}{8}$$
$$\left(-\frac{4}{13}\right)\left(-\frac{13}{4}\right)x = \frac{39}{8} \cdot \left(-\frac{4}{13}\right)$$
$$x = \textbf{-\frac{3}{2}}$$

19. $2^3 + 2^3[2^2(\sqrt{16} - \sqrt{9})(\sqrt{9} + 2^2)]$
$$= 8 + 8[4(4 - 3)(3 + 4)]$$
$$= 8 + 8[4(1)(7)] = 8 + 8[28] = 8 + 224$$
$$= \textbf{232}$$

20. $-2(-2 - 3 \cdot 5) - [2(3 - 5) + 2]$
$$= -2(-17) - [2(-2) + 2] = 34 - [-2] = \textbf{36}$$

21. $-3(-2 - 6 \cdot 2) - [4(2 - 4) - 2]$
$$= -3(-14) - [4(-2) - 2] = 42 - [-10] = \textbf{52}$$

22. $\dfrac{5[(6 - 3) + 2] - 1}{2(5 - 1)} = \dfrac{5[3 + 2] - 1}{2(4)}$
$$= \frac{5[5] - 1}{2(4)} = \frac{25 - 1}{8} = \frac{24}{8} = \textbf{3}$$

23. $\dfrac{2(-1 - 6) + 4(2 - 5)}{3(-1 - 3)} = \dfrac{2(-7) + 4(-3)}{3(-4)}$
$$= \frac{-14 + (-12)}{-12} = \frac{-26}{-12} = \textbf{\frac{13}{6}}$$

24. $\dfrac{(7 + 3)(-7 + 3)}{-3[(11 - 2) + 5]} = \dfrac{10(-4)}{-3[9 + 5]} = \dfrac{10(-4)}{-3[14]}$
$$= \frac{-40}{-42} = \textbf{\frac{20}{21}}$$

25. $\dfrac{1}{3}\left(2\dfrac{1}{4} \cdot \dfrac{2}{3} - \dfrac{1}{6} \cdot \dfrac{1}{2}\right)$
$$= \frac{1}{3}\left(\frac{9}{4} \cdot \frac{2}{3} - \frac{1}{6} \cdot \frac{1}{2}\right) = \frac{1}{3}\left(\frac{18}{12} - \frac{1}{12}\right)$$
$$= \frac{1}{3}\left(\frac{17}{12}\right) = \textbf{\frac{17}{36}}$$

26. $2\dfrac{1}{3} \times 1\dfrac{1}{4} \div \dfrac{3}{8} \div \dfrac{1}{2}$
$$= \frac{7}{3} \times \frac{5}{4} \times \frac{8}{3} \times \frac{2}{1} = \textbf{\frac{140}{9}}$$

27. $y = x + 3$

x	y
-3	0
0	3
1	4

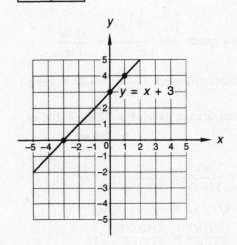

28. $yy - xy = (-2)(-2) - (-1)(-2) = 4 - 2 = \textbf{2}$

29. $abc + ab = (-2)(-1)(3) + (-2)(-1)$
$\qquad = 6 + 2$
$\qquad = \mathbf{8}$

30. $a^2 + b^2 + c^2 + 2ab$
$\qquad = (1)^2 + (2)^2 + (3)^2 + 2(1)(2)$
$\qquad = 1 + 4 + 9 + 4$
$\qquad = 5 + 9 + 4$
$\qquad = 14 + 4$
$\qquad = \mathbf{18}$

PROBLEM SET 94

1. $-4N + 12 = 36$
$\qquad -4N = 24$
$\qquad N = \mathbf{-6}$

2. $8N + 12 = 84$
$\qquad 8N = 72$
$\qquad N = \mathbf{9}$

3. $\dfrac{4\frac{1}{2}}{6} = \dfrac{675}{O}$

$\dfrac{9}{2}O = 4050$

$\dfrac{2}{9} \cdot \dfrac{9}{2}O = 4050 \cdot \dfrac{2}{9}$

$\qquad O = \mathbf{900 \ were \ opaque}$

4. Original speed: $\dfrac{18 \text{ mi}}{1 \text{ hr}}$

Distance traveled: $\dfrac{18 \text{ mi}}{1 \text{ hr}} \times 4 \text{ hr} = 72 \text{ mi}$

New speed: $\dfrac{\left(\frac{3}{2} \cdot 18\right) \text{ mi}}{1 \text{ hr}} = \dfrac{27 \text{ mi}}{1 \text{ hr}}$

Distance traveled: $\dfrac{27 \text{ mi}}{1 \text{ hr}} \times 4 \text{ hr} = 108 \text{ mi}$

Total distance traveled $= 72 \text{ mi} + 108 \text{ mi}$
$\qquad\qquad\qquad\qquad = \mathbf{180 \ mi}$

5. $\dfrac{40}{\$5400} = \dfrac{C}{\$13,500}$

$\$5400 \cdot C = 40 \cdot \$13,500$

$\dfrac{\$5400\,C}{\$5400} = \dfrac{\$540,000}{\$5400}$

$\qquad C = \mathbf{100 \ collectibles}$

6. $215\% \times 400{,}000{,}000 = B$
$\quad 2.15 \times 400{,}000{,}000 = B$
$\qquad\qquad 860{,}000{,}000 = B$

$400{,}000{,}000 + 860{,}000{,}000$
$= \mathbf{1{,}260{,}000{,}000 \ automobiles}$

7. $x \nleq 0$
$\quad x > 0$

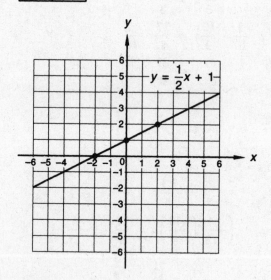

8. $12^2 = 144$
$\quad 13^2 = 169$
So, $\mathbf{12 < \sqrt{150} < 13}$

9. $y = \dfrac{1}{2}x + 1$

x	y
-2	0
0	1
2	2

10. Increase $= 0.1 \cdot \$90 = \$9$
Price $= \$90 + \$9 = \$99$

Decrease $= 0.1 \cdot \$99 = \$9.90$
Price $= \$99 - \$9.90 = \mathbf{\$89.10}$

11. $P \cdot 600 = 2280$

$\qquad P = \dfrac{2280}{600}$

$\qquad P = 3.8$

$\qquad P = \mathbf{380\%}$

12.

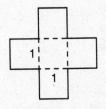

$V = A_{\text{Base}} \times \text{height}$
$= [5A_{\text{Square}}] \times \text{height}$
$= [5(1)(1)] \text{ yd}^2 \times \left(\frac{3}{3}\right) \text{yd}$
$= \textbf{5 yd}^3$

13. $L.A. = \text{Perimeter} \times \text{height}$
$= 12(1) \text{ yd} \times 3\left(\frac{1}{3}\right) \text{yd}$
$= \textbf{12 yd}^2$

14. $V = A_{\text{Base}} \times \text{height}$
$= \pi(10)^2 \text{ m}^2 \times 6 \text{ m}$
$= 100\pi \text{ m}^2 \times 6 \text{ m}$
$\approx \textbf{1884 m}^3$

15. $-x + 3y - 2x + y + 4 + 5x - 1$
$= (-x - 2x + 5x) + (3y + y) + (4 - 1)$
$= \textbf{2x + 4y + 3}$

16. $5a - 3b + 12 - a + 3 - 14b$
$= (5a - a) + (-3b - 14b) + (12 + 3)$
$= \textbf{4a - 17b + 15}$

17. $-9x - 4 = 5$
$\underline{\quad +4 \quad +4}$
$-9x \quad = 9$
$\frac{-9x}{-9} = \frac{9}{-9}$
$x = \textbf{-1}$

18.
$\frac{2\frac{1}{11}}{5\frac{6}{8}} = \frac{\frac{1}{11}}{p}$
$\frac{\frac{23}{11}}{\frac{46}{8}} = \frac{\frac{1}{11}}{p}$
$\frac{23}{11}p = \frac{46}{88}$
$\frac{11}{23} \cdot \frac{23}{11}p = \frac{46}{88} \cdot \frac{11}{23}$
$p = \textbf{\frac{1}{4}}$

19.
$-5\frac{2}{3}x - \frac{1}{5} = 4\frac{7}{15}$
$-5\frac{2}{3}x - \frac{1}{5} + \frac{1}{5} = 4\frac{7}{15} + \frac{1}{5}$
$-\frac{17}{3}x = \frac{67}{15} + \frac{3}{15}$
$\left(-\frac{3}{17}\right)\left(-\frac{17}{3}\right)x = \frac{70}{15}\left(-\frac{3}{17}\right)$
$x = -\textbf{\frac{14}{17}}$

20. $(3 \times 10^{-17}) \times (3 \times 10^{-19}) = \textbf{9} \times \textbf{10}^{-36}$

21. $3^2\left[\sqrt{25}(4^2 - 5 \cdot 2)\sqrt{\frac{1}{169}}\right]$
$= 9\left[5(6)\left(\frac{1}{13}\right)\right] = \textbf{\frac{270}{13}}$

22. $5\frac{3}{4} \cdot 1\frac{7}{9} - \frac{35}{36} = \frac{23}{4} \cdot \frac{16}{9} - \frac{35}{36}$
$= \frac{368}{36} - \frac{35}{36} = \frac{333}{36} = \textbf{\frac{37}{4}}$

23. $-5(-3 - 4 \cdot 3) - 4 \cdot 6 - (-1)$
$= -5(-3 - 12) - 4 \cdot 6 - (-1)$
$= -5(-15) - 4 \cdot 6 - (-1)$
$= 75 - 24 + 1 = 51 + 1 = \textbf{52}$

24. $-(-4) - (-9) + 7 + (-9) = 4 + 9 + 7 - 9$
$= 13 + 7 - 9 = 20 - 9 = \textbf{11}$

25. $-[-(+3)] - (-2) - 4 + (-6)$
$= 3 + 2 - 4 - 6 = 5 - 4 - 6$
$= 1 - 6 = \textbf{-5}$

26. $\frac{1}{4}\left(3\frac{1}{2} \cdot 5\frac{1}{5} - \frac{11}{20}\right) = \frac{1}{4}\left(\frac{7}{2} \cdot \frac{26}{5} - \frac{11}{20}\right)$
$= \frac{1}{4}\left(\frac{364}{20} - \frac{11}{20}\right) = \textbf{\frac{353}{80}}$

27. $\frac{2[(23 - 19) + (6 + 2)3]}{5 - 7 + 6} = \frac{2[4 + (8)3]}{5 - 7 + 6}$
$= \frac{2[28]}{-2 + 6} = \frac{56}{4} = \textbf{14}$

28. $\frac{18 - 3[-1(2 - 4) + 2]}{3(-4 + 2)} = \frac{18 - 3[-1(-2) + 2]}{3(-2)}$
$= \frac{18 - 3[4]}{-6} = \frac{6}{-6} = \textbf{-1}$

29. $-zm - (-m)(-z) = -(3)(5) - (-5)(-3)$
$$= -15 - 15$$
$$= \textbf{-30}$$

30. $zp - mp + z^p = (2)(4) - (5)(4) + (2^4)$
$$= 8 - 20 + 16$$
$$= -12 + 16$$
$$= \textbf{4}$$

PROBLEM SET 95

1. $4N + 4 = 24$
$$4N = 20$$
$$N = \textbf{5}$$

2. $-4N - 4 = -24$
$$-4N = -20$$
$$N = \textbf{5}$$

3. $30N - 75 = -225$
$$30N = -150$$
$$N = \textbf{-5}$$

4. $3\frac{3}{5} \cdot A = 288$
$$\frac{5}{18} \cdot \frac{18}{5} \cdot A = 288 \cdot \frac{5}{18}$$
$$A = \textbf{80 gallons}$$

5. $x \not< 2$
$$x \geq 2$$

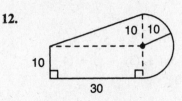

6. $\frac{2}{3} + \frac{3}{4} + 3\frac{1}{3} + 2\frac{3}{4}$
$$= \left(\frac{2}{3} + 3\frac{1}{3}\right) + \left(\frac{3}{4} + 2\frac{3}{4}\right)$$
$$= 4 + 3\frac{2}{4} = \textbf{7}\frac{\textbf{1}}{\textbf{2}}$$

7. $B = 130\% \cdot 220$
$$B = 1.3 \cdot 220$$
$$B = \textbf{286}$$

8. Increase $= 99¢ - 45¢ = 54¢$
$$54 = P \cdot 45$$
$$P = \frac{54}{45}$$
$$P = 1.2$$
$$P = \textbf{120\%}$$

$99 = P \cdot 45$
$$P = \frac{99}{45}$$
$$P = 2.2$$
$$P = \textbf{220\%}$$

9. $76 = P \cdot 95$
$$P = \frac{76}{95}$$
$$P = 0.8$$
$$P = \textbf{80\%}$$

10.

FRACTION	DECIMAL	PERCENT
$\frac{51}{100}$	0.51	51%
(a) $\frac{53}{100}$	0.53	(b) **53%**

11. $14^2 = 196$
$15^2 = 225$

So, $\textbf{14} < \sqrt{\textbf{201}} < \textbf{15}$

12.

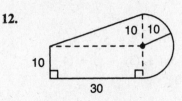

$V = A_{\text{Base}} \times \text{height}$
$$= [A_{\text{Rect}} + A_{\text{Triangle}} + A_{\text{Semicircle}}](\text{height})$$
$$= \left[10(30) + \frac{30(10)}{2} + \frac{\pi(10)^2}{2}\right](2)(12) \text{ in.}^3$$
$$= [300 + 150 + 50\pi](24) \text{ in.}^3$$
$$\approx \textbf{14,568 in.}^3$$

13. $P = [2(38) + 2(30) \text{ yd}] = \textbf{136 yd}$

14. $L.A. = \text{Perimeter} \times \text{height}$
$$= 136 \text{ yd} \times 12 \text{ yd}$$
$$= \textbf{1632 yd}^2$$

15. $4p - m + 2p + 3m - 5 - 2m$
$$= (4p + 2p) + (-m + 3m - 2m) - 5$$
$$= \textbf{6p - 5}$$

16. $-3c + 8 - 2b + c - 4 + 4b - c$
$$= (-3c + c - c) + (-2b + 4b) + (8 - 4)$$
$$= \textbf{2b - 3c + 4}$$

17. $3a + 4 - 2b + 6 - a - 3b + 1$
$$= (3a - a) + (-2b - 3b) + (4 + 6 + 1)$$
$$= \textbf{2a - 5b + 11}$$

18.
$$-2x - 12 = -2$$
$$\underline{+12 \quad +12}$$
$$-2x = 10$$
$$\frac{-2x}{-2} = \frac{10}{-2}$$
$$x = -5$$

19.
$$-3x + 6 = 4 - 6x$$
$$\underline{+6x +6x}$$
$$3x + 6 = 4$$
$$ \underline{-6 \quad -6}$$
$$3x = -2$$
$$\frac{3x}{3} = \frac{-2}{3}$$
$$x = -\frac{2}{3}$$

20.
$$2x - 5 = 9x$$
$$\underline{-2x -2x}$$
$$-5 = 7x$$
$$\frac{-5}{7} = \frac{7x}{7}$$
$$-\frac{5}{7} = x$$

21.
$$6x + 6 = -4 + 4x$$
$$\underline{-4x -4x}$$
$$2x + 6 = -4$$
$$ \underline{-6 \quad -6}$$
$$2x = -10$$
$$\frac{2x}{2} = \frac{-10}{2}$$
$$x = -5$$

22.
$$\frac{4\frac{1}{5}}{2\frac{1}{3}} = \frac{1\frac{1}{2}}{x}$$
$$\frac{\frac{21}{5}}{\frac{7}{3}} = \frac{\frac{3}{2}}{x}$$
$$\frac{21}{5}x = \frac{7}{2}$$
$$\frac{5}{21} \cdot \frac{21}{5} \cdot x = \frac{7}{2} \cdot \frac{5}{21}$$
$$x = \frac{5}{6}$$

23.
$$3^2 + 2^3[2^2(\sqrt{16} - \sqrt{9}) - (\sqrt{25} - \sqrt{4})]$$
$$= 9 + 8[4(1) - (3)] = 9 + 8[1] = 17$$

24.
$$-2(-6 - 2 \cdot 5) + 2(-2 - 1)$$
$$= -2(-16) + 2(-3) = 32 - 6 = 26$$

25.
$$\frac{5 - (8 - 6) + 3 \cdot 4 + 6}{5(4 - 5)}$$
$$= \frac{5 - 2 + 12 + 6}{-5} = -\frac{21}{5}$$

26.
$$\frac{5 - (4 - 6)2 + 4 \cdot 2}{3(-2 - 1)} = \frac{5 - (-2)2 + 8}{3(-3)}$$
$$= \frac{5 + 4 + 8}{-9} = -\frac{17}{9}$$

27.
$$\frac{1}{4}\left(2\frac{1}{3} \cdot \frac{1}{4} - \frac{5}{6} \cdot \frac{1}{2}\right)$$
$$= \frac{1}{4}\left(\frac{7}{3} \cdot \frac{1}{4} - \frac{5}{6} \cdot \frac{1}{2}\right) = \frac{1}{4}\left(\frac{7}{12} - \frac{5}{12}\right)$$
$$= \frac{1}{4}\left(\frac{2}{12}\right) = \frac{1}{24}$$

28. $y = -x + 3$

x	y
-1	4
0	3
1	2

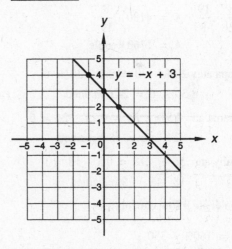

29.
$$-bc - ac = -(-3)(-5) - (-1)(-5)$$
$$= -15 - 5$$
$$= -20$$

30.
$$a^2 + b^2 + c^2 + 2abc + ab$$
$$= (3)^2 + (2)^2 + (4)^2 + 2(3)(2)(4) + (3)(2)$$
$$= 9 + 4 + 16 + 48 + 6$$
$$= 13 + 16 + 48 + 6$$
$$= 29 + 48 + 6$$
$$= 77 + 6$$
$$= 83$$

PROBLEM SET 96

1. $6N + 3 = 33$

$\quad 6N = 30$

$\quad N = \mathbf{5}$

2. $15N - 63 = 72$

$\quad 15N = 135$

$\quad N = \mathbf{9}$

3. If $\frac{5}{8}$ were equine, $\frac{3}{8}$ were not equine.

$$\frac{3}{8} \cdot A = 330$$

$$\frac{8}{3} \cdot \frac{3}{8} \cdot A = 330 \cdot \frac{8}{3}$$

$$A = \mathbf{880\ ranch\ animals}$$

4. $\quad \dfrac{2}{29} = \dfrac{I}{8410}$

$16{,}820 = 29I$

$\quad I = \mathbf{580\ idle\ onlookers}$

5. $\quad 2\frac{3}{8} \cdot A = 4180$

$$\frac{8}{19} \cdot \frac{19}{8} \cdot A = 4180 \cdot \frac{8}{19}$$

$$A = \mathbf{1760\ people}$$

6. Papa ate: $20\% \cdot 30 = 0.2 \cdot 30 = 6$

$\quad 30 - 6 = 24$ remained

Mama ate: $25\% \cdot 24 = 0.25 \cdot 24 = 6$

$\quad 24 - 6 = 18$ remained

Baby ate: $50\% \cdot 18 = 0.5 \cdot 18 = 9$

$\quad 18 - 9 = 9$ remained

Goldi ate **9 fine chocolates.**

7. $B = 160\% \cdot 350$

$B = 1.6 \cdot 350$

$B = \mathbf{560}$

8. $B = 40\% \cdot 95$

$B = 0.4 \cdot 95$

$B = 38$

$95 + 38 = \mathbf{133}$

9. $\left(\dfrac{3}{2}\right)^3 - \left(\dfrac{2}{5}\right)^2 = \dfrac{3^3}{2^3} - \dfrac{2^2}{5^2} = \dfrac{27}{8} - \dfrac{4}{25}$

$= \dfrac{675}{200} - \dfrac{32}{200} = \dfrac{\mathbf{643}}{\mathbf{200}}$

10. Rate Method:

$$\frac{30\ \text{corn dogs}}{48\ \text{min}} = \frac{5\ \text{corn dogs}}{8\ \text{min}}, \frac{8\ \text{min}}{5\ \text{corn dogs}}$$

$$\frac{8\ \text{min}}{5\ \text{corn dogs}} \times 50\ \text{corn dogs} = \mathbf{80\ min}$$

Proportion Method:

$$\frac{30}{48} = \frac{50}{M}$$

$30M = 48 \cdot 50$

$\quad M = \dfrac{2400}{30} = \mathbf{80\ min}$

11. $V = \pi r^2 h$

$= \pi(100)^2\ \text{cm}^2 \times 100\ \text{cm}$

$= 10{,}000\pi\ \text{cm}^2 \times 100\ \text{cm}$

$\approx \mathbf{3{,}140{,}000\ cm^3}$

$S.A. = 2\pi r^2 + \pi dh$

$= 2(10{,}000\pi)\ \text{cm}^2 + \pi(200)(100)\ \text{cm}^2$

$= 20{,}000\pi\ \text{cm}^2 + 20{,}000\pi\ \text{cm}^2$

$= 40{,}000\pi\ \text{cm}^2$

$\approx \mathbf{125{,}600\ cm^2}$

12.

$V = A_{\text{Base}} \times \text{height}$

$= [A_{\text{Semicircle}} + A_{\text{Rect}} + A_{\text{Triangle}}] \times \text{height}$

$= \left[\dfrac{\pi(2)^2}{2} + 4(6) + \dfrac{3(4)}{2}\right] \text{ft}^2 \times 5\ \text{ft}$

$= [2\pi + 24 + 6]\ \text{ft}^2 \times 5\ \text{ft}$

$\approx \mathbf{181.4\ ft^3}$

13. $L.A. = \text{Perimeter} \times \text{height}$

$= \left[2 + 4 + 2 + 5 + 7 + \dfrac{\pi(4)}{2}\right] \text{ft} \times 5\ \text{ft}$

$= [20 + 2\pi](5)\ \text{ft}^2$

$\approx \mathbf{131.4\ ft^2}$

14. $-6m + 2m + 8 - p - m = \mathbf{-5m - p + 8}$

15. $5x + 5 - 11x - 9 + 2x = \mathbf{-4x - 4}$

16. $-2x + 3 - 2x = 2 - 5x$

$\quad -4x + 3 = 2 - 5x$

$\quad \underline{+5x \qquad\qquad +5x}$

$\quad x + 3 = 2$

$\quad \underline{\quad -3 \quad -3}$

$\quad x = \mathbf{-1}$

17. $2x - 5 = -x + 10x$

$2x - 5 = 9x$

$\underline{-2x \qquad -2x}$

$-5 = 7x$

$\dfrac{-5}{7} = \dfrac{7x}{7}$

$-\dfrac{5}{7} = x$

18. $6x + 6 + 2x = 2x - 4 - 4x$

$8x + 6 = -2x - 4$

$\underline{+2x \qquad\quad +2x}$

$10x + 6 = \qquad -4$

$\underline{\qquad -6 \qquad -6}$

$10x = \qquad -10$

$\dfrac{10x}{10} = \dfrac{-10}{10}$

$x = -1$

19. $4x + 6 = x - 5$

$\underline{-x \qquad\quad -x}$

$3x + 6 = \qquad -5$

$\underline{\qquad -6 \qquad -6}$

$3x = \qquad -11$

$\dfrac{3x}{3} = \dfrac{-11}{3}$

$x = -\dfrac{11}{3}$

20. $-5x + 9 = 6$

$\underline{\quad -9 \quad -9}$

$-5x = -3$

$\dfrac{-5x}{-5} = \dfrac{-3}{-5}$

$x = \dfrac{3}{5}$

21. $\dfrac{2\frac{1}{4}}{5\frac{2}{5}} = \dfrac{x}{4}$

$\dfrac{\frac{9}{4}}{\frac{27}{5}} = \dfrac{x}{4}$

$9 = \dfrac{27}{5}x$

$\dfrac{5}{27} \cdot \dfrac{27}{5}x = 9 \cdot \dfrac{5}{27}$

$x = \dfrac{5}{3}$

22. $\sqrt{81} + 4^3[7(2^3 - 3^2) - (\sqrt{4} - 1^{10})]$

$= 9 + 64[7(-1) - (1)] = 9 + 64[-8]$

$= 9 - 512 = -503$

23. $-3(-5 - 2 \cdot 3) + 3[5 + 2(-3)]$

$= -3(-11) + 3[-1] = 33 - 3 = 30$

24. $\dfrac{-6 - (3 - 4)3 + 2(-2)}{3(-2 - 1)} = \dfrac{-6 - (-1)3 - 4}{3(-3)}$

$= \dfrac{-6 + 3 - 4}{-9} = \dfrac{7}{9}$

25. $\dfrac{6 - (2 - 3)3 + 4 \cdot 1}{3(-1 + 2)} = \dfrac{6 - (-1)3 + 4}{3(1)}$

$= \dfrac{6 + 3 + 4}{3} = \dfrac{13}{3}$

26. $x = -\dfrac{y}{2}$

x	y
-1	2
0	0
1	-2

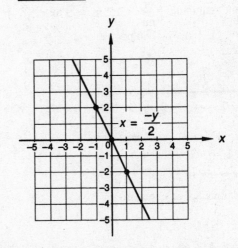

27. $-[bp - (-b)] = -[(-3)(2) - (-(-3))]$

$= -[-6 - 3]$

$= 9$

28. $a^a b + \sqrt[a]{ab} = (2^2)(8) + (\sqrt[2]{(2)(8)})$

$= 4(8) + \sqrt{16}$

$= 32 + 4$

$= 36$

29. $4,800,000 \times 40,000 = (4.8 \times 10^6) \times (4 \times 10^4)$
$= 19.2 \times 10^{10} = \mathbf{1.92 \times 10^{11}}$

30. $4^3 = 64$
$5^3 = 125$

So, $4 < \sqrt[3]{100} < 5$

PROBLEM SET 97

1. $7N + 42 = -98$
$\qquad 7N = -140$
$\qquad\ N = \mathbf{-20}$

2. $-4N - 6 = -34$
$\qquad -4N = -28$
$\qquad\ \ N = \mathbf{7}$

3. $\dfrac{140}{\$980} = \dfrac{200}{C}$
$140C = \$196,000$
$\qquad C = \mathbf{\$1400}$

4. Starting speed $= \dfrac{14 \text{ mi}}{2 \text{ hr}} = \dfrac{7 \text{ mi}}{1 \text{ hr}}$

Reduced speed $= \dfrac{7 \text{ mi}}{1 \text{ hr}} - \dfrac{2 \text{ mi}}{1 \text{ hr}} = \dfrac{5 \text{ mi}}{1 \text{ hr}}$

Time required: $\dfrac{1 \text{ hr}}{5 \text{ mi}} \times 15 \text{ mi} = \mathbf{3 \text{ hr}}$

5. $x \ngeq 3$
$x < 3$

6. $3x - 4y - 4x + 2x + 5 - y$
$= (3x - 4x + 2x) + (-4y - y) + 5$
$= \mathbf{x - 5y + 5}$

7. $3^3 = 3 \cdot 3 \cdot 3 = 27$
$4^3 = 4 \cdot 4 \cdot 4 = 64$

So, $3 < \sqrt[3]{40} < 4$

8. $170\% \cdot 260 = B$
$1.7 \cdot 260 = B$
$\qquad\quad 442 = B$

$260 + 442 = \mathbf{702}$

9. $P \cdot 80 = 116$

$P = \dfrac{116}{80}$

$P = 1.45$

$P = \mathbf{145\%}$

10. $30\% \cdot A = 180$
$\qquad 0.3A = 180$

$\qquad \dfrac{0.3A}{0.3} = \dfrac{180}{0.3}$

$\qquad\quad\ A = \mathbf{600}$

11.

FRACTION	DECIMAL	PERCENT
$\frac{51}{100}$	0.51	51%
$\frac{9}{10}$	(a) **0.9**	(b) **90%**

12.

$V = A_{\text{Base}} \times \text{height}$
$\ = [A_{\text{Rectangle}} + A_{\text{Triangle}}] \times \text{height}$
$\ = \left[(10)(30) + \dfrac{(30)(12)}{2}\right] \text{cm}^2 \times 12 \text{ cm}$
$\ = [300 + 180](12) \text{ cm}^3$
$\ = \mathbf{5760 \text{ cm}^3}$

13. $L.A. = \text{Perimeter} \times \text{height}$
$\quad = [30 + 10 + 18 + 20.47 + 10] \text{ cm}$
$\qquad \times 12 \text{ cm}$
$\quad = [88.47](12) \text{ cm}^2$
$\quad = \mathbf{1061.64 \text{ cm}^2}$

14. $1287.321 \text{ cm} \times \dfrac{1 \text{ m}}{100 \text{ cm}} \times \dfrac{1 \text{ km}}{1000 \text{ m}}$

$= \dfrac{1287.321}{(100)(1000)} \text{ km} = \mathbf{0.01287321 \text{ km}}$

15. $3x - 9 = 6$
$\quad\ 3x = 15$
$\qquad x = 5$

$2x - 6 = 2(5) - 6 = 10 - 6 = \mathbf{4}$

16. $6x - 2 = 10$
$\qquad 6x = 12$
$\qquad\ x = 2$

$\dfrac{1}{2}x + 3 = \dfrac{1}{2}(2) + 3 = 1 + 3 = \mathbf{4}$

17. $\dfrac{3}{4}x - 2 = 1$

$\dfrac{3}{4}x = 3$

$\qquad x = 4$

$3x - 4 = 3(4) - 4 = 12 - 4 = \mathbf{8}$

18. $14x - 1 + 5x = 10 - 2x + 5 + 4x$

$$19x - 1 = 15 + 2x$$
$$\underline{-2x \qquad\qquad -2x}$$
$$17x - 1 = 15$$
$$\underline{\quad +1 \quad +1}$$
$$17x \quad = 16$$
$$\frac{17x}{17} = \frac{16}{17}$$
$$x = \frac{16}{17}$$

19. $6x + 6 + 2x = 2x - 4 + 4x$

$$8x + 6 = 6x - 4$$
$$\underline{-6x \qquad -6x}$$
$$2x + 6 = \quad -4$$
$$\underline{\quad -6 \qquad -6}$$
$$2x \quad = \quad -10$$
$$\frac{2x}{2} = \frac{-10}{2}$$
$$x = -5$$

20. $-4x - 12 = -1$
$$\underline{\quad +12 \ +12}$$
$$-4x \quad = 11$$
$$\frac{-4x}{-4} = \frac{11}{-4}$$
$$x = -\frac{11}{4}$$

21. $4x + 2 = 2x - 4$
$$\underline{\quad -2 \qquad -2}$$
$$4x \quad = 2x - 6$$
$$\underline{-2x \qquad -2x}$$
$$2x \quad = \quad -6$$
$$\frac{2x}{2} = \frac{-6}{2}$$
$$x = -3$$

22. $$-2\frac{1}{3}x + \frac{2}{3} = 1\frac{4}{6}$$
$$-2\frac{1}{3}x + \frac{2}{3} - \frac{2}{3} = 1\frac{4}{6} - \frac{2}{3}$$
$$-\frac{7}{3}x = 1\frac{4}{6} - \frac{4}{6}$$
$$\left(-\frac{3}{7}\right)\left(-\frac{7}{3}\right)x = 1\left(-\frac{3}{7}\right)$$
$$x = -\frac{3}{7}$$

23. $$\frac{3\frac{1}{2}}{2\frac{1}{3}} = \frac{\frac{1}{2}}{x}$$
$$\frac{\frac{7}{2}}{\frac{7}{3}} = \frac{\frac{1}{2}}{x}$$
$$\frac{7}{2}x = \frac{7}{6}$$
$$\frac{2}{7} \cdot \frac{7}{2}x = \frac{7}{6} \cdot \frac{2}{7}$$
$$x = \frac{1}{3}$$

24. $-2(8 - 2 \cdot 3) - 4(3 \cdot 1 - 2) = -2(2) - 4(1)$
$$= -4 - 4 = -8$$

25. $$\frac{7 - (2 - 4)2 + 5 \cdot 1}{4(-2 + 3)} = \frac{7 - (-2)2 + 5}{4(1)}$$
$$= \frac{16}{4} = 4$$

26. $$\frac{8 - (4 - 7)3 + 4 \cdot 2}{5(-3 + 5)} = \frac{8 - (-3)3 + 8}{5(2)}$$
$$= \frac{25}{10} = \frac{5}{2}$$

27. $3^2 + 3[2^3(\sqrt{49} - 2^2)(3^2 - 2^3) - 2^2]$
$$= 9 + 3[8(3)(1) - 4] = 9 + 3[20] = 69$$

28. $3\frac{1}{3} \times 2\frac{1}{5} \div 1\frac{2}{3} \div \frac{1}{4}$
$$= \frac{10}{3} \times \frac{11}{5} \times \frac{3}{5} \times \frac{4}{1} = \frac{88}{5}$$

29. $bcb - ac = (-2)(-5)(-2) - (-1)(-5)$
$$= -20 - 5$$
$$= -25$$

30. $y = 2x - 4$

x	y
-1	-6
0	-4
2	0
4	4

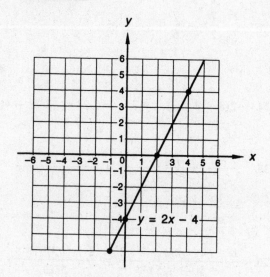

PROBLEM SET 98

1. $6N = -N - 14$
$7N = -14$
$N = -2$

2. $7N = (2N + 5) + 30$
$7N = 2N + 35$
$5N = 35$
$N = 7$

3. $x \not\leq -1$
$x > -1$

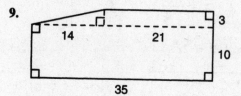

4. $\dfrac{1}{29} = \dfrac{R}{3509}$
$29R = 3509$
$R = $ **121 ratio problems**

5. Starting rate: $\dfrac{90 \text{ mi}}{15 \text{ hr}} = \dfrac{6 \text{ mi}}{1 \text{ hr}}$

Faster rate: $\dfrac{(2 \times 6) \text{ mi}}{1 \text{ hr}} = \dfrac{12 \text{ mi}}{1 \text{ hr}}$

Time required: $180 \text{ mi} \times \dfrac{1 \text{ hr}}{12 \text{ mi}} = 15 \text{ hr}$

Total time required: $15 \text{ hr} + 15 \text{ hr} = $ **30 hr**

6. $190\% \cdot 250 = B$
$1.9 \cdot 250 = B$
$475 = B$
$250 + 475 = $ **725**

7. $P \cdot 90 = 162$
$P = \dfrac{162}{90}$
$P = 1.8$
$P = $ **180%**

8. $40\% \cdot A = 62$
$0.4A = 62$
$\dfrac{0.4A}{0.4} = \dfrac{62}{0.4}$
$A = $ **155**

9.

$V = A_{\text{Base}} \times \text{height}$
$= [A_{\text{Rect 1}} + A_{\text{Rect 2}} + A_{\text{Triangle}}] \times \text{height}$
$= \left[10(35) + 21(3) + \dfrac{14(3)}{2} \right] \text{cm}^2 \times 10 \text{ cm}$
$= [350 + 63 + 21](10) \text{ cm}^3$
$= $ **4340 cm^3**

10. $L.A. = \text{Perimeter} \times \text{height}$
$= [10 + 14.32 + 21 + 13 + 35] \text{ cm}$
$\times 10 \text{ cm}$
$= [93.32](10) \text{ cm}^2$
$= $ **933.2 cm^2**

11. The measure of $\angle A$ is $180° - 38° = $ **142°**.

The measure of $\angle C$ is $90° - 38° = $ **52°**.

12. (a) $\angle A$ is acute, **43°**

  (b) $\angle B$ is right, **90°**

  (c) $\angle C$ is obtuse, **120°**

13. Any three of the following: $\angle R$, $\angle C$, $\angle QRS$, or $\angle SRQ$

14. $4^2 = 16$
  $5^2 = 25$
  So, $\mathbf{4 < \sqrt{17.4} < 5}$

15. $3p - 2q + 5q - 5p + 3 - q$
  $= (3p - 5p) + (-2q + 5q - q) + 3$
  $= \mathbf{-2p + 2q + 3}$

16. $6x - 2 = 16$
  $6x = 18$
  $x = 3$
  $x - 3 = (3 - 3) = \mathbf{0}$

17. $\dfrac{3}{2}x - 3 = 1$

  $\dfrac{3}{2}x = 4$

  $x = \dfrac{8}{3}$

  $2x + 3 = 2\left(\dfrac{8}{3}\right) + 3 = \dfrac{16}{3} + \dfrac{9}{3} = \mathbf{\dfrac{25}{3}}$

18. $-3x + 12 = 1 + 5x + 10 - 6x$
  $-3x + 12 = 11 - x$
  $\underline{ +x \qquad\qquad +x}$
  $-2x + 12 = 11$
  $\underline{ -12 \quad -12}$
  $-2x = -1$
  $\dfrac{-2x}{-2} = \dfrac{-1}{-2}$
  $x = \mathbf{\dfrac{1}{2}}$

19. $-2x + 10 - 3x = 5x - 3 - 6x$
  $-5x + 10 = -x - 3$
  $\underline{+5x \qquad\qquad +5x}$
  $10 = 4x - 3$
  $\underline{ +3 \qquad +3}$
  $13 = 4x$
  $\dfrac{13}{4} = \dfrac{4x}{4}$
  $\mathbf{\dfrac{13}{4}} = x$

20. $6x + 2 = 3x - 5$
  $\underline{ -2 \qquad\quad -2}$
  $6x = 3x - 7$
  $\underline{-3x \qquad\quad -3x}$
  $3x = -7$
  $\dfrac{3x}{3} = \dfrac{-7}{3}$
  $x = \mathbf{-\dfrac{7}{3}}$

21. $-6x + 2 = 3x - 6$
  $\underline{ -2 \qquad\quad -2}$
  $-6x = 3x - 8$
  $\underline{ -3x \qquad -3x}$
  $-9x = -8$
  $\dfrac{-9x}{-9} = \dfrac{-8}{-9}$
  $x = \mathbf{\dfrac{8}{9}}$

22. $-3\dfrac{1}{4}x + \dfrac{3}{4} = 1\dfrac{3}{8}$

  $-3\dfrac{1}{4}x + \dfrac{3}{4} - \dfrac{3}{4} = 1\dfrac{3}{8} - \dfrac{3}{4}$

  $-\dfrac{13}{4}x = \dfrac{11}{8} - \dfrac{6}{8}$

  $\left(-\dfrac{4}{13}\right)\left(-\dfrac{13}{4}\right)x = \dfrac{5}{8}\left(-\dfrac{4}{13}\right)$

  $x = \mathbf{-\dfrac{5}{26}}$

23. $-2(3 \cdot 2 - 7) - 3(2 \cdot 1 - 5) + \sqrt[4]{\dfrac{16}{81}}$

  $= -2(-1) - 3(-3) + \dfrac{2}{3} = 2 + 9 + \dfrac{2}{3} = \mathbf{11\dfrac{2}{3}}$

24. $\dfrac{6 - 3(2 - 1) + 4 \cdot 3}{5(-3 + 6)} = \dfrac{6 - 3(1) + 12}{5(3)}$

  $= \dfrac{15}{15} = \mathbf{1}$

25. $\dfrac{8 - 2(3 - 2) + 4 \cdot 2}{4(2 - 3)} = \dfrac{8 - 2(1) + 8}{4(-1)}$

  $= \dfrac{14}{-4} = \mathbf{-\dfrac{7}{2}}$

26. $2^3 + 3\left[2^2(\sqrt{36} + \sqrt{4})(\sqrt[3]{8} - 1) + 1\right]$
  $= 8 + 3[4(8)(1) + 1] = 8 + 3[33] = 8 + 99$
  $= \mathbf{107}$

27. $2\frac{1}{4} \times 3\frac{1}{3} \div 2\frac{1}{3} \times 1\frac{1}{4}$

$= \frac{9}{4} \times \frac{10}{3} \times \frac{3}{7} \times \frac{5}{4} = \frac{225}{56}$

28. $x - 4 = 2y$

x	y
-2	-3
0	-2
4	0

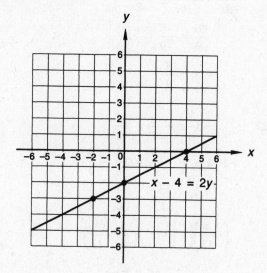

29. $d + c - ac = (1) + (-1) - (-1)(-1)$

$ = 1 - 1 - 1$

$ = -1$

30. $abc - a = (-2)(-1)(-3) - (-2)$

$ = -6 + 2$

$ = -4$

PROBLEM SET 99

1. $5N = -2N + 35$

$7N = 35$

$N = 5$

2. $-7N = 3N - 50$

$-10N = -50$

$N = 5$

3. $\frac{3}{5} \cdot 1200 = B$

720 students $= B$

4. $\dfrac{2\frac{1}{2}}{7} = \dfrac{1400}{D}$

$\frac{5}{2}D = 9800$

$\frac{2}{5} \cdot \frac{5}{2}D = 9800 \cdot \frac{2}{5}$

$D = 3920$

5. $x \not< -2$

$x \geq -2$

$$\overset{}{\underset{\begin{matrix}-3 & -2 & -1 & 0 & 1\end{matrix}}{\longleftarrow\!\!\!-\!\!\!-\!\!\!-\!\!\!-\!\!\!-\!\!\!\rightarrow}}$$

6. (a) **90°, right**

(b) **120°, obtuse**

(c) **40°, acute**

7. $B = 160\% \cdot 230$

$B = 1.6 \cdot 230$

$B = 368$

8. $180\% \cdot 180 = B$

$1.8 \cdot 180 = B$

$324 = B$

$180 + 324 = 504$

9. $25\% \cdot A = 61$

$0.25A = 61$

$\dfrac{0.25A}{0.25} = \dfrac{61}{0.25}$

$A = 244$

10. $7.34 \times 10^{-5} = 0.0000734$

11. $\dfrac{3121}{7} = 445\frac{6}{7}$

$$
\begin{array}{r}
445 \\
7\overline{)3121} \\
\underline{28} \\
32 \\
\underline{28} \\
41 \\
\underline{35} \\
6
\end{array}
$$

12. $-\left\{-[-(-3)]\right\} + 3 - [-(-3)] = 3 + 3 - 3 = 3$

13.

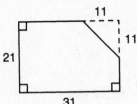

$V = A_{\text{Base}} \times \text{height}$

$\quad = [A_{\text{Rectangle}} - A_{\text{Missing}}] \times \text{height}$

$\quad = \left[31(21) - \dfrac{11(11)}{2} \right] \text{ft}^2 \times 6 \text{ ft}$

$\quad = [651 - 60.5] \text{ ft}^2 \times 6 \text{ ft}$

$\quad = 590.5 \text{ ft}^2 \times 6 \text{ ft}$

$\quad = \mathbf{3543 \text{ ft}^3}$

14. $2^5 = 32$

$3^5 = 243$

So, $\mathbf{2 < \sqrt[5]{40} < 3}$

15. $3x - 4a + 5a - 3x + a$

$= (3x - 3x) + (-4a + 5a + a) = \mathbf{2a}$

16. $2x + 10 = 7$

$\qquad 2x = -3$

$\qquad x = -\dfrac{3}{2}$

$6x + 12 = 6\left(-\dfrac{3}{2}\right) + 12 = -9 + 12 = \mathbf{3}$

17. $\dfrac{2}{3}x + 5 = 7$

$\qquad \dfrac{2}{3}x = 2$

$\qquad \dfrac{3}{2} \cdot \dfrac{2}{3}x = 2 \cdot \dfrac{3}{2}$

$\qquad x = 3$

$3 - \dfrac{4}{3}x = 3 - \dfrac{4}{3}(3) = 3 - 4 = \mathbf{-1}$

18. $-4x + 3 = x + 4$

$\qquad \dfrac{-3 \qquad\qquad -3}{-4x \qquad = x + 1}$

$\qquad \dfrac{-x \qquad\qquad -x}{-5x \qquad = \qquad 1}$

$\qquad \dfrac{-5x}{-5} = \dfrac{1}{-5}$

$\qquad x = -\dfrac{1}{5}$

19. $3x - 4 = -x + 6 + 11x + 4x - 3$

$\qquad 3x - 4 = 14x + 3$

$\qquad \dfrac{-3x \qquad\qquad -3x}{-4 = 11x + 3}$

$\qquad \dfrac{-3 \qquad\qquad -3}{-7 = 11x}$

$\qquad \dfrac{-7}{11} = \dfrac{11x}{11}$

$\qquad -\dfrac{7}{11} = x$

20. $-3\dfrac{1}{3}x - \dfrac{1}{6} = 1\dfrac{1}{3}$

$-3\dfrac{1}{3}x - \dfrac{1}{6} + \dfrac{1}{6} = 1\dfrac{1}{3} + \dfrac{1}{6}$

$\qquad -\dfrac{10}{3}x = \dfrac{8}{6} + \dfrac{1}{6}$

$\left(-\dfrac{3}{10}\right)\left(-\dfrac{10}{3}\right)x = \dfrac{3}{2}\left(-\dfrac{3}{10}\right)$

$\qquad x = -\dfrac{9}{20}$

21. $\dfrac{6\dfrac{1}{2}}{2\dfrac{1}{3}} = \dfrac{\dfrac{1}{2}}{x}$

$\qquad \dfrac{\dfrac{13}{2}}{\dfrac{7}{3}} = \dfrac{\dfrac{1}{2}}{x}$

$\qquad \dfrac{13}{2}x = \dfrac{7}{6}$

$\qquad \dfrac{2}{13} \cdot \dfrac{13}{2}x = \dfrac{7}{6} \cdot \dfrac{2}{13}$

$\qquad x = \dfrac{7}{39}$

22. $-2^2 - (-3)^2 - 3^4 = -4 - 9 - 81 = \mathbf{-94}$

23. $-3^2 + (-2)^2 = -9 + 4 = \mathbf{-5}$

24. $-2^2(2 \cdot 2 - 3 \cdot 2) - 3(2 \cdot 3 - 2^2)$

$= -4(-2) - 3(2) = 8 - 6 = \mathbf{2}$

25. $\dfrac{3 - 2(3 - 2) + 3 \cdot 2}{4(-2 + 5)} = \dfrac{3 - 2(1) + 6}{4(3)} = \dfrac{\mathbf{7}}{\mathbf{12}}$

26. $\dfrac{-3 - 2(3 \cdot 2 - 1) + 2^2}{2^2(-3 - 2^2)} = \dfrac{-3 - 2(5) + 4}{4(-7)}$

$= \dfrac{-9}{-28} = \dfrac{\mathbf{9}}{\mathbf{28}}$

27. $3\frac{1}{2} \times 2\frac{1}{3} \div \frac{1}{4} \div \frac{1}{6} = \frac{7}{2} \times \frac{7}{3} \times \frac{4}{1} \times \frac{6}{1}$

$\quad = \textbf{196}$

28. $x = 2y$

x	y
-2	-1
0	0
4	2

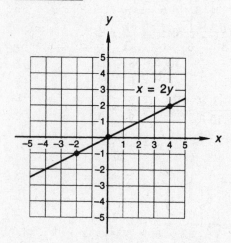

29. $abc + ab = (-2)(-3)(-4) + (-2)(-3)$

$\quad\quad\quad\quad = -24 + 6$

$\quad\quad\quad\quad = \textbf{-18}$

30. $-ab + bc = -(2)(-3) + (-3)(-4)$

$\quad\quad\quad\quad = 6 + 12$

$\quad\quad\quad\quad = \textbf{18}$

PROBLEM SET 100

1. $\frac{R}{B} = \frac{7}{5}, \quad \frac{R}{T} = \frac{7}{12}, \quad \frac{B}{T} = \frac{5}{12}$

$\quad \frac{7}{12} = \frac{R}{240}$

$\quad 1680 = 12R$

$\quad\quad R = \textbf{140 acres}$

$\quad B = 240 - 140 = \textbf{100 acres}$

2. $\frac{G}{P} = \frac{3}{17}, \quad \frac{G}{T} = \frac{3}{20}, \quad \frac{P}{T} = \frac{17}{20}$

$\quad \frac{3}{20} = \frac{G}{400}$

$\quad 1200 = 20G$

$\quad\quad G = \textbf{60 greens}$

3. $4N = 2N + 22$

$\quad 2N = 22$

$\quad\quad N = \textbf{11}$

4. If $\frac{5}{16}$ swooned, $\frac{11}{16}$ did not swoon.

$\quad \frac{11}{16} \cdot A = 2200$

$\quad \frac{16}{11} \cdot \frac{11}{16} \cdot A = 2200 \cdot \frac{16}{11}$

$\quad\quad\quad A = \textbf{3200 teenies}$

5. $x > -1$

$$\overset{\oplus}{\underset{-2 \quad -1 \quad 0 \quad 1 \quad 2}{\longleftrightarrow}}$$

6. $B = 140\% \cdot 75$

$\quad B = 1.4 \cdot 75$

$\quad B = \textbf{105}$

7. $30\% \cdot 150 = B$

$\quad 0.3 \cdot 150 = B$

$\quad\quad\quad\quad 45 = B$

$\quad 150 + 45 = \textbf{195}$

8. $P \cdot 70 = 91$

$\quad P = \frac{91}{70}$

$\quad P = 1.3$

$\quad P = \textbf{130\%}$

9. $\quad 35 = P \cdot 140$

$\quad \frac{35}{140} = P$

$\quad 0.25 = P$

$\quad \textbf{25\%} = P$

10. $3^4 = 81$

$\quad 4^4 = 196$

So, $3 < \sqrt[4]{180} < 4$

11. $85\% = \frac{85}{100} = \frac{17}{20}$

FRACTION	DECIMAL	PERCENT
$\frac{51}{100}$	0.51	51%
(a) $\frac{17}{20}$	(b) **0.85**	85%

12. $V = A_{\text{Base}} \times \text{height}$

$= [A_{\text{Large Semicircle}} + A_{\text{Small Semicircle}}] \times \text{height}$

$= \left[\dfrac{\pi(10)^2}{2} + \dfrac{\pi(5)^2}{2}\right] \text{ft}^2 \times 2(3) \text{ ft}$

$= [50\pi + 12.5\pi](6) \text{ ft}^3$

$\approx \mathbf{1177.5 \text{ ft}^3}$

13. $S.A. = 2A_{\text{Base}} + (\text{Perimeter} \times \text{height})$

$= 2[62.5\pi]\text{ft}^2 + [5\pi + 10\pi + 10](2)(3) \text{ ft}^2$

$= (125\pi + 90\pi + 60) \text{ ft}^2$

$\approx \mathbf{735.1 \text{ ft}^2}$

14. $61.131121 \text{ km} \times \dfrac{1000 \text{ m}}{1 \text{ km}} \times \dfrac{100 \text{ cm}}{1 \text{ m}}$

$= (63.131121)(1000)(100) \text{ cm} = \mathbf{6,113,112.1 \text{ cm}}$

15. $4x - 3a + 2y - 5x + x + 3a - y - y$

$= (4x - 5x + x) + (-3a + 3a)$

$+ (2y - y - y) = \mathbf{0}$

16. The measure of $\angle A$ is $180° - 47° = \mathbf{133°}$.

The measure of $\angle C$ is $90° - 47° = \mathbf{43°}$.

17. $6x - 3 = 12$

$6x = 15$

$x = \dfrac{15}{6} = \dfrac{5}{2}$

$\dfrac{2}{3}x + 5 = \dfrac{2}{3}\left(\dfrac{5}{2}\right) + 5 = \dfrac{5}{3} + 5 = \dfrac{20}{3} = \mathbf{6\dfrac{2}{3}}$

18.
$$7 - 6x = 2x + 13$$
$$\underline{-7 \qquad\qquad -7}$$
$$-6x = 2x + 6$$
$$\underline{-2x \quad -2x}$$
$$-8x = \qquad 6$$
$$\dfrac{-8x}{-8} = \dfrac{6}{-8}$$
$$x = -\dfrac{6}{8} = -\dfrac{3}{4}$$
$$22x = 22\left(-\dfrac{3}{4}\right) = -\dfrac{33}{2}$$

19.
$$-3x + 6 + 4x = 2x + 1 - 9 + 2x$$
$$6 + x = 4x - 8$$
$$\underline{-x \qquad -x}$$
$$6 = 3x - 8$$
$$\underline{+8 \qquad\qquad +8}$$
$$14 = 3x$$
$$\dfrac{14}{3} = \dfrac{3x}{3}$$
$$\dfrac{14}{3} = x$$

20.
$$2x - 4 - 5x = -x + 5 + 7x - 3$$
$$-3x - 4 = 6x + 2$$
$$\underline{+3x \qquad\quad +3x}$$
$$-4 = 9x + 2$$
$$\underline{-2 \qquad -2}$$
$$-6 = 9x$$
$$\dfrac{-6}{9} = \dfrac{9x}{9}$$
$$x = -\dfrac{6}{9} = -\dfrac{2}{3}$$

21.
$$-2\dfrac{1}{2}x - \dfrac{1}{3} = 2\dfrac{2}{3}$$
$$-2\dfrac{1}{2}x - \dfrac{1}{3} + \dfrac{1}{3} = 2\dfrac{2}{3} + \dfrac{1}{3}$$
$$-\dfrac{5}{2}x = 3$$
$$\left(-\dfrac{2}{5}\right)\left(-\dfrac{5}{2}\right)x = 3\left(-\dfrac{2}{5}\right)$$
$$x = -\dfrac{6}{5}$$

22.
$$\dfrac{\dfrac{3}{5}}{\dfrac{6}{25}} = \dfrac{\dfrac{1}{3}}{x}$$
$$\dfrac{3}{5}x = \dfrac{2}{25}$$
$$\dfrac{5}{3} \cdot \dfrac{3}{5}x = \dfrac{2}{25} \cdot \dfrac{5}{3}$$
$$x = \dfrac{2}{15}$$

23. $-2^2 - (-2)^3 = -4 + 8 = \mathbf{4}$

24. $-3^2 - (-3)^2 = -9 - 9 = \mathbf{-18}$

25. $-8(2 \cdot 3 - 4) - 3(8 - 9) = -8(2) - 3(-1)$

$= -16 + 3 = \mathbf{-13}$

26. $\dfrac{4 - 2(6 - 4) + 2^2 \cdot 3}{3(-3 + 5)} = \dfrac{4 - 2(2) + 12}{3(2)}$

$= \dfrac{12}{6} = \mathbf{2}$

27. $\dfrac{-2 + 3(2 \cdot 4 - 3) + 2^2}{3(2^2 - 1)}$

$= \dfrac{-2 + 3(5) + 4}{3(3)} = \dfrac{\mathbf{17}}{\mathbf{9}}$

28. $2\dfrac{1}{2} \div 3\dfrac{1}{2} \times \sqrt{\dfrac{9}{16}} \div \dfrac{1}{3}$

$= \dfrac{5}{2} \times \dfrac{2}{7} \times \dfrac{3}{4} \times \dfrac{3}{1} = \dfrac{\mathbf{45}}{\mathbf{28}}$

29. $-bc - ab = -(-2)(-5) - (-2)(-2)$

$= -10 - 4$

$= \mathbf{-14}$

30. $ab - ac = (-1)(-2) - (-1)(-1) = 2 - 1 = \mathbf{1}$

Problem Set 101

1. $\dfrac{H}{C} = \dfrac{2}{5}, \qquad \dfrac{H}{T} = \dfrac{2}{7}, \qquad \dfrac{C}{T} = \dfrac{5}{7}$

$\dfrac{2}{7} = \dfrac{H}{4900}$

$7H = 9800$

$H = \mathbf{1400\ hats}$

2. $-5N = 2N - 56$

$-7N = -56$

$N = \mathbf{8}$

3. $\dfrac{400}{\$200} = \dfrac{140}{C}$

$400C = \$28,000$

$C = \mathbf{\$70}$

4. $\text{Avg} = \dfrac{(60 \times \$40) + (40 \times \$150)}{60 + 40}$

$= \dfrac{\$8400}{100} = \mathbf{\$84}$

5. $x \not\le -2$

$x > -2$

6. $B = 130\% \cdot 80$

$B = 1.3 \cdot 80$

$B = \mathbf{104}$

7. $40\% \cdot 160 = B$

$0.4 \cdot 160 = B$

$64 = B$

$160 + 64 = \mathbf{224}$

8. $0.65 \cdot A = 78$

$\dfrac{0.65A}{0.65} = \dfrac{78}{0.65}$

$A = \mathbf{120\ discriminating\ gourmets}$

9. $\dfrac{40}{100} \times A = 240$

$A = 240 \cdot \dfrac{100}{40}$

$A = \mathbf{600}$

10. $V = \pi r^2 h$

$= \pi(2\text{ m})^2(1\text{ m})$

$= 4\pi\text{ m}^3$

$\approx 12.56\text{ m}^3 \left(\dfrac{100\text{ cm}}{1\text{ m}}\right)\left(\dfrac{100\text{ cm}}{1\text{ m}}\right)\left(\dfrac{100\text{ cm}}{1\text{ m}}\right)$

$\approx \mathbf{12,560,000\ cm^3}$

11. $P = (25 + 29 + 28 + 29 + 18 + 15)\text{ in.}$

$= \mathbf{144\ in.}$

12. $-2x + 5 + 4x + 8 = 4x - 2x + 5 + 8$

$= \mathbf{2x + 13}$

13. $y = -2x + 1$

x	y
$-\dfrac{1}{2}$	2
0	1
$\dfrac{1}{2}$	0

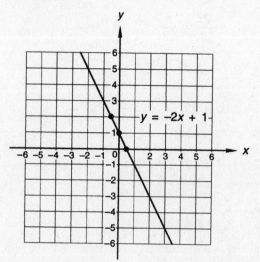

14. (a) obtuse, **130°**

(b) acute, **35°**

(c) obtuse, **100°**

15. $6x - 5 = 31$

$6x = 36$

$x = 6$

$2x - 3 = 2(6) - 3 = 12 - 3 = \mathbf{9}$

16.

$$-2x + 4 = 3x + 2$$
$$\underline{-2 -2}$$
$$-2x + 2 = 3x$$
$$\underline{+2x +2x}$$
$$2 = 5x$$
$$\frac{2}{5} = \frac{5x}{5}$$
$$\frac{2}{5} = x$$

17. $3x - 2 = -2x + 5 + 4x + 8$

$3x - 2 = 2x + 13$

$\underline{+2 +2}$

$3x = 2x + 15$

$\underline{-2x -2x}$

$x = + 15$

$x = \mathbf{15}$

18.

$$-3\frac{1}{2}x - \frac{1}{2} = 2\frac{1}{4}$$
$$-3\frac{1}{2}x - \frac{1}{2} + \frac{1}{2} = 2\frac{1}{4} + \frac{1}{2}$$
$$-\frac{7}{2}x = \frac{11}{4}$$
$$\left(-\frac{2}{7}\right)\left(-\frac{7}{2}\right)x = \left(-\frac{2}{7}\right)\left(\frac{11}{4}\right)$$
$$x = -\frac{11}{14}$$

19.

$$\frac{-\dfrac{2}{5}}{\dfrac{4}{25}} = \frac{-\dfrac{1}{2}}{x}$$
$$-\frac{2}{5}x = -\frac{2}{25}$$
$$\left(-\frac{5}{2}\right)\left(-\frac{2}{5}\right)x = \left(-\frac{2}{25}\right)\left(-\frac{5}{2}\right)$$
$$x = \frac{1}{5}$$

20. $-2^3 - (-2)^3 = -8 - (-8) = \mathbf{0}$

21. $-(-3)^2 = -(9) = \mathbf{-9}$

22. $(-3)^2 + (-2)^3 = 9 + (-8) = \mathbf{1}$

23. $x^3 \cdot x^2 \cdot y^{15} \cdot x^6 \cdot y^6 = x^3 x^2 x^6 y^{15} y^6 = \mathbf{x^{11}y^{21}}$

24. $5^{15} \cdot 5^{17} = \mathbf{5^{32}}$

25. $aa^3mm^3am^4a^2 = aa^3aa^2mm^3m^4 = \mathbf{a^7m^8}$

26. $xx^3x^2yy^5y^6xy^2 = xx^3x^2xyy^5y^6y^2 = \mathbf{x^7y^{14}}$

27. $\dfrac{2^3 - 2(3 - 1)(2^3 - 3)}{2^2(2^2 - 1)} = \dfrac{8 - 2(2)(5)}{4(3)}$

$= \dfrac{-12}{12} = \mathbf{-1}$

28. $\dfrac{3 - 2^2(2^2 - 1)(2^3 - 1)}{3^2(3^2 - 2^3)} = \dfrac{3 - 4(3)(7)}{9(1)}$

$= \dfrac{-81}{9} = \mathbf{-9}$

29. $3\frac{1}{2} \div \frac{1}{4} \times 2\frac{1}{3} \div \frac{1}{6}$

$= \frac{7}{2} \times \frac{4}{1} \times \frac{7}{3} \times \frac{6}{1} = \mathbf{196}$

30. $-ab + bc = -(-2)(-1) + (-1)(1)$

$= -2 - 1$

$= \mathbf{-3}$

PROBLEM SET 102

1. $\dfrac{J}{M} = \dfrac{2}{5}, \qquad \dfrac{J}{T} = \dfrac{2}{7}, \qquad \dfrac{M}{T} = \dfrac{5}{7}$

$\dfrac{5}{7} = \dfrac{M}{350}$

$7M = 1750$

$M = \mathbf{250 \text{ marbles}}$

2. $\dfrac{F}{D} = \dfrac{2}{9}, \qquad \dfrac{F}{T} = \dfrac{2}{11}, \qquad \dfrac{D}{T} = \dfrac{9}{11}$

$\dfrac{2}{11} = \dfrac{F}{1210}$

$11F = 2420$

$F = \mathbf{220 \text{ acres}}$

3. $2N = -3N - 20$

$5N = -20$

$N = \mathbf{-4}$

4. $2\frac{3}{4} \times 140{,}000 = S$

$\frac{11}{4} \times 140{,}000 = S$

$S = \mathbf{385{,}000 \text{ sprouts}}$

5. $x \geq 1$

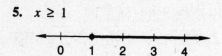

6. $B = 75\% \cdot 220$
   $B = 0.75 \cdot 220$
   $B = \mathbf{165}$

7. $60\% \cdot 150 = B$
   $0.6 \cdot 150 = B$
   $90 = B$

   $150 + 90 = \mathbf{240}$

8. Math test: English test:
   $P \cdot 20 = 17$ $P \cdot 25 = 21$

   $P = \dfrac{17}{20}$ $P = \dfrac{21}{25}$

   $P = 0.85$ $P = 0.84$

   $P = \mathbf{85\%}$ $P = \mathbf{84\%}$

The score on the **math test** was better.

9. $14^2 = 196$
   $15^2 = 225$

   So, $\mathbf{14 < \sqrt{211} < 15}$

10.

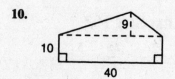

$A_{\text{Figure}} = A_{\text{Rectangle}} + A_{\text{Triangle}}$

$= (10)(40) \text{ ft}^2 + \dfrac{(40)(9)}{2} \text{ ft}^2$

$= 400 \text{ ft}^2 + 180 \text{ ft}^2$

$= \mathbf{580 \text{ ft}^2}$

11. $12 = 2 \cdot 2 \cdot 3$
   $30 = 2 \cdot 3 \cdot 5$
   $36 = 2 \cdot 2 \cdot 3 \cdot 3$

   $\text{LCM } (12, 30, 36) = 2 \cdot 2 \cdot 3 \cdot 3 \cdot 5 = \mathbf{180}$

12. $c + c = 90°$ $s + s = 180°$
   $2c = 90°$ $2s = 180°$

   $c = \dfrac{90°}{2}$ $s = \dfrac{180°}{2}$

   $c = 45°$ $s = 90°$

The two complementary angles measure **45°**, and the two supplementary angles measure **90°**.

13. $x = \dfrac{y}{3} - 3$

x	y
-4	-3
-3	0
-2	3

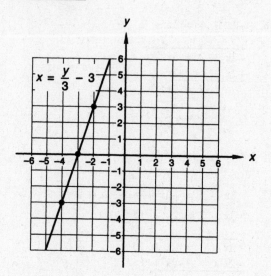

14. $2x - 5 = 5x - 6$
   $\underline{+6 \qquad\quad +6}$
   $2x + 1 = 5x$
   $\underline{-2x \qquad\quad -2x}$
   $+ 1 = 3x$

   $\dfrac{1}{3} = \dfrac{3x}{3}$

   $\dfrac{1}{3} = x$

15. $6x - 2 + 2x - 5 = 2x + 4 + 12 + 3x$
   $8x - 7 = 5x + 16$
   $\underline{+7 \qquad\qquad +7}$
   $8x = 5x + 23$
   $\underline{-5x \qquad\quad -5x}$
   $3x = 23$

   $\dfrac{3x}{3} = \dfrac{23}{3}$

   $x = \dfrac{23}{3}$

16.
$$-4\frac{1}{3}x - \frac{1}{4} = 2\frac{1}{3}$$
$$-4\frac{1}{3}x - \frac{1}{4} + \frac{1}{4} = 2\frac{1}{3} + \frac{1}{4}$$
$$-\frac{13}{3}x = \frac{31}{12}$$
$$\left(-\frac{3}{13}\right)\left(-\frac{13}{3}\right)x = \left(-\frac{3}{13}\right)\left(\frac{31}{12}\right)$$
$$x = -\frac{31}{52}$$

17.
$$\frac{-\frac{1}{4}}{\frac{1}{2}} = \frac{\frac{1}{3}}{x}$$
$$-\frac{1}{4}x = \frac{1}{6}$$
$$\left(-\frac{4}{1}\right)\left(-\frac{1}{4}\right)x = \left(\frac{1}{6}\right)\left(-\frac{4}{1}\right)$$
$$x = -\frac{2}{3}$$

18. $6x + 8 + 6x + 5 + 2xy + 9yx + 2$
$= \mathbf{12x + 11xy + 15}$

19. $3 + 2ab^2 + 3a + 2abb - 4a$
$= 3 + 2ab^2 + 3a + 2ab^2 - 4a$
$= \mathbf{4ab^2 - a + 3}$

20. $6x + 2y + 3xy + 2 + 3x + 4y - xy$
$= \mathbf{2xy + 9x + 6y + 2}$

21. $-2^3 - (-3)^2 = -8 - 9 = \mathbf{-17}$

22. $-(-3)^2 - [-(-1)] = -9 - 1 = \mathbf{-10}$

23. $2^3 - 2[3(2 - 3) + 2(2 - 4)]$
$= 8 - 2[3(-1) + 2(-2)]$
$= 8 - 2[-7] = 8 + 14 = \mathbf{22}$

24. $aa^3ba^2a^4b^3b^2 = \mathbf{a^{10}b^6}$

25. $xxyyx^2y^3x^3y^4 = \mathbf{x^7y^9}$

26. $ababa^2b^2a^2b^2a^3b^3 = \mathbf{a^9b^9}$

27. $\dfrac{2^4 - 3(2^2 - 1)(3^2 - 2^3)}{2^3(2^2 - 1)}$
$= \dfrac{16 - 3(3)(1)}{8(3)} = \mathbf{\dfrac{7}{24}}$

28. $\dfrac{2^3 - 3^2(2^2 - 1)(3 - 2^2)}{2(3^2 - 2^3)}$
$= \dfrac{8 - 9(3)(-1)}{2(1)} = \mathbf{\dfrac{35}{2}}$

29. $3\frac{1}{3} \div \frac{1}{4} \times 2\frac{1}{3} \div \frac{1}{3}$
$= \frac{10}{3} \times \frac{4}{1} \times \frac{7}{3} \times \frac{3}{1} = \mathbf{\dfrac{280}{3}}$

30. $p^x + \frac{1}{3}p + \sqrt[x]{p} = (9^2) + \frac{1}{3}(9) + (\sqrt{9})$
$= 81 + 3 + 3$
$= \mathbf{87}$

PROBLEM SET 103

1. $\dfrac{D}{I} = \dfrac{3}{7}, \qquad \dfrac{D}{T} = \dfrac{3}{10}, \qquad \dfrac{I}{T} = \dfrac{7}{10}$
$$\dfrac{3}{10} = \dfrac{D}{140}$$
$$10D = 420$$
$$D = \mathbf{42 \ players}$$

2. $\dfrac{E}{D} = \dfrac{2}{9}, \qquad \dfrac{E}{T} = \dfrac{2}{11}, \qquad \dfrac{D}{T} = \dfrac{9}{11}$
$$\dfrac{2}{11} = \dfrac{420}{T}$$
$$2T = 4620$$
$$T = \mathbf{2310 \ students}$$

3. Initial Speed $= \dfrac{24 \text{ mi}}{6 \text{ hr}} = \dfrac{4 \text{ mi}}{1 \text{ hr}}$
Reduced speed $= 4 \text{ mph} - 1 \text{ mph} = 3 \text{ mph}$
$$\dfrac{1 \text{ hr}}{3 \text{ mi}} \times 18 \text{ mi} = \mathbf{6 \ hr}$$

4. $4N = -2N + 36$
$6N = 36$
$N = \mathbf{6}$

5. $0.56 \cdot 21{,}000 = T$
$T = \mathbf{11{,}760 \ Trojans}$

6. $80\% \cdot 80 = B$
$0.8 \cdot 80 = B$
$64 = B$

$80 + 64 = \mathbf{144}$

7. $P \times 60 = 93$

$$P = \frac{93}{60}$$

$$P = 1.55$$

$$P = \mathbf{155\%}$$

8. $20 = 2 \cdot 2 \cdot 5$

$25 = 5 \cdot 5$

$30 = 2 \cdot 3 \cdot 5$

LCM $(20, 25, 30) = 2 \cdot 2 \cdot 3 \cdot 5 \cdot 5 = \mathbf{300}$

9. $S.A. = 2A_{End} + 2A_{Side} + A_{Bottom}$

$$= \left[2\left(\frac{(12)(8)}{2}\right) + 2(10)(20) + (20)(12) \right] cm^2$$

$$= [96 + 400 + 240] \, cm^2$$

$$= \mathbf{736 \ cm^2}$$

10. $5^3 = 125$

$6^3 = 216$

So, $\mathbf{5 < \sqrt[3]{155} < 6}$

11. $5x(x^2 - 3 + 3x) = \mathbf{5x^3 + 15x^2 - 15x}$

12. $2ab(a + b) = \mathbf{2a^2b + 2ab^2}$

13. $\quad y = x + y$

$y - y = x$

$\quad\quad 0 = x$

Regardless of the value of y, x will always be zero.

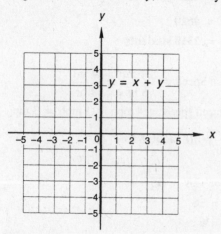

Note: The line $x = 0$ represents the y-axis. Though it is difficult to notice in the graph above, if you look carefully you will see that we have made the y-axis thicker than usual in an attempt to indicate that it is the line we are graphing.

14. $3x + 5 + 3y + 2xy + 6x + 5xy$

$= \mathbf{7xy + 9x + 3y + 5}$

15. $a^2 + 5a^2 + 3b^2 + 2a^2 - 5b^2 = \mathbf{8a^2 - 2b^2}$

16. $-4yxx + 5xyx + 7x - 6y + 5x$

$= (-4x^2y + 5x^2y) + (7x + 5x) - 6y$

$= \mathbf{x^2y + 12x - 6y}$

17. $2x - 6 - 4 = 3x - 4 + 5x$

$2x - 10 = 8x - 4$

$\underline{\quad +4 \quad\quad\quad +4}$

$2x - 6 = 8x$

$\underline{-2x \quad\quad\quad -2x}$

$-6 = 6x$

$\dfrac{-6}{6} = \dfrac{6x}{6}$

$\mathbf{-1} = x$

18. $6x - 10 = -3 + 2x + 3(5 - 2x)$

$6x - 10 = -3 + 2x + 15 - 6x$

$6x - 10 = 12 - 4x$

$\underline{+4x \quad\quad\quad\quad +4x}$

$10x - 10 = 12$

$\underline{\quad\quad +10 \ +10}$

$10x \quad\quad = 22$

$\dfrac{10x}{10} = \dfrac{22}{10}$

$x = \dfrac{\mathbf{11}}{\mathbf{5}}$

19. $-2x + 6 = -4x + 6$

$-2x = -4x$

$2x = 0$

$x = \mathbf{0}$

20. $\dfrac{-\dfrac{2}{3}}{\dfrac{1}{4}} = \dfrac{\dfrac{1}{6}}{x}$

$-\dfrac{2}{3}x = \dfrac{1}{24}$

$\left(-\dfrac{3}{2}\right)\left(-\dfrac{2}{3}\right)x = \left(\dfrac{1}{24}\right)\left(-\dfrac{3}{2}\right)$

$x = -\dfrac{\mathbf{1}}{\mathbf{16}}$

21. $-3^2 - (-2)^3 = -9 + 8 = \mathbf{-1}$

22. $-[-(-3)] + (-2)^3 = -3 - 8 = \mathbf{-11}$

23. $2^3 - 2^4[2(3 - 5) + 2^2(3^2 - 2^2)]$

$= 8 - 16[2(-2) + 4(5)] = 8 - 16[16]$

$= 8 - 256 = \mathbf{-248}$

24. $aa^2ba^3b^2a^2b^3 = \mathbf{a^8b^6}$

25. $xy^2x^2x^3y^3y^2 = \mathbf{x^6y^7}$

26. $m^2nn^3m^3n^2n^3 = m^5n^9$

27. $\dfrac{2^3 - 2(1 - 2^2)(3 - 2^2)}{2(2^3 - 3^2)}$

$= \dfrac{8 - 2(-3)(-1)}{2(-1)} = \dfrac{2}{-2} = \mathbf{-1}$

28. $\dfrac{1 - 3(2^2 - 3) + (3^3 - 2 \cdot 3)}{6(2 \cdot 3 - 2)}$

$= \dfrac{1 - 3(1) + 21}{6(4)} = \dfrac{\mathbf{19}}{\mathbf{24}}$

29. $\dfrac{1}{2} + \dfrac{1}{3} + \dfrac{1}{4} + \dfrac{1}{5} + \dfrac{1}{6}$

$= \dfrac{30}{60} + \dfrac{20}{60} + \dfrac{15}{60} + \dfrac{12}{60} + \dfrac{10}{60}$

$= \dfrac{87}{60} = 1\dfrac{27}{60} = \mathbf{1\dfrac{9}{20}}$

30. (a) **90°, right**

(b) **30°, acute**

(c) **15°, acute**

PROBLEM SET 104

1. If $\frac{7}{12}$ mispronounced *chic*, $\frac{5}{12}$ pronounced *chic* correctly.

$$\dfrac{5}{12} \cdot A = 1440$$

$$\dfrac{12}{5} \cdot \dfrac{5}{12} \cdot A = 1440 \cdot \dfrac{12}{5}$$

$$A = 3456 \text{ students in all}$$

$M = 3456 - 1440$

$= \mathbf{2016\ students}$ mispronounced *chic*

2. $\dfrac{P}{N} = \dfrac{12}{5}, \qquad \dfrac{P}{T} = \dfrac{12}{17}, \qquad \dfrac{N}{T} = \dfrac{5}{17}$

$\dfrac{12}{17} = \dfrac{P}{170}$

$17P = 2040$

$P = \mathbf{120}$ paid the piper

3. $\dfrac{S}{N} = \dfrac{14}{1}, \qquad \dfrac{S}{T} = \dfrac{14}{15}, \qquad \dfrac{N}{T} = \dfrac{1}{15}$

$\dfrac{14}{15} = \dfrac{S}{750}$

$15S = 10{,}500$

$S = \mathbf{700\ sycophants}$

4. $[500 + 3(500)]\text{ m} = \mathbf{2000\ m}$

5. $3N = 5N - 12$

$-2N = -12$

$N = \mathbf{6}$

6. $B = 170\% \cdot 120$

$B = 1.7 \cdot 120$

$B = \mathbf{204}$

7. $B = 60\% \cdot 80$

$B = 0.6 \cdot 80$

$B = 48$

$80 + 48 = \mathbf{128}$

8. If 60% were in the club, then 40% were not in the club.

$0.40 \cdot 30 = N$

$N = \mathbf{12\ students}$

9. $V = A_{\text{Base}} \times \text{height}$

$= \pi(2)^2 \text{ cm}^2 \times 100 \text{ cm}$

$= 400\pi \text{ cm}^3$

$\approx \mathbf{1256\ cm^3}$

10. $\text{S.A.} = 2A_{\text{Base}} + (\text{Perimeter} \times \text{height})$

$= 2\left[(20)(40) + \dfrac{(40)(10)}{2}\right] \text{cm}^2$

$+ (40 + 20 + 41.23 + 30)(100) \text{ cm}^2$

$= 2[1000]\text{cm}^2 + 13{,}123 \text{ cm}^2$

$= \mathbf{15{,}123\ cm^2}$

11.

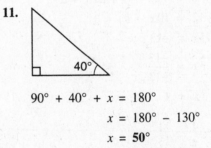

$90° + 40° + x = 180°$

$x = 180° - 130°$

$x = \mathbf{50°}$

12. The measure of $\angle B = 180° - 179\dfrac{1}{2}° = \dfrac{\mathbf{1}}{\mathbf{2}}\mathbf{°}$

13. The measure of $\angle A = 180° - (59° + 59°)$

$= 180° - 118° = \mathbf{62°}$

Since two sides have the same length, the triangle is **isosceles.** Since all three angles are acute, the triangle is **acute.**

14. Since each side has a different length, $\triangle XYZ$ is **scalene**. Since $\angle Z$ is greater than 90°, $\triangle XYZ$ is **obtuse**.

15. $2bc(a + b + c) = \mathbf{2abc + 2b^2c + 2bc^2}$

16. $ab(a + b + ab) = \mathbf{a^2b + ab^2 + a^2b^2}$

17. $a^3 + 2abb + 3aa^2 + 2ab^2$
$= a^3 + 2ab^2 + 3a^3 + 2ab^2 = \mathbf{4a^3 + 4ab^2}$

18. $yx^2 + xyx + 3y^2x + 2yxy$
$= yx^2 + x^2y + 3y^2x + 2y^2x = \mathbf{2x^2y + 5xy^2}$

19. $4x - 3 = -x + 6$
$4x = -x + 9$
$5x = 9$
$x = \dfrac{9}{5}$

20. $-5x - 2 = -x + 7 + 4(2x - 1)$
$-5x - 2 = -x + 7 + 8x - 4$
$-5x - 2 = 7x + 3$

$$\begin{array}{rcl} +2 & & +2 \\ \hline -5x & = & 7x + 5 \\ -7x & & -7x \\ \hline -12x & = & + 5 \end{array}$$

$\dfrac{-12x}{-12} = \dfrac{5}{-12}$

$x = -\dfrac{5}{12}$

21. $7 - 3x + 5 = -2x - 3 + 12x + 2$
$-3x + 12 = 10x - 1$
$-13x = -13$
$x = \dfrac{-13}{-13}$
$x = 1$

22.
$\dfrac{-\dfrac{1}{3}}{2\dfrac{1}{3}} = \dfrac{\dfrac{1}{2}}{x}$

$\dfrac{-\dfrac{1}{3}}{\dfrac{7}{3}} = \dfrac{\dfrac{1}{2}}{x}$

$-\dfrac{1}{3}x = \dfrac{7}{6}$

$\left(-\dfrac{3}{1}\right)\left(-\dfrac{1}{3}\right)x = \left(\dfrac{7}{6}\right)\left(-\dfrac{3}{1}\right)$

$x = -\dfrac{7}{2}$

23. $-[-(-6)] + (-2)(-3) = -6 + 6 = \mathbf{0}$

24. $2^3 - 2^2\left[2(3 - 2^2) + 2^2(2 \cdot 3 - 2^2)\right]$
$= 8 - 4[2(-1) + 4(2)] = 8 - 4[6] = \mathbf{-16}$

25. $aba^2b^2a^3 = \mathbf{a^6b^3}$

26. $xy^3x^2yy^2x = \mathbf{x^4y^6}$

27. $\dfrac{2^2 - 2^3(3^2 - 2^2) + 2^3}{6(5 - 2^2)}$

$= \dfrac{4 - 8(5) + 8}{6(1)} = \dfrac{-28}{6} = \mathbf{-\dfrac{14}{3}}$

28. $\dfrac{1}{2} - 0.3 = 0.5 - 0.3 = \mathbf{0.2}$

or

$\dfrac{1}{2} - 0.3 = \dfrac{1}{2} - \dfrac{3}{10} = \dfrac{5}{10} - \dfrac{3}{10} = \dfrac{2}{10}$

$= \mathbf{\dfrac{1}{5}}$

29. $ab^2 + \dfrac{b}{4} = (-2)(8)^2 + \dfrac{(8)}{4}$
$= (-2)(64) + 2$
$= -128 + 2$
$= \mathbf{-126}$

30. $a + b - ab = (-1) + (-2) - (-1)(-2)$
$= -1 - 2 - 2$
$= \mathbf{-5}$

PROBLEM SET 105

1. If $\frac{6}{13}$ could discriminate, $\frac{7}{13}$ could not discriminate.

$\dfrac{7}{13} \times A = 28$

$\dfrac{13}{7} \cdot \dfrac{7}{13} \cdot A = 28 \cdot \dfrac{13}{7}$

$A = 52$ audiophiles total

$52 - 28 = \mathbf{24\ audiophiles}$ could discriminate

2. If $\frac{6}{7}$ were proclaiming, $\frac{1}{7}$ were not proclaiming.

$\dfrac{1}{7} \times A = 1100$

$\dfrac{7}{1} \cdot \dfrac{1}{7} \cdot A = 1100 \cdot \dfrac{7}{1}$

$A = \mathbf{7700\ delegates}$

3. $\dfrac{C}{A} = \dfrac{2}{17}$, $\qquad \dfrac{C}{T} = \dfrac{2}{19}$, $\qquad \dfrac{A}{T} = \dfrac{17}{19}$

$$\dfrac{2}{19} = \dfrac{C}{7600}$$

$$19C = 15{,}200$$

$$C = \textbf{800 contumacious}$$

$$A = 7600 - 800 = \textbf{6800 affable}$$

4. If $\dfrac{1}{4}$ had white eyes, $\dfrac{3}{4}$ had red eyes.

$$\dfrac{3}{4} \cdot 848 = F$$

$$F = \textbf{636 flies}$$

5. $7N = -20N - 9$

$27N = -9$

$N = -\dfrac{1}{3}$

6. Original speed $= \dfrac{16 \text{ mi}}{2 \text{ hr}} = 8\dfrac{\text{mi}}{\text{hr}}$

Reduced speed $= \dfrac{1}{2} \times 8\dfrac{\text{mi}}{\text{hr}} = 4\dfrac{\text{mi}}{\text{hr}}$

$\dfrac{1 \text{ hr}}{4 \text{ mi}} \times 20 \text{ mi} = \textbf{5 hr}$

7. $20\% \cdot A = 28$

$0.20 \cdot A = 28$

$A = \dfrac{28}{0.2}$

$A = 140$ total

$5\% \cdot 140 = B$

$0.05 \cdot 140 = B$

$B = \textbf{7}$ chose Bubble Gum Delight

8. $80\% \cdot 90 = B$

$0.8 \cdot 90 = B$

$72 = B$

$90 + 72 = \textbf{162}$

9. $P \times 60 = 78$

$P = \dfrac{78}{60}$

$P = 1.3$

$P = \textbf{130\%}$

10. $V = A_{\text{Base}} \times \text{height}$

$= \left[(4)(4) + \dfrac{(8)(6)}{2} + \dfrac{\pi(5)^2}{2} \right](2) \text{ m}^3$

$= [16 + 24 + 12.5\pi](2) \text{ m}^3$

$\approx \textbf{158.5 m}^3$

11. $L.A. = \text{Perimeter} \times \text{height}$

$= [4 + 4 + 2 + \pi(5) + 12](2) \text{ m}^2$

$= [22 + 5\pi](2) \text{ m}^2$

$\approx \textbf{75.4 m}^2$

12. $3^4 = 81$

$4^4 = 256$

So, $3 < \sqrt[4]{95} < 4$

13. Since it has one right angle, $\triangle ABC$ is a **right** triangle. Since 2 sides have equal lengths, $\triangle ABC$ is an **isosceles** triangle.

Since $\angle C$ measures $90°$, the sum of the measures of $\angle A$ and $\angle B$ must be $180° - 90° = 90°$. Also, $\angle A$ and $\angle B$ must have equal measures, since the sides opposite them have equal lengths. Thus, the measure of $\angle A$ is $\frac{90°}{2} = \textbf{45}°$, and the measure of $\angle B$ is also $\textbf{45}°$.

14. $3px(p + x + 2px) = \textbf{6}\boldsymbol{p^2x^2} + \textbf{3}\boldsymbol{p^2x} + \textbf{3}\boldsymbol{px^2}$

15. $mn^2(m + mny + 3m^2)$

$= \textbf{3}\boldsymbol{m^3n^2} + \boldsymbol{m^2n^3y} + \boldsymbol{m^2n^2}$

16. $m^2n^3 + 3mn^2nn - mnmn^2 + 2n^3mm$

$= m^2n^3 + 3m^2n^3 - m^2n^3 + 2m^2n^3 = \textbf{5}\boldsymbol{m^2n^3}$

17. $ap^3 + pap^2 - 5p^2ap = ap^3 + ap^3 - 5ap^3$

$= \boldsymbol{-3ap^3}$

18. $2x + 5 = 85$

$2x = 80$

$x = \dfrac{80}{2}$

$x = 40$

$3x \div (10 - x) = 3(40) \div (10 - (40))$

$= 120 \div (-30) = \textbf{-4}$

19. $5x - 4 = -3x + 20 + 2(x - 1)$
    $5x - 4 = -3x + 20 + 2x - 2$
    $5x - 4 = -x + 18$
    $6x = 22$
    $x = \dfrac{22}{6}$
    $x = \dfrac{11}{3}$

20. $3 - 4x - 3 = -10x + 7 + 2x + 5(1 - x)$
    $-4x = -8x + 7 + 5 - 5x$
    $-4x = -13x + 12$
    $9x = 12$
    $x = \dfrac{12}{9}$
    $x = \dfrac{4}{3}$

21. $\dfrac{-\frac{2}{5}}{2\frac{1}{2}} = \dfrac{-\frac{11}{13}}{x}$

    $\dfrac{-\frac{2}{5}}{\frac{5}{2}} = \dfrac{-\frac{11}{13}}{x}$

    $-\dfrac{2}{5}x = -\dfrac{55}{26}$

    $\left(-\dfrac{5}{2}\right)\left(-\dfrac{2}{5}\right)x = \left(-\dfrac{55}{26}\right)\left(-\dfrac{5}{2}\right)$

    $x = \dfrac{275}{52}$

22. $-[-(-6)^2] + (-3)(-4) = 36 + 12 = \mathbf{48}$

23. $a^3 p^2 a p^2 a^4 aa = a^{10} p^4$

24. $y^8 z^2 yzyyz^4 = y^{11} z^7$

25. $2\frac{1}{4} - 0.315 = 2.25 - 0.315 = \mathbf{1.935}$

26. $m^2 + \dfrac{n^2}{m} = (-8)^2 + \dfrac{(4)^2}{(-8)} = 64 - 2 = \mathbf{62}$

27. $p^3 + c^3 = (-3)^3 + (-1)^3 = -27 - 1 = \mathbf{-28}$

28. $\dfrac{a^2}{b^3} - a = \dfrac{(4)^2}{(-2)^3} - (4)$
    $= \dfrac{16}{-8} - 4$
    $= -2 - 4$
    $= \mathbf{-6}$

29. $p^q + \sqrt[q]{p} = 4^2 + \sqrt{4} = 16 + 2 = \mathbf{18}$

30. $y = \dfrac{1}{2}x - 1$

x	y
-2	-2
0	-1
2	0

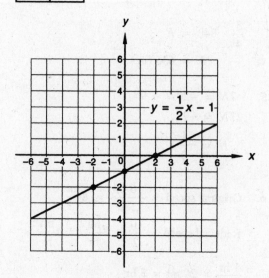

PROBLEM SET 106

1. $5N = -2N + 28$
   $7N = 28$
   $N = \mathbf{4}$

2. $\dfrac{A}{H} = \dfrac{4}{7}, \quad \dfrac{A}{T} = \dfrac{4}{11}, \quad \dfrac{H}{T} = \dfrac{7}{11}$

   $\dfrac{4}{11} = \dfrac{A}{4400}$

   $11A = 17{,}600$

   $A = \mathbf{1600\ ascetics}$

3. $2\dfrac{3}{4} \cdot A = 2200$

   $\dfrac{4}{11} \cdot \dfrac{11}{4} \cdot A = 2200 \cdot \dfrac{4}{11}$

   $A = \mathbf{800\ customers}$

4. $\dfrac{50}{\$3150} = \dfrac{78}{D}$

   $50D = \$245{,}700$

   $D = \mathbf{\$4914}$

5. Initial speed $= \dfrac{1600 \text{ mi}}{4 \text{ hr}} = \dfrac{400 \text{ mi}}{1 \text{ hr}}$

Reduced speed $= \dfrac{400 \text{ mi}}{1 \text{ hr}} - \dfrac{45 \text{ mi}}{1 \text{ hr}} = \dfrac{355 \text{ mi}}{1 \text{ hr}}$

$\dfrac{1 \text{ hr}}{355 \text{ mi}} \times 2485 \text{ mi} = \mathbf{7 \text{ hr}}$

6. $x \geq -3$

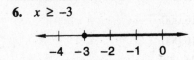

7. If 22% voted for Josh, 78% voted for Moe.

$\dfrac{78}{100} \times 900 = B$

$B = \mathbf{702 \text{ students}}$ voted for Moe.

8. $90\% \cdot 350 = B$

$0.9 \cdot 350 = B$

$315 = B$

$350 + 315 = \mathbf{665}$

9. $P \cdot 75 = 135$

$P = \dfrac{135}{75}$

$P = 1.8$

$P = \mathbf{180\%}$

10.

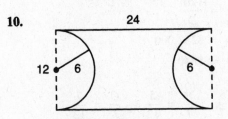

$V = A_{\text{Base}} \times \text{height}$

$= \left[A_{\text{Rectangle}} - 2A_{\text{Semicircle}} \right] \times \text{height}$

$= \left[(24)(12) - 2\left(\dfrac{\pi(6)^2}{2} \right) \right](1)(3) \text{ ft}^3$

$= [288 - 36\pi](3) \text{ ft}^3$

$\approx [174.96](3) \text{ ft}^3$

$\approx \mathbf{524.88 \text{ ft}^3}$

11. $S.A. = 2A_{\text{Base}} + (\text{Perimeter} \times \text{height})$

$\approx 2[174.96] \text{ ft}^2 + [2(24) + 2\pi(6)](3) \text{ ft}^2$

$\approx 349.92 \text{ ft}^2 + 257.04 \text{ ft}^2$

$\approx \mathbf{606.96 \text{ ft}^2}$

12. Mean $= \left(\dfrac{5}{4} + \dfrac{2}{5} + \dfrac{3}{10} \right) \div 3$

$= \left(\dfrac{25}{20} + \dfrac{8}{20} + \dfrac{6}{20} \right) \cdot \dfrac{1}{3}$

$= \dfrac{39}{20} \cdot \dfrac{1}{3}$

$= \dfrac{13}{20}$

13. $x = \dfrac{y}{2^{-2}} = (2^2)y = 4y$

x	y
-4	-1
0	0
4	1

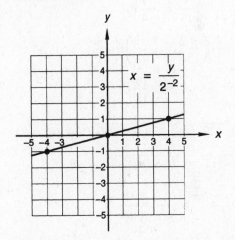

14. Since $\angle ACB$ is the supplement of $110°$, it measures $180° - 110° = 70°$. The measure of $\angle B$ must also equal $70°$, since the sides opposite $\angle B$ and $\angle C$ have equal lengths. Thus, $\angle A$ measures $180° - 70° - 70° = \mathbf{40°}$.

15. $4ax(3a + 3z + 9ax)$

$= \mathbf{36a^2x^2 + 12a^2x + 12axz}$

16. $m^3 + \dfrac{x}{y} - x^2 = (-3)^3 + \dfrac{(-4)}{(-2)} - (-4)^2$

$= -27 + 2 - 16$

$= \mathbf{-41}$

17. $a^3 + b^3 + c^3 = (-1)^3 + (-2)^3 + (-3)^3$

$= -1 - 8 - 27$

$= \mathbf{-36}$

18. $10^a + 10^b + ba = 10^0 + 10^3 + (3)(0)$
$$= 1 + 1000 + 0$$
$$= \mathbf{1001}$$

19. $3m^2 + 2c^2 - 4cm + 6c^2 - 9m^2$
$$= \mathbf{-6m^2 + 8c^2 - 4cm}$$

20.
$$-7x + 4 = -x - 8 + 5(2 - 3x)$$
$$-7x + 4 = -x - 8 + 10 - 15x$$
$$-7x + x + 15x = -4 - 8 + 10$$
$$9x = -2$$
$$x = \mathbf{-\frac{2}{9}}$$

21.
$$\frac{\frac{2}{3}}{-\frac{1}{4}} = \frac{x}{\frac{2}{5}}$$
$$\frac{4}{15} = -\frac{1}{4}x$$
$$\left(-\frac{4}{1}\right)\left(\frac{4}{15}\right) = \left(-\frac{1}{4}\right)x\left(-\frac{4}{1}\right)$$
$$x = \mathbf{-\frac{16}{15}}$$

22. $\sqrt[7]{-128} = \sqrt[7]{(-2)(-2)(-2)(-2)(-2)(-2)(-2)} = \mathbf{-2}$

23. $\sqrt[3]{-512} = \sqrt[3]{(-8)(-8)(-8)} = \mathbf{-8}$

24. (a) $4^{-3} = \frac{1}{4^3} = \mathbf{\frac{1}{64}}$

(b) $\frac{1}{5^{-3}} = 5^3 = \mathbf{125}$

25. $-5^2 - (-3)^3 + \sqrt[3]{-64} = -25 + 27 - 4 = \mathbf{-2}$

26. $-[-(-3)^2] - (4)^2 - \sqrt[3]{-216} = 9 - 16 + 6$
$$= \mathbf{-1}$$

27. $a^3m^3aa^2m^2a = \mathbf{a^7m^5}$

28. $\frac{4^3 - 2^2(1^5 - 3^2)(3 - 2^2)}{2^2(2^3 - 3^2)}$
$$= \frac{64 - 4(-8)(-1)}{4(-1)} = \frac{32}{-4} = \mathbf{-8}$$

29. $5^{-2} - 0.04 = \frac{1}{5^2} - 0.04 = \frac{1}{25} - 0.04$
$$= 0.04 - 0.04 = \mathbf{0}$$

30. $3^{-3} + \frac{1}{9} = \frac{1}{3^3} + \frac{1}{9} = \frac{1}{27} + \frac{1}{9}$
$$= \frac{1}{27} + \frac{3}{27} = \mathbf{\frac{4}{27}}$$

PROBLEM SET 107

1. If 60% were iconoclasts, 40% were traditionalists.
$$40\% \cdot A = 16,000$$
$$0.4 \cdot A = 16,000$$
$$\frac{0.4A}{0.4} = \frac{16,000}{0.4}$$
$$A = 40,000$$
$$40,000 - 16,000 = \mathbf{24,000 \text{ iconoclasts}}$$

2. $360\% \cdot 22,000 = B$
$$3.6(22,000) = B$$
$$79,200 = B$$
$$22,000 + 79,200 = \mathbf{101,200 \text{ flowers}}$$

3. $2\frac{2}{13} \times 39,000 = B$
$$\frac{28}{13} \times 39,000 = B$$
$$B = \mathbf{84,000 \text{ people}}$$

4. $\frac{300}{\$2100} = \frac{120}{D}$
$$300D = \$252,000$$
$$D = \mathbf{\$840}$$

5. $\frac{A}{O} = \frac{5}{7}, \quad \frac{A}{T} = \frac{5}{12}, \quad \frac{O}{T} = \frac{7}{12}$
$$\frac{5}{12} = \frac{A}{4800}$$
$$12A = 24,000$$
$$A = \mathbf{2000 \text{ advocates}}$$
$$O = 4800 - 2000 = \mathbf{2800 \text{ opponents}}$$

6. $6N = -3N + 45$
$$9N = 45$$
$$N = \mathbf{5}$$

7. $x > 0$

8. $84\% \times 1600 = B$

$\dfrac{84}{100} \times 1600 = B$

$\mathbf{1344 = B}$

9. $35\% \cdot 900 = B$

$0.35(900) = B$

$315 = B$

$900 + 315 = \mathbf{1215}$

10. $P \cdot 89 = 445$

$\dfrac{P \cdot 89}{89} = \dfrac{445}{89}$

$P = 5$

$P = \mathbf{500\%}$

11.

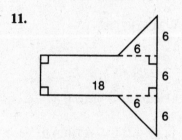

$V = A_{\text{Base}} \times \text{height}$

$= [A_{\text{Rectangle}} + 2A_{\text{Triangle}}](\text{height})$

$= \left[(6)(18) + 2\left(\dfrac{(6)(6)}{2}\right)\right](2) \text{ m}^3$

$= [108 + 36](2) \text{ m}^3$

$= (288)(100)(100)(100) \text{ cm}^3$

$= \mathbf{288{,}000{,}000 \text{ cm}^3}$

12. $L.A. = \text{Perimeter} \times \text{height}$

$= [2(6 + 12 + 8.49 + 6)](2) \text{ m}^2$

$= [64.98](2) \text{ m}^2$

$= (129.96)(100)(100) \text{ cm}^2$

$= \mathbf{1{,}299{,}600 \text{ cm}^2}$

13. $17 = \mathbf{XVII}$

14. $939 = \mathbf{CMXXXIX}$

15. $\text{VI} = 5 + 1 = \mathbf{6}$

16. $\text{CDXLIV} = 400 + 40 + 4 = \mathbf{444}$

17. $13.05 = 13\dfrac{5}{100} = \mathbf{13\dfrac{1}{20}}$

18. Since $\angle B$ is the complement of $42°$, it must measure $90° - 42° = 48°$. Three-fourths of the measure of $\angle B$, then, is $\frac{3}{4} \cdot 48° = \mathbf{36°}$.

19. $5c(3m + 2p + c) = \mathbf{5c^2 + 15cm + 10cp}$

20. $7mp(m + p + x + 2mp)$

$= \mathbf{14m^2p^2 + 7mpx + 7m^2p + 7mp^2}$

21. $m^2 + c^3 + 3mm + c^2c$

$= m^2 + c^3 + 3m^2 + c^3 = \mathbf{4m^2 + 2c^3}$

22. $p^2 + 3aam - 2p^2p + a^2m$

$= p^2 + 3a^2m - 2p^3 + a^2m$

$= \mathbf{4a^2m + p^2 - 2p^3}$

23. $(-5)^2 - 5^2 - (-5)^4 + (-5)^4$

$= 25 - 25 - 625 + 625 = \mathbf{0}$

24. $-4x - 3 = -11x + 16$

$7x = 19$

$x = \mathbf{\dfrac{19}{7}}$

25. $-7x - 4 = 3x + 3$

$-10x = 7$

$x = \mathbf{-\dfrac{7}{10}}$

26. $\dfrac{x}{\dfrac{1}{2}} = \dfrac{2\dfrac{1}{4}}{\dfrac{3}{8}}$

$\dfrac{3}{8}x = \dfrac{1}{2} \cdot \dfrac{9}{4}$

$\dfrac{8}{3} \cdot \dfrac{3}{8} \cdot x = \dfrac{9}{8} \cdot \dfrac{8}{3}$

$x = \mathbf{3}$

27. $-[-(-4)] + (2)^2 - 3^2 + \sqrt[3]{-64} - \sqrt[3]{\dfrac{125}{1000}}$

$= -4 + 4 - 9 + (-4) - \dfrac{5}{10} = -13 - \dfrac{1}{2}$

$= \mathbf{-13\dfrac{1}{2}}$

28. Let x be the measure of $\angle A$.

$x + 93° + 61° = 180°$

$x = 180° - 93° - 61°$

$x = \mathbf{26°}$

$\angle A$ is an **acute** angle.

29.
$$m^2p + p^2 = (-4)^2(5) + (5)^2$$
$$= (16)5 + 25$$
$$= \mathbf{105}$$

30.
$$x^2yx^4y + a^2a^{10} - b^3b^3b + 5^2c(5)$$
$$= \mathbf{x^6y^2 + a^{12} - b^7 + 125c}$$

PROBLEM SET 108

1.
$$110\% \cdot 500{,}000 = B$$
$$1.1 \cdot 500{,}000 = B$$
$$550{,}000 = B$$
$$500{,}000 + 550{,}000 = \mathbf{1{,}050{,}000 \text{ units}}$$

2.
$$2\frac{1}{2} \cdot R = 8500$$
$$\frac{2}{5} \cdot \frac{5}{2} \cdot R = 8500 \cdot \frac{2}{5}$$
$$R = \mathbf{3400 \text{ items}}$$

3.
$$\frac{11}{\$74{,}800} = \frac{6}{D}$$
$$11D = \$448{,}800$$
$$D = \mathbf{\$40{,}800}$$

4.
$$\frac{B}{P} = \frac{7}{13}, \quad \frac{B}{T} = \frac{7}{20}, \quad \frac{P}{T} = \frac{13}{20}$$
$$\frac{7}{20} = \frac{B}{2600}$$
$$20B = 18{,}200$$
$$B = \mathbf{910 \text{ bears}}$$
$$P = 2600 - 910 = \mathbf{1690 \text{ pumas}}$$

5. $x \geq 1$

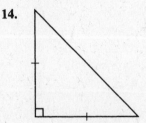

6.
$$62\frac{1}{4}\% \cdot 750 = B$$
$$0.6225 \cdot 750 = B$$
$$466.875 = B$$
$$750 + 466.875 = \mathbf{1216.875}$$

7.
$$P \cdot 95 = 228$$
$$P = \frac{228}{95}$$
$$P = 2.4$$
$$P = \mathbf{240\%}$$

8.
$$15\frac{1}{2}\% \cdot 9000 = B$$
$$0.155 \cdot 9000 = B$$
$$\mathbf{1395 = B}$$

9.

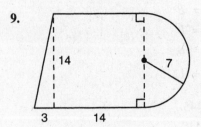

$$V = A_{\text{Base}} \times \text{height}$$
$$= [A_{\text{Triangle}} + A_{\text{Square}} + A_{\text{Semicircle}}](\text{height})$$
$$= \left[\frac{(3)(14)}{2} + 14(14) + \frac{\pi(7)^2}{2}\right](2) \text{ ft}^3$$
$$= [21 + 196 + 24.5\pi](2) \text{ ft}^3$$
$$\approx [293.93](2) \text{ ft}^3$$
$$\approx \mathbf{587.86 \text{ ft}^3}$$

10.
$$S.A. = 2A_{\text{Base}} + (\text{Perimeter} \times \text{height})$$
$$\approx 2[293.93] \text{ ft}^2$$
$$\quad + [17 + 14.32 + 14 + \pi(7)](2) \text{ ft}^2$$
$$\approx 587.86 \text{ ft}^2 + (67.3)(2) \text{ ft}^2$$
$$\approx \mathbf{722.46 \text{ ft}^2}$$

11. (a) $19 = \mathbf{XIX}$

(b) $1644 = \mathbf{MDCXLIV}$

12. $XXVIII + V = 28 + 5 = \mathbf{33}$

13. (a) **45°**

(b) **90°**

(c) **170°**

14.

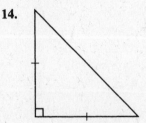

Right angle = **90°**.

The remaining 2 angles are equal, and their sum must be $180° - 90° = 90°$. Therefore, both of them measure **45°**.

15. $\left(\dfrac{2}{3}\right)^{-2} + (4)^{-2} = \left(\dfrac{3}{2}\right)^2 + \dfrac{1}{4^2} = \dfrac{9}{4} + \dfrac{1}{16}$

$\qquad = \dfrac{36}{16} + \dfrac{1}{16} = \mathbf{\dfrac{37}{16}}$

16. $4x(3a + b + ab) = \mathbf{12xa + 4xb + 4xab}$

17. $11p(11 + p + 2ab) = \mathbf{11p^2 + 121p + 22pab}$

18. $mp^2 + mpm + 2a^2b^2 + mp + abab$

$\qquad = mp^2 + m^2p + 2a^2b^2 + mp + a^2b^2$

$\qquad = \mathbf{mp^2 + m^2p + 3a^2b^2 + mp}$

19. $p^2ap + m^2nx + ap^3 + nmxm$

$\qquad = ap^3 + m^2nx + ap^3 + m^2nx$

$\qquad = \mathbf{2ap^3 + 2m^2nx}$

20. $\dfrac{2}{5} + 0.13 + \dfrac{3}{25} = \dfrac{40}{100} + \dfrac{13}{100} + \dfrac{12}{100}$

$\qquad = \dfrac{65}{100} = \mathbf{0.65}$

21. $\dfrac{8}{9}$

$$\begin{array}{r} 0.888 \\ 9\overline{)8.000} \\ 7\,2 \\ \hline 80 \\ 72 \\ \hline 80 \\ 72 \\ \hline 8 \end{array}$$

The rounded number is **0.89.**

22. $9\dfrac{1}{5}\% = 9.2\% = \mathbf{0.092}$

23. $86\dfrac{1}{4}\% = 86.25\% = \mathbf{0.8625}$

24. $6x + 3 = x - 22$

$\qquad 5x = -25$

$\qquad x = \mathbf{-5}$

25. $10 - x - 9 = -11x + 91 + 4(3 - x)$

$\qquad 1 - x = -11x + 91 + 12 - 4x$

$\qquad 11x + 4x - x = 91 + 12 - 1$

$\qquad 14x = 102$

$\qquad \dfrac{14x}{14} = \dfrac{102}{14}$

$\qquad x = \dfrac{51}{7}$

$\qquad x = \mathbf{7\dfrac{2}{7}}$

26. $\dfrac{\dfrac{4}{5}}{3\dfrac{1}{2}} = \dfrac{\dfrac{6}{7}}{x}$

$\qquad \dfrac{4}{5}x = 3\dfrac{1}{2} \cdot \dfrac{6}{7}$

$\qquad \dfrac{5}{4} \cdot \dfrac{4}{5}x = 3 \cdot \dfrac{5}{4}$

$\qquad x = \mathbf{\dfrac{15}{4}}$

27. $-[-(-4)(-4)] - [-(3^2)] = 16 + 9 = \mathbf{25}$

28. $-[-(6)(3)] + (-9) - [-(-4)] - \sqrt[3]{-27}$

$\qquad = 18 - 9 - 4 + 3 = \mathbf{8}$

29. $t^3m^{-4}t^{-1}t^{-2}m^{-1}t^5am^5 = t^3t^{-1}t^{-2}t^5m^{-4}m^{-1}m^5a$

$\qquad = t^5m^0a = \mathbf{at^5}$

30. $a^2b^2 + b^2 + \dfrac{b}{a} = (-3)^2(-9)^2 + (-9)^2 + \dfrac{(-9)}{(-3)}$

$\qquad = 729 + 81 + 3$

$\qquad = \mathbf{813}$

PROBLEM SET 109

1. $\dfrac{G}{R} = \dfrac{2}{5}, \qquad \dfrac{G}{T} = \dfrac{2}{7}, \qquad \dfrac{R}{T} = \dfrac{5}{7}$

$\dfrac{2}{7} = \dfrac{G}{4900}$

$7G = 9800$

$G = \mathbf{1400\ people}$

2. $\dfrac{500}{\$3300} = \dfrac{350}{D}$

$500D = 350 \cdot \$3300$

$D = \mathbf{\$2310}$

3. $\dfrac{230}{100} \times 11{,}000{,}000 = B$

$25{,}300{,}000 = B$

$11{,}000{,}000 + 25{,}300{,}000 = \mathbf{36{,}300{,}000\ bushels}$

4. $3\dfrac{3}{4} \cdot A = 12{,}000$

$\dfrac{4}{15} \cdot \dfrac{15}{4} \cdot A = 12{,}000 \cdot \dfrac{4}{15}$

$A = \mathbf{3200\ people}$

5. 1st yr's interest = 0.07($5000) = $350

Amount after 1 yr = $5000 + $350 = $5350

2nd yr's interest = 0.07($5350) = $374.50

Amount after 2 yr = $5350 + $374.50
= **$5724.50**

6. Interest each yr = 0.10($15,000) = $1500

$15,000 + $1500 + $1500 + $1500 = **$19,500**

7. Interest 1st yr = 0.10($15,000) = $1500

Total after 1st yr = $15,000 + $1500 = $16,500

Interest 2nd yr = 0.10($16,500) = $1650

Total after 2nd yr = $16,500 + $1650
= $18,150

Interest 3rd yr = 0.10($18,150) = $1815

Total after 3rd yr = $18,150 + $1815
= **$19,965**

8. $x \leq -2$

9. $90\% \cdot 90 = A$

$0.9 \cdot 90 = A$

$81 = A$

$90 + 81 = 171$

$0.9 \cdot 171 = B$

$153.9 = B$

$171 - 153.9 = \textbf{17.1}$

10. $P \cdot 68 = 170$

$P = \dfrac{170}{68}$

$P = 2.5$

$P = \textbf{250\%}$

11. $18\dfrac{1}{20}\% \cdot 12,000 = B$

$0.1805 \cdot 12,000 = B$

$\textbf{2166} = B$

12.

$V = A_{\text{Base}} \times \text{height}$

$= [A_{\text{Rect 1}} + A_{\text{Rect 2}} + A_{\text{Semicircle}}](\text{height})$

$= \left[5(0.5) + 2.5(2) + \dfrac{\pi(1)^2}{2} \right](0.5) \text{ mi}^3$

$= [7.5 + 0.5\pi](0.5) \text{ mi}^3$

$\approx \textbf{4.535 mi}^3$

13. $V = A_{\text{Base}} \times \text{height}$

$= \pi(0.1)^2 \text{ ft}^2 \times 0.5 \text{ ft}$

$\approx \textbf{0.0157 ft}^3$

$L.A. = \text{Circumference} \times \text{height}$

$= 2\pi(0.1) \text{ ft} \times (0.5) \text{ ft}$

$= (0.1)\pi \text{ ft}^2$

$\approx \textbf{0.314 ft}^2$

14. (a) $26 = \textbf{XXVI}$

(b) $825 = \textbf{DCCCXXV}$

15. $\dfrac{\sqrt[3]{-27}}{\sqrt[5]{-32}} - \left(\dfrac{2}{5}\right)^{-3} = \dfrac{-3}{-2} - \left(\dfrac{5}{2}\right)^3 = \dfrac{3}{2} - \dfrac{125}{8}$

$= \dfrac{12}{8} - \dfrac{125}{8} = -\dfrac{113}{8} = \textbf{-14}\dfrac{\textbf{1}}{\textbf{8}}$

16. $x + 14\dfrac{3}{4}° + 90.31° = 180°$

$x + 14.75° + 90.31° = 180°$

$x = 180° - 90.31° - 14.75°$

$x = \textbf{74.94}°$

17. $3c(m + mn + n) = \textbf{3cm + 3cmn + 3cn}$

18. $a^2(a + ab + b^2) = \textbf{a}^3 + \textbf{a}^3\textbf{b} + \textbf{a}^2\textbf{b}^2$

19. $a^3p + mb^2m + apa^2 + m^2bb$

$= a^3p + b^2m^2 + a^3p + b^2m^2 = \textbf{2a}^3\textbf{p} + \textbf{2m}^2\textbf{b}^2$

20. $p^2bc^2 + pbcpc + mx^3 + mp$

$= p^2bc^2 + p^2bc^2 + mx^3 + mp$

$= \textbf{2p}^2\textbf{c}^2\textbf{b} + \textbf{mx}^3 + \textbf{pm}$

21. $301\frac{1}{40} = \mathbf{301.025}$

$$40\overline{)1.000} \quad \begin{array}{r} 0.025 \\ \hline 80 \\ \hline 200 \\ 200 \\ \hline 0 \end{array}$$

22. $\frac{3}{8} = \mathbf{0.375}$

$$8\overline{)3.000} \quad \begin{array}{r} 0.375 \\ \hline 24 \\ \hline 60 \\ 56 \\ \hline 40 \\ 40 \\ \hline 0 \end{array}$$

23. $8\frac{1}{5}\% = 8.2\% = \mathbf{0.082}$

24. $9\frac{1}{8}\% = 9.125\% = \mathbf{0.09125}$

25. $2x - 11 = -8x + 109$

$10x = 120$

$x = \mathbf{12}$

26. $6 - x - 4 = -13x - 100 + 3(2 - x)$

$2 - x = -13x - 100 + 6 - 3x$

$2 - x = -16x - 94$

$$\begin{array}{r} +16x \qquad +16x \\ \hline 2 + 15x = \qquad -94 \\ -2 \qquad\qquad -2 \\ \hline 15x = \qquad -96 \end{array}$$

$\dfrac{15x}{15} = \dfrac{-96}{15}$

$x = -\dfrac{32}{5}$

27. $\dfrac{\frac{1}{4}}{-3\frac{1}{6}} = \dfrac{-1\frac{1}{2}}{x}$

$\dfrac{\frac{1}{4}}{-\frac{19}{6}} = \dfrac{-\frac{3}{2}}{x}$

$\dfrac{1}{4}x = \dfrac{19}{4}$

$x = \mathbf{19}$

28. $-[-(-3)(2)] - [-(-4)] - (2^2) + \sqrt[3]{-64}$
$= -6 - 4 - 4 - 4 = \mathbf{-18}$

29. $mx^3 + \dfrac{x}{m} + \dfrac{m}{x} = (-2)(-4)^3 + \dfrac{(-4)}{(-2)} + \dfrac{(-2)}{(-4)}$

$= 128 + 2 + \dfrac{1}{2}$

$= \mathbf{130\frac{1}{2}}$

30. $a^c b - a^2 b - a^2 bc$
$= (-2)^0(-3) - (-2)^2(-3) - (-2)^2(-3)(0)$
$= 1(-3) - 4(-3) - 0$
$= -3 + 12$
$= \mathbf{9}$

PROBLEM SET 110

1. If 20% suffered from agoraphobia, 80% did not suffer.

$80\% \cdot A = 1600$

$0.8 \cdot A = 1600$

$\dfrac{0.8A}{0.8} = \dfrac{1600}{0.8}$

$A = \mathbf{2000\ little\ people}$ lived in the forest

2. $\dfrac{R}{G} = \dfrac{3}{14}, \quad \dfrac{R}{T} = \dfrac{3}{17}, \quad \dfrac{G}{T} = \dfrac{14}{17}$

$\dfrac{14}{17} = \dfrac{G}{15,300}$

$214,200 = 17G$

$\mathbf{12,600} = G$

3. $-2N + 7 = 3N + 27$

$-5N = 20$

$N = \mathbf{-4}$

4. Initial speed $= \dfrac{100\text{ mi}}{4\text{ hr}} = 25$ mph

Increased speed $= 25$ mph $+ 5$ mph $= 30$ mph

$\dfrac{1\text{ hr}}{30\text{ mi}} \times 120\text{ mi} = \mathbf{4\ hr}$

5. Markup $= 60\%$ of $20
$= 0.6 \cdot \$20$
$= \$12$

Selling price $= \$20 + \$12 = \mathbf{\$32}$

6. If we have a 20% markdown, then 80% of the original price is equal to the sale price.

$$0.80 \cdot A = \$42$$

$$A = \frac{\$42}{0.80}$$

$$A = \mathbf{\$52.50}$$

7. $4^3 = 64$

$5^3 = 125$

So, $\mathbf{4 < \sqrt[3]{80} < 5}$

8. $V = A_{\text{Base}} \times \text{height}$

$= [A_{\text{Triangle}} + A_{\text{Semicircle}}] \times \text{height}$

$= \left[\frac{(3)(4)}{2} + \frac{\pi(1.25)^2}{2} \right] \text{ft}^2 \times 4 \text{ ft}$

$\approx \mathbf{33.8125 \text{ ft}^3}$

9. $L.A. = \text{Perimeter} \times \text{height}$

$= [3 + 4 + 1.25 + \pi(1.25) + 1.25](4) \text{ ft}^2$

$= [9.5 + 1.25\pi](4) \text{ ft}^2$

$\approx \mathbf{53.7 \text{ ft}^2}$

10. $524 = \mathbf{DXXIV}$

11. $CDXIX = 400 + 10 + 9 = \mathbf{419}$

12. $80\% \cdot 120 = B$

$0.8 \cdot 120 = B$

$96 = B$

$120 + 96 = \mathbf{216}$

13. $P \cdot 50 = 60$

$P = \frac{60}{50}$

$P = \frac{120}{100}$

$P = \mathbf{120\%}$

14. (a) $61\frac{3}{5}\% = 61.6\% = \mathbf{0.616}$

(b) $7\frac{4}{5}\% = 7.8\% = \mathbf{0.078}$

15. $12 \text{ mi} \times \frac{5280 \text{ ft}}{1 \text{ mi}} \times \frac{12 \text{ in.}}{1 \text{ ft}}$

$= 12(5280)(12) \text{ in.} = \mathbf{760{,}320 \text{ in.}}$

16. $\text{Interest} = 0.045 \cdot \$200 = \mathbf{\$9}$

17. $2ac(ab + a - b) = \mathbf{2a^2bc + 2a^2c - 2abc}$

18. $-am(a^2 + m - am) = \mathbf{-a^3m - am^2 + a^2m^2}$

19. $3xy^2 - 2xy + 3xyy + 3xy - y^2x$

$= 3xy^2 - 2xy + 3xy^2 + 3xy - xy^2$

$= \mathbf{5xy^2 + xy}$

20. $3x + 6 = x + 7$

$2x = 1$

$x = \frac{1}{2}$

21. $-2\frac{1}{3}x - \frac{3}{4} = \frac{1}{2}$

$-2\frac{1}{3}x - \frac{3}{4} + \frac{3}{4} = \frac{1}{2} + \frac{3}{4}$

$-\frac{7}{3}x = \frac{5}{4}$

$x = -\frac{15}{28}$

22. $\dfrac{-\frac{1}{3}}{\frac{4}{9}} = \dfrac{\frac{1}{4}}{x}$

$-\frac{1}{3}x = \frac{1}{9}$

$x = -\frac{1}{3}$

23. $-[-(-4)^2] + (-3)(4) + (-2)^3 + \sqrt[7]{-128}$

$= 16 - 12 - 8 - 2 = \mathbf{-6}$

24. $2^2 + 2^3[-2(-3 + 2^2)(2^2 - 1) + 2]$

$= 4 + 8[-2(1)(3) + 2] = 4 - 32 = \mathbf{-28}$

25. $6a^2bab^2b^3 = \mathbf{6a^3b^6}$

26. $\dfrac{-(-2)^2 + 3^2(2^2 - 5) + 3}{2(2^3 - 4)}$

$= \dfrac{-4 + 9(-1) + 3}{2(4)} = \dfrac{-10}{8} = -\dfrac{5}{4}$

27. $\frac{1}{4}\left(2\frac{1}{3} \cdot \frac{1}{4} - \frac{7}{12} \right) = \frac{1}{4}\left(\frac{7}{12} - \frac{7}{12} \right) = \mathbf{0}$

28. $(-2)^{-2} - \left(\frac{1}{3}\right)^{-3} + \sqrt[3]{-64}$

$= \dfrac{1}{(-2)^2} - \left(\frac{3}{1}\right)^3 + (-4)$

$= \dfrac{1}{4} - 27 - 4 = \dfrac{1}{4} - 31$

$= \mathbf{-30\dfrac{3}{4}}$

29. $a^a - 2a = (-2)^{-2} - 2(-2)$

$$= \frac{1}{(-2)^2} + 4$$

$$= \frac{1}{4} + 4$$

$$= 4\frac{1}{4}$$

30. $a^2 + 3ab^2 = (-1)^2 + 3(-1)(-3)^2$

$$= 1 - 27$$

$$= \mathbf{-26}$$

PROBLEM SET 111

1. $\dfrac{R}{G} = \dfrac{2}{19}, \qquad \dfrac{R}{T} = \dfrac{2}{21}, \qquad \dfrac{G}{T} = \dfrac{19}{21}$

$$\frac{2}{21} = \frac{R}{84,000}$$

$$168,000 = 21R$$

$$R = \textbf{8000 red marbles}$$

2. $30\% \cdot A = 120$

$$0.3A = 120$$

$$\frac{0.3A}{0.3} = \frac{120}{0.3}$$

$$A = 400 \text{ airplanes total}$$

Not biplanes: $400 - 120 = \textbf{280 airplanes}$

3. $3\dfrac{2}{5} \times 8500 = W$

$$\frac{17}{5} \times 8500 = W$$

$$\textbf{28,900 wet ones} = W$$

4. Commission = $(0.1)(\$45,000) = \$4500$

$$\begin{array}{ll} \$4500 & \text{Commission} \\ \underline{+\ \$\ 350} & \text{Base salary} \\ \$4850 & \text{Paycheck} \end{array}$$

5. Markup = $(0.6)(\$16,000) = \$9600$

$$\begin{array}{ll} \$9600 & \text{Markup} \\ \underline{-\ \$7000} & \text{Expenses} \\ \$2600 & \text{Profit} \end{array}$$

6. Interest 1st yr = $0.08(\$8000) = \$640$

Total after 1 yr = $\$8000 + \$640 = \$8640$

Interest 2nd yr = $0.08(\$8640) = \$691.20$

Total after 2 yr = $\$8640 + \$691.20 = \textbf{\$9331.20}$

7. $6\dfrac{1}{4}\% = 6.25\% = \textbf{0.0625}$

8. MMMCDXXXIV = $1000 + 1000 + 1000$
$+ 400 + 10 + 10 + 10 + 4 = \textbf{3434}$

9. $3000 = \text{MMM}$
$400 = \text{CD}$
$60 = \text{LX}$
$5 = \text{V}$
$3465 = \textbf{MMMCDLXV}$

10. $45 = P \cdot 225$

$$\frac{45}{225} = P$$

$$\mathbf{20\%} = P$$

11. $42\% \times A = 126$

$$0.42A = 126$$

$$A = \frac{126}{0.42}$$

$$A = \mathbf{300}$$

12. Commission = $0.35(\$12.50 + \$15.50)$

$$= 0.35(\$28)$$

$$= \textbf{\$9.80}$$

13. 15% off means the sale price is 85% of the original price.

$$\$4.25 = 0.85 \times A$$

$$\frac{\$4.25}{0.85} = \frac{0.85A}{0.85}$$

$$\textbf{\$5.00} = A$$

14. $5x = 2\dfrac{1}{2}$

$$\frac{1}{5} \cdot 5x = \frac{5}{2} \cdot \frac{1}{5}$$

$$x = \frac{1}{2}$$

$$4x - 12 = 4\left(\frac{1}{2}\right) - 12 = 2 - 12 = \mathbf{-10}$$

15. $3^{-4} - (-2)^3 + \left(\dfrac{3}{2}\right)^{-5} = \dfrac{1}{3^4} - (-2)^3 + \left(\dfrac{2}{3}\right)^5$

$$= \frac{1}{81} - (-8) + \frac{32}{243} = \frac{3}{243} + 8 + \frac{32}{243}$$

$$= 8\frac{35}{243}$$

16. $2ac(a + c - ac + a^2)$
$= 2a^2c + 2ac^2 - 2a^2c^2 + 2a^3c.$

17. $2xyy + 3xyx - 6xy^2 + 4x^2y$

$= 2xy^2 + 3x^2y - 6xy^2 + 4x^2y$

$= \mathbf{-4xy^2 + 7x^2y}$

18. Since two sides have equal lengths, $\triangle QRS$ is **isosceles**. Since $\angle S$ measures more than 90°, $\triangle QRS$ is **obtuse**.

19. $V = A_{\text{Base}} \times \text{height}$

$= [A_{\text{Big Semicircle}} + 2A_{\text{Small Semicircle}}] \times \text{height}$

$= \left[\dfrac{\pi(12)^2}{2} + 2\left(\dfrac{\pi(6)^2}{2}\right)\right] \text{cm}^2 \times 3 \text{ cm}$

$= [108\pi](3) \text{ cm}^3$

$\approx 1017.36 \text{ cm}^3$

$\approx (1017.36)\left(\dfrac{1}{100}\right)\left(\dfrac{1}{100}\right)\left(\dfrac{1}{100}\right) \text{m}^3$

$\approx \mathbf{1.01736 \times 10^{-3} \text{ m}^3}$

20. $5 - 3x + 6 = 2x - 21 + 4(2x - 3)$

$11 - 3x = 2x - 21 + 8x - 12$

$11 - 3x = 10x - 33$

$\dfrac{\begin{array}{r}+3x \qquad +3x\end{array}}{11 \qquad = 13x - 33}$

$\dfrac{\begin{array}{r}+33 \qquad\qquad +33\end{array}}{44 \qquad = 13x}$

$\dfrac{44}{13} = \dfrac{13x}{13}$

$\dfrac{44}{13} = x$

21. $-2\dfrac{1}{3}x - \dfrac{5}{6} = \dfrac{1}{3}$

$-2\dfrac{1}{3}x - \dfrac{5}{6} + \dfrac{5}{6} = \dfrac{1}{3} + \dfrac{5}{6}$

$-\dfrac{7}{3}x = \dfrac{7}{6}$

$x = -\dfrac{1}{2}$

22. $-[-(-1)^3] + (-4) + \sqrt[3]{-125} = -1 - 4 - 5$
$= \mathbf{-10}$

23. $2^2aa^3ba^2b^3 = \mathbf{4a^6b^4}$

24. $\dfrac{2^2(2^2 - 2 \cdot 4)}{2^3(3^2 - 2^2)} = \dfrac{4(-4)}{8(5)} = \mathbf{-\dfrac{2}{5}}$

25. $2\dfrac{1}{3}\left(2\dfrac{1}{2} \cdot \dfrac{1}{2} - \dfrac{1}{3} \cdot \dfrac{4}{5}\right) = \dfrac{7}{3}\left(\dfrac{5}{4} - \dfrac{4}{15}\right)$

$= \dfrac{7}{3}\left(\dfrac{75}{60} - \dfrac{16}{60}\right) = \dfrac{7}{3}\left(\dfrac{59}{60}\right) = \mathbf{\dfrac{413}{180}}$

26. $a^2b + ab^2 = (-1)^2(-2) + (-1)(-2)^2$

$= -2 - 4$

$= \mathbf{-6}$

27. $a^2 - b^2 = (-2)^2 - (-3)^2 = 4 - 9 = \mathbf{-5}$

28. $x \nleq 3$

$x > 3$

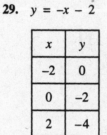

29. $y = -x - 2$

x	y
-2	0
0	-2
2	-4

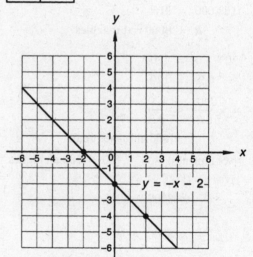

30.

$62,9\textcircled{8}7,134.32$

The rounded number is **62,990,000**.

PROBLEM SET 112

1. $P(B \text{ or } R) = \dfrac{7 + 5}{22} = \dfrac{12}{22} = \mathbf{\dfrac{6}{11}}$

2. $P(D) = \dfrac{13}{52} = \mathbf{\dfrac{1}{4}}$

3. (a) $P(3) = \mathbf{\dfrac{1}{6}}$

(b) $P(4) = \mathbf{\dfrac{1}{6}}$

4. $40\% \cdot A = 660$

$0.4A = 660$

$\dfrac{0.4A}{0.4} = \dfrac{660}{0.4}$

$A = 1650$ airplanes total

Not hydroplanes: $1650 - 660 = \textbf{990 airplanes}$

5.

$2\dfrac{2}{3} \cdot A = 3000$

$\dfrac{8}{3} \cdot A = 3000$

$\dfrac{3}{8} \cdot \dfrac{8}{3} \cdot A = 3000 \cdot \dfrac{3}{8}$

$A = \textbf{1125 were aloof}$

6. $3N + 16 = -4N - 5$

$7N = -21$

$N = \textbf{-3}$

7. $0.09 \times \$3000 = \$270$

After 1 yr: $\$3000 + \$270 = \$3270$

$0.09 \times \$3270 = \$294.30$

After 2 yr: $\$3270 + \$294.30 = \$3564.30$

$0.09 \times \$3564.30 = \$320.79$

After 3 yr: $\$3564.30 + \$320.79 = \$3885.09$

$0.09 \times \$3885.09 = \$349.66$

After 4 yr: $\$3885.09 + \$349.66 = \textbf{\$4234.75}$

8. Commission $= 0.06(\$260,000) = \$15,600$

Base pay $= \dfrac{\$510}{1\ \text{mo}} \times 2\ \text{mo} = \$1020$

Total pay $= \$15,600 + \$1020 = \textbf{\$16,620}$

9. Commission $= 0.12(\$550,000) = \$66,000$

Base pay $= \dfrac{\$400}{1\ \text{mo}} \times 12\ \text{mo} = \$4800$

Total pay $= \$66,000 + \$4800 = \textbf{\$70,800}$

10. $12\dfrac{1}{2}\% \times \$20 = B$

$0.125(\$20) = B$

$\$2.50 = B$

Sale price: $\$20 - \$2.50 = \textbf{\$17.50}$

11. MMMXV $= 1000 + 1000 + 1000 + 10 + 5$

$= \textbf{3015}$

12. (a) $\angle AOB$ measures **45°**

(b) $\angle AOC$ measures **100°**

(c) $\angle AOD$ measures **135°**

13. (a) $5\dfrac{1}{5}\% = 5.2\% = \textbf{0.052}$

(b) $17\dfrac{3}{4}\% = 17.75\% = \textbf{0.1775}$

(c) $11\dfrac{3}{8}\% = 11.375\% = \textbf{0.11375}$

14. $S.A. = 2A_{\text{Base}} + L.A.$

$= 2\pi r^2 + \pi dh$

$= [2\pi(10)^2 + \pi(20)(7)]\ \text{cm}^2$

$= [340\pi]\ \text{cm}^2$

$\approx \textbf{1067.6 cm}^2$

15. $78\% \cdot A = 156$ million

$0.78A = 156$ million

$\dfrac{0.78A}{0.78} = \dfrac{156}{0.78}$ million

$A = 200$ million vehicles

200 million $- 156$ million

$= \textbf{44 million commercial vehicles}$

16. $3mp\left(2m + p + \dfrac{m^2}{p} + \dfrac{p^4}{m}\right)$

$= \textbf{6}\boldsymbol{m^2p} \textbf{ + 3}\boldsymbol{mp^2} \textbf{ + 3}\boldsymbol{m^3} \textbf{ + 3}\boldsymbol{p^5}$

17. $7d^2z + 7zdz + 7ddz + 7dz^2$

$= 7d^2z + 7dz^2 + 7d^2z + 7dz^2$

$= \textbf{14}\boldsymbol{d^2z} \textbf{ + 14}\boldsymbol{dz^2}$

18.

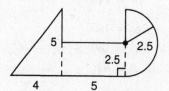

$V = A_{\text{Base}} \times \text{height}$

$= \left[\dfrac{(4)(5)}{2} + (2.5)(5) + \dfrac{\pi(2.5)^2}{2}\right](0.5)\ \text{mi}^3$

$= [10 + 12.5 + 3.125\pi](0.5)\ \text{mi}^3$

$\approx \textbf{16.15625 mi}^3$

19. $S.A. = 6A_{\text{Side}} = 6(3)(3)\ \text{in.}^2 = \textbf{54 in.}^2$

20. $2 - 5x + 6 = 14x - 32 + 10(2 - x)$

$\qquad 8 - 5x = 14x - 32 + 20 - 10x$

$\qquad 8 - 5x = 4x - 12$

$\qquad \dfrac{+5x \quad +5x}{8 \qquad = 9x - 12}$

$\qquad \dfrac{+12 \qquad\qquad +12}{20 \qquad = 9x}$

$\qquad\qquad \dfrac{20}{9} = \dfrac{9x}{9}$

$\qquad\qquad \mathbf{\dfrac{20}{9}} = x$

21. $\qquad -5\dfrac{5}{6}x - \dfrac{1}{2} = 1\dfrac{1}{4}$

$\qquad -5\dfrac{5}{6}x - \dfrac{1}{2} + \dfrac{1}{2} = 1\dfrac{1}{4} + \dfrac{1}{2}$

$\qquad\qquad -\dfrac{35}{6}x = \dfrac{7}{4}$

$\qquad \left(-\dfrac{6}{35}\right)\left(-\dfrac{35}{6}\right)x = \dfrac{7}{4}\left(-\dfrac{6}{35}\right)$

$\qquad\qquad\qquad x = \mathbf{-\dfrac{3}{10}}$

22. $\qquad \dfrac{-\dfrac{5}{6}}{1\dfrac{2}{3}} = \dfrac{\dfrac{3}{4}}{x}$

$\qquad\qquad -\dfrac{5}{6}x = \dfrac{5}{4}$

$\qquad \left(-\dfrac{6}{5}\right)\left(-\dfrac{5}{6}\right)x = \dfrac{5}{4}\left(-\dfrac{6}{5}\right)$

$\qquad\qquad\qquad x = \mathbf{-\dfrac{3}{2}}$

23. (a) $A + 21° + 41° = 180°$

$\qquad\qquad A = 180° - 21° - 41°$

$\qquad\qquad A = \mathbf{118°}$

(b) $B + 90° + 26° = 180°$

$\qquad\qquad B = 180° - 90° - 26°$

$\qquad\qquad B = \mathbf{64°}$

(c) $2C + 42° = 180°$

$\qquad\qquad 2C = 180° - 42°$

$\qquad\qquad C = \dfrac{138°}{2}$

$\qquad\qquad C = \mathbf{69°}$

24. $x \not\geq -1$

$\qquad x < -1$

25. $-[-(-2)^5] + (-9) - \sqrt[5]{-243}$

$= -[-(-32)] + (-9) - (-3)$

$= -32 - 9 + 3 = \mathbf{-38}$

26. $3^2 mn^5 mn^2 m^6 (-1)^5 = \mathbf{-9m^8 n^7}$

27. $\dfrac{-(-3)^2 - 1^9(3^3 - 3 \cdot 2)}{3^3(3^2 - 2^3)}$

$= \dfrac{-9 - 1(27 - 6)}{27(9 - 8)} = \dfrac{-9 - (21)}{27(1)} = -\dfrac{10}{9}$

28. $3\dfrac{1}{2}\left(2\dfrac{1}{3} \cdot \dfrac{1}{3} - \dfrac{1}{2} \cdot \dfrac{3}{7}\right) = \dfrac{7}{2}\left(\dfrac{7}{9} - \dfrac{3}{14}\right)$

$= \dfrac{7}{2}\left(\dfrac{98}{126} - \dfrac{27}{126}\right) = \dfrac{7}{2}\left(\dfrac{71}{126}\right) = \mathbf{\dfrac{71}{36}}$

29. (a) $2^{-5} = \dfrac{1}{2^5} = \mathbf{\dfrac{1}{32}}$

(b) $\dfrac{1}{3^{-3}} = 3^3 = \mathbf{27}$

30. $m^3 n^{-2} = (-2)^3(-3)^{-2} = \dfrac{(-2)^3}{(-3)^2} = -\dfrac{8}{9}$

PROBLEM SET 113

1. $P(R) = \dfrac{1}{5}$

2. $P(>2) = \dfrac{4}{6} = \dfrac{2}{3}$

3. $260\% \cdot 230{,}000 = B$

$\qquad 2.6(230{,}000) = B$

$\qquad\qquad 598{,}000 = B$

$230{,}000 + 598{,}000 = \mathbf{828{,}000 \text{ tons}}$

4. $\dfrac{P}{D} = \dfrac{2}{5}, \qquad \dfrac{P}{T} = \dfrac{2}{7}, \qquad \dfrac{D}{T} = \dfrac{5}{7}$

$\qquad \dfrac{2}{7} = \dfrac{P}{2450}$

$\qquad 4900 = 7P$

$\qquad \mathbf{700} = P$

5. Simple interest: $4[0.06(\$11{,}000)] = \$2640$

Compound interest:

$\qquad 0.06(\$11{,}000) = \$660$

$\qquad 0.06(\$11{,}660) = \$699.60$

$\qquad 0.06(\$12{,}359.60) = \$741.58$

$\qquad 0.06(\$13{,}101.18) = \$786.07$

$\qquad\qquad \text{Total} = \$2887.25$

Difference = $\$2887.25 - \$2640 = \mathbf{\$247.25}$

6. $2^4 = 16$

$3^4 = 81$

So, $2 < \sqrt[4]{63} < 3$

7. $\dfrac{6}{8} = \dfrac{9}{x}$ $\dfrac{6}{8} = \dfrac{12}{y}$

$6x = 72$ $6y = 96$

$x = \mathbf{12}$ $y = \mathbf{16}$

8. **3 in., 7.6 cm**

9. (a) $1\dfrac{8}{16}$ in. $= \mathbf{1\dfrac{1}{2}}$ **in.**

    (b) **47 mm**

10. (a) $A + 29° + 70° = 180°$

$A = 180° - 29° - 70°$

$A = \mathbf{81°}$

    (b) $B + 90° + 25° = 180°$

$B = 180° - 90° - 25°$

$B = \mathbf{65°}$

    (c) $C = \mathbf{60°}$ because the triangle is an equilateral triangle.

11. (a) $2(4^{-2}) = \dfrac{2}{4^2} = \dfrac{2}{16} = \mathbf{\dfrac{1}{8}}$

    (b) $2\left(\dfrac{1}{4^{-2}}\right) = 2(4^2) = 2(16) = \mathbf{32}$

12. (a) $6\dfrac{7}{8}\% = 6.875\% = \mathbf{0.06875}$

    (b) $132\dfrac{3}{4}\% = 132.75\% = \mathbf{1.3275}$

13. $P \cdot 90 = 162$

$\dfrac{P \cdot 90}{90} = \dfrac{162}{90}$

$P = 1.8$

$P = \mathbf{180\%}$

14. Markup = 150% of $20

$= 1.5 \times \$20 = \$30$

Retail price = Wholesale price + Markup

$= \$20 + \$30 = \$50$

Discount = 20% of $50

$= 0.20 \times \$50 = \$10$

Sale price = $50 - $10 = **$40**

15. MDXCIV $= 1000 + 500 + 90 + 4 = \mathbf{1594}$

16. Commission $= 0.2(\$14{,}000) = \$2800$

Paycheck $= \$2800 + \$350 = \mathbf{\$3150}$

17. $2ac(ab + bc - c) = \mathbf{2a^2bc + 2abc^2 - 2ac^2}$

18. $2xyx + 3yxy^2 - 3x^2y + 5y^3x - 6yx^2$

$= 2x^2y + 3xy^3 - 3x^2y + 5xy^3 - 6x^2y$

$= \mathbf{-7x^2y + 8xy^3}$

19.

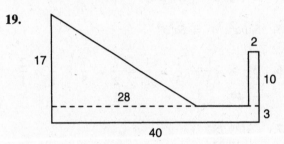

$V = A_{\text{Base}} \times \text{height}$

$= \left[A_{\text{Triangle}} + A_{\text{Rect 1}} + A_{\text{Rect 2}}\right] \times \text{height}$

$= \left[\dfrac{17(28)}{2} + 40(3) + 2(10)\right] \text{ft}^2 \times (2)(3) \text{ ft}$

$= [238 + 120 + 20](6) \text{ ft}^3$

$= \mathbf{2268 \text{ ft}^3}$

20. $S.A. = 2A_{\text{Front}} + 2A_{\text{Top}} + 2A_{\text{Side}}$

$= 2(2)(10) \text{ ft}^2 + 2(2)(4) \text{ ft}^2 + 2(4)(10) \text{ ft}^2$

$= 40 \text{ ft}^2 + 16 \text{ ft}^2 + 80 \text{ ft}^2$

$= \mathbf{136 \text{ ft}^2}$

21. $3x - 3 + 2(x - 1) = -8x + 5 + 5(2 - 3x)$

$3x - 3 + 2x - 2 = -8x + 5 + 10 - 15x$

$5x - 5 = -23x + 15$

$28x = 20$

$x = \dfrac{20}{28}$

$x = \mathbf{\dfrac{5}{7}}$

22. $-\dfrac{1}{4}x + \dfrac{2}{3} = 3\dfrac{5}{12}$

$-\dfrac{1}{4}x + \dfrac{2}{3} - \dfrac{2}{3} = \dfrac{41}{12} - \dfrac{2}{3}$

$-\dfrac{1}{4}x = \dfrac{11}{4}$

$-\dfrac{4}{1}\left(-\dfrac{1}{4}\right)x = \dfrac{11}{4}\left(-\dfrac{4}{1}\right)$

$x = \mathbf{-11}$

23. $\dfrac{-\dfrac{2}{3}}{\dfrac{4}{9}} = \dfrac{\dfrac{1}{3}}{x}$

$-\dfrac{2}{3}x = \dfrac{4}{27}$

$x = -\dfrac{2}{9}$

24. $-[-(-1)^4] + (-3)(-2) - \sqrt[5]{-1}$
$= 1 + 6 - (-1) = \mathbf{8}$

25. $5aba^2b^3a^2 = \mathbf{5a^5b^4}$

26. $3\dfrac{1}{2} \times 2\dfrac{1}{3} \div \dfrac{1}{3} \div \dfrac{1}{2} = \dfrac{7}{2} \times \dfrac{7}{3} \times \dfrac{3}{1} \times \dfrac{2}{1}$
$= \mathbf{49}$

27. $\begin{array}{r} 139.287 \\ -\ 19.876 \\ \hline \mathbf{119.411} \end{array}$

28. $-xy + y = -(-2)(-3) + (-3) = -6 - 3 = \mathbf{-9}$

29. $ab^2 - a^2b = (2)(-1)^2 - (2)^2(-1) = 2 + 4 = \mathbf{6}$

30. (a) The measure of $\angle AEX$ is greater than 90°, so $\angle AEX$ is **obtuse.**

(b) The measure of $\angle XYZ$ is 90°, so $\angle XYZ$ is **right.**

(c) The measure of $\angle BOZ$ is less than 90°, so $\angle BOZ$ is **acute.**

PROBLEM SET 114

1. $P(R, R) = \dfrac{1}{7} \cdot \dfrac{3}{12} = \dfrac{\mathbf{1}}{\mathbf{28}}$

2. $\dfrac{P}{A} = \dfrac{21}{2}, \qquad \dfrac{P}{T} = \dfrac{21}{23}, \qquad \dfrac{A}{T} = \dfrac{2}{23}$

$\dfrac{2}{23} = \dfrac{A}{92}$

$184 = 23A$

8 aesthetes $= A$

3. A 50% increase means the new number is 150% of the old number.

$150\% \cdot A = 3600$

$1.5A = 3600$

$\dfrac{1.5A}{1.5} = \dfrac{3600}{1.5}$

$A = \mathbf{2400\ incidents}$

4. $7N = -9N + 32$
$16N = 32$
$N = \mathbf{2}$

5. $P(B_J, B_J) = \dfrac{2}{52} \cdot \dfrac{2}{52} = \dfrac{1}{26} \cdot \dfrac{1}{26} = \dfrac{\mathbf{1}}{\mathbf{676}}$

6. $P(H, H, T, H, H) = \dfrac{1}{2} \cdot \dfrac{1}{2} \cdot \dfrac{1}{2} \cdot \dfrac{1}{2} \cdot \dfrac{1}{2}$
$= \dfrac{\mathbf{1}}{\mathbf{32}}$

7. Normal rates: $\dfrac{2500 \text{ letters}}{1 \text{ hr}}, \dfrac{1 \text{ hr}}{2500 \text{ letters}}$

Holiday rates:

$3\left(\dfrac{2500 \text{ letters}}{1 \text{ hr}}\right) = \dfrac{7500 \text{ letters}}{1 \text{ hr}}, \dfrac{1 \text{ hr}}{7500 \text{ letters}}$

Time needed:

$\dfrac{1 \text{ hr}}{7500 \text{ letters}} \times 60{,}000 \text{ letters} = \mathbf{8\ hr}$

8. Interest 1st yr: $0.1(\$10{,}000) = \$1000$
Amount after 1 yr $= \$10{,}000 + \$1000$
$= \$11{,}000$

Interest 2nd yr: $0.1(\$11{,}000) = \$1100$
Amount after 2 yr $= \$11{,}000 + \$1100$
$= \$12{,}100$

Interest 3rd yr $= 0.1(\$12{,}100) = \$1210$
Amount after 3 yr $= \$12{,}100 + \$1210$
$= \$13{,}310$

Interest 4th yr $= 0.1(\$13{,}310) = \$1331$
Amount after 4 yr $= \$13{,}310 + \$1331$
$= \mathbf{\$14{,}641}$

9. (a) $180° - 83° = \mathbf{97°}$

(b) $90° - 83° = \mathbf{7°}$

10. $\dfrac{6}{2} = \dfrac{9}{x} \qquad \dfrac{6}{12} = \dfrac{2}{y}$

$6x = 18 \qquad 6y = 24$

$x = \mathbf{3} \qquad y = \mathbf{4}$

11. (a) $A + 17° + 22° = 180°$
$A = 180° - 17° - 22°$
$A = \mathbf{141°}$

(b) $B + 90° + 44° = 180°$
$B = 180° - 90° - 44°$
$B = \mathbf{46°}$

(c) $2C + 50° = 180°$

$\qquad 2C = 180° - 50°$

$\qquad C = \dfrac{130°}{2}$

$\qquad C = \mathbf{65°}$

12. **152 mm**

13. $\qquad 87 = P \cdot 8.7$

$\qquad \dfrac{87}{8.7} = \dfrac{8.7P}{8.7}$

$\qquad 10 = P$

$\qquad \mathbf{1000\%} = P$

14. $\qquad 0.61A = 183$

$\qquad \dfrac{0.61A}{0.61} = \dfrac{183}{0.61}$

$\qquad A = \mathbf{300}$

15. $4az(z^2 + a - 3a^2z)$

$= \mathbf{4az^3 + 4a^2z - 12a^3z^2}$

16. $7mpm + 2mpm^2 - 4m^2p + 3pm^3 - 9pm^2$

$= 7m^2p + 2m^3p - 4m^2p + 3m^3p - 9m^2p$

$= \mathbf{-6m^2p + 5m^3p}$

17. $L.A. = \pi dh$

$\qquad = \pi(2\text{ cm})(2\text{ cm})$

$\qquad = 4\pi\text{ cm}^2$

$\qquad \approx \mathbf{12.56\text{ cm}^2}$

18. Markup $= 100\%$ of $30¢ = 1 \times 30¢ = 30¢$

Total Markup $= 325 \times \$0.30 = \$97.50$

Profit $= \$97.50 - \$95.00 = \mathbf{\$2.50}$

19. $1000 = M$

$900 = CM$

$90 = XC$

$9 = IX$

$1999 = \mathbf{MCMXCIX}$

20. $79\dfrac{1}{8}\% = 79.125\% = \mathbf{0.79125}$

21. $4x - 4 + 5x - 3 = -4x - 46$

$\qquad 9x - 7 = -4x - 46$

$\qquad 13x = -39$

$\qquad x = \mathbf{-3}$

22. $\qquad -\dfrac{1}{3}x + \dfrac{2}{4} = 1\dfrac{5}{6}$

$\qquad -\dfrac{1}{3}x + \dfrac{2}{4} - \dfrac{2}{4} = \dfrac{11}{6} - \dfrac{2}{4}$

$\qquad -\dfrac{1}{3}x = \dfrac{11}{6} - \dfrac{3}{6}$

$\qquad -\dfrac{1}{3}x = \dfrac{8}{6}$

$\qquad x = \mathbf{-4}$

23. $\qquad \dfrac{-\dfrac{1}{4}}{\dfrac{7}{11}} = \dfrac{\dfrac{22}{35}}{x}$

$\qquad -\dfrac{1}{4}x = \dfrac{2}{5}$

$\qquad x = -\dfrac{\mathbf{8}}{\mathbf{5}}$

24. $(-2)^3(-1) - \sqrt[6]{64} = 8 - 2 = \mathbf{6}$

25. $6mp^2m^3p^5m = \mathbf{6m^5p^7}$

26. $4\dfrac{1}{5} \times 2\dfrac{1}{2} \div \dfrac{1}{5} \div \dfrac{1}{4}$

$= \dfrac{21}{5} \times \dfrac{5}{2} \times \dfrac{5}{1} \times \dfrac{4}{1} = \mathbf{210}$

27. $\dfrac{3}{4} - 0.2 + (2.1)\left(\dfrac{1}{2}\right)$

$= 0.75 - 0.2 + (2.1)(0.5)$

$= 0.75 - 0.2 + 1.05 = \mathbf{1.6}$

28. $y = \dfrac{3}{2}x - 1$

x	y
-2	-4
0	-1
2	2

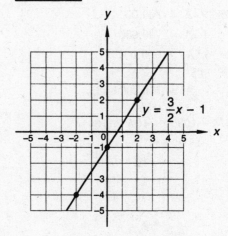

29. $-x^y - y^x = -(-2)^{-3} - (-3)^{-2}$

$$= -\frac{1}{(-2)^3} - \frac{1}{(-3)^2}$$

$$= \frac{1}{8} - \frac{1}{9}$$

$$= \frac{9}{72} - \frac{8}{72}$$

$$= \mathbf{\frac{1}{72}}$$

30. $a^2b + \dfrac{b}{a^2} = (-4)^2(2) + \dfrac{2}{(-4)^2}$

$$= 16(2) + \frac{2}{16}$$

$$= 32 + \frac{1}{8}$$

$$= \mathbf{32\frac{1}{8}}$$

PROBLEM SET 115

1. $\dfrac{E}{B} = \dfrac{8}{7}, \qquad \dfrac{E}{T} = \dfrac{8}{15}, \qquad \dfrac{B}{T} = \dfrac{7}{15}$

$$\frac{7}{15} = \frac{B}{600}$$
$$4200 = 15B$$
$$\mathbf{280} = B$$

2. $P(H, H, T, T, H, T) = \left(\dfrac{1}{2}\right)^6 = \mathbf{\dfrac{1}{64}}$

3. $P(W, W) = \dfrac{\overset{1}{\cancel{2}}}{\underset{3}{\cancel{6}}} \cdot \dfrac{11}{15} = \mathbf{\dfrac{11}{45}}$

4. An increase of 350% means the new number is 450% of the old number.

$$450\% \times A = 900,000$$
$$4.5A = 900,000$$
$$\frac{4.5A}{4.5} = \frac{900,000}{4.5}$$
$$A = \mathbf{200,000\ bulbs}$$

5. $-25N - 108 = -10N - 16$

$$-15N = 92$$
$$N = -\mathbf{\frac{92}{15}}$$

6. $\qquad 2\dfrac{3}{5} \cdot A = 6500$

$$\frac{5}{13} \cdot \frac{13}{5} \cdot A = 6500 \cdot \frac{5}{13}$$
$$A = \mathbf{2500}$$

7. (a) $180° - 65° = \mathbf{115°}$

(b) $90° - 65° = \mathbf{25°}$

8. (a) $A + 49° + 51° = 180°$

$$A = 180° - 49° - 51°$$
$$A = \mathbf{80°}$$

(b) $B + 90° + 51° = 180°$

$$B = 180° - 90° - 51°$$
$$B = \mathbf{39°}$$

(c) $2C + 58° = 180°$

$$2C = 180° - 58°$$
$$C = \frac{122°}{2}$$
$$C = \mathbf{61°}$$

9. (a) **reflection**   (b) **rotation**

10. (a) **concave polygon, hexagon**

(b) **convex polygon, right triangle, scalene triangle**

(c) **convex polygon, parallelogram, rhombus, quadrilateral**

11. $5\dfrac{14}{16}$ **in.** $= \mathbf{5\dfrac{7}{8}}$ **in.**

12. Commission $= 11\dfrac{1}{2}\% \times \$2350.00$

$$= 0.115(\$2350.00)$$
$$= \mathbf{\$270.25}$$

13. Interest $= 4\dfrac{1}{4}\% \times \$550.00$

$$= 0.0425(\$550)$$
$$= \mathbf{\$23.38}$$

14. $P \cdot 28 = 70$

$$P = \frac{70}{28}$$
$$P = \frac{5}{2}$$
$$P = 2.5$$
$$P = \mathbf{250\%}$$

15. $1000 = M$

$900 = CM$

$80 = LXXX$

$4 = IV$

$1984 = \textbf{MCMLXXXIV}$

16. $2xp(x^3 + 4x^2p - 2xp)$

$= \textbf{2px}^4 + \textbf{8p}^2\textbf{x}^3 - \textbf{4p}^2\textbf{x}^2$

17. $3x^2am^4 - xamxmm^2 + 2am^2xmxm$

$= 3x^2am^4 - x^2am^4 + 2x^2am^4 = \textbf{4x}^2\textbf{am}^4$

18.

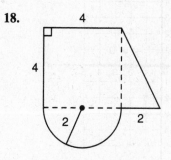

$V = A_{\text{Base}} \times \text{height}$

$= \left[A_{\text{Square}} + A_{\text{Triangle}} + A_{\text{Semicircle}}\right] \times \text{height}$

$= \left[4(4) + \dfrac{2(4)}{2} + \dfrac{\pi(2)^2}{2}\right] \text{km}^2 \times 0.6 \text{ km}$

$= [16 + 4 + 2\pi](0.6) \text{ km}^3$

$\approx \textbf{15.768 km}^3$

19. $L.A. = \text{Perimeter} \times \text{height}$

$= [4 + 4 + 4.47 + 2 + \pi(2)] \text{ km}$

$\times 0.6 \text{ km}$

$= [14.47 + 2\pi](0.6) \text{ km}^2$

$\approx \textbf{12.45 km}^2$

20. $7x + 4 - 9x = 12x - 48$

$4 - 2x = 12x - 48$

$52 = 14x$

$\dfrac{\textbf{26}}{\textbf{7}} = x$

21. $-11\dfrac{1}{2}x - \dfrac{1}{4} = 5\dfrac{1}{2}$

$-\dfrac{23}{2}x - \dfrac{1}{4} + \dfrac{1}{4} = \dfrac{11}{2} + \dfrac{1}{4}$

$-\dfrac{23}{2}x = \dfrac{23}{4}$

$x = -\dfrac{\textbf{1}}{\textbf{2}}$

22. $\dfrac{-\dfrac{1}{5}}{3\dfrac{1}{12}} = \dfrac{x}{\dfrac{1}{2}}$

$-\dfrac{1}{10} = \dfrac{37}{12}x$

$-\dfrac{\textbf{6}}{\textbf{185}} = x$

23. $x \not\leq 0$

$x > 0$

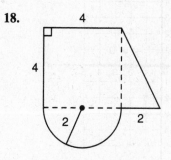

24. $-\left[-(-3^3)\right] + \sqrt[3]{-27} = -27 - 3 = \textbf{-30}$

25. $-6^2mp^5m^2pm^4(-1)^5 = \textbf{36m}^7\textbf{p}^6$

26. $\dfrac{-(-4^2) - 2^2(3 - 4 \cdot 2)}{2^3(2^3 - 3^2)} = \dfrac{16 - 4(-5)}{8(-1)}$

$= \dfrac{36}{-8} = -\dfrac{\textbf{9}}{\textbf{2}}$

27. $2\dfrac{1}{5}\left(3\dfrac{1}{2} \cdot \dfrac{1}{4} - \dfrac{1}{3} \cdot \dfrac{3}{5}\right)$

$= 2\dfrac{1}{5}\left(\dfrac{7}{2} \cdot \dfrac{1}{4} - \dfrac{1}{3} \cdot \dfrac{3}{5}\right)$

$= \dfrac{11}{5}\left(\dfrac{7}{8} - \dfrac{1}{5}\right) = \dfrac{11}{5}\left(\dfrac{35}{40} - \dfrac{8}{40}\right)$

$= \dfrac{11}{5}\left(\dfrac{27}{40}\right) = \dfrac{\textbf{297}}{\textbf{200}}$

28. (a) $3^{-3} = \dfrac{1}{3^3} = \dfrac{\textbf{1}}{\textbf{27}}$

(b) $\dfrac{1}{3^{-3}} = 3^3 = \textbf{27}$

29. $a^{-3}b^3 = (-2)^{-3}(-3)^3 = \dfrac{(-3)^3}{(-2)^3} = \dfrac{\textbf{27}}{\textbf{8}}$

30. $ab^2 + ab = (-3)(3)^2 + (-3)(3)$

$= -27 - 9$

$= \textbf{-36}$

PROBLEM SET 116

1. $P(R_8, B_{10}) = \dfrac{2}{52} \cdot \dfrac{2}{52} = \dfrac{1}{676}$

2. $\dfrac{D}{S} = \dfrac{3}{2}, \qquad \dfrac{D}{T} = \dfrac{3}{5}, \qquad \dfrac{S}{T} = \dfrac{2}{5}$

$\dfrac{3}{5} = \dfrac{D}{25,000}$

$75,000 = 5D$

$D = \mathbf{15,000 \ customers}$

3. $9N + 5 = 2N - 16$

$7N = -21$

$N = \mathbf{-3}$

4. A 92% decrease means 8% remains.

$8\% \cdot A = 6$

$0.08A = 6$

$\dfrac{0.08A}{0.08} = \dfrac{6}{0.08}$

$A = \mathbf{75}$

5. $9\dfrac{1}{2}\% \times \$30,000 = B$

$0.095 \times \$30,000 = B$

$\$2850 = B$

$\$30,000 + \$2850 = \mathbf{\$32,850}$

6. (a) $1\dfrac{8}{16}$ in. $= \mathbf{1\dfrac{1}{2} \ in.}$

(b) **3.8 cm**

7. $6\% \times A = \$6540$

$0.06A = \$6540$

$\dfrac{0.06A}{0.06} = \dfrac{\$6540}{0.06}$

$A = \mathbf{\$109,000}$

8. $P \cdot 90 = 162$

$P = \dfrac{162}{90}$

$P = 1.8$

$P = \mathbf{180\%}$

9. $35\% \cdot A = 2450$

$0.35A = 2450$

$\dfrac{0.35A}{0.35} = \dfrac{2450}{0.35}$

$A = \mathbf{7000}$

10. $y = 0$

x	y
-2	0
0	0
2	0

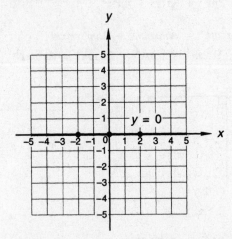

Note: The line $y = 0$ represents the x-axis. Though it is difficult to notice in the graph above, if you look carefully you will see that we have made the x-axis thicker than usual in an attempt to indicate that it is the line we are graphing.

11. MMM = 3000

CM = 900

XC = 90

IV = 4

MMMCMXCIV = **3994**

12. $7x - 23 = 5$

$7x = 28$

$x = 4$

$1.2x + 100 = 1.2(4) + 100 = \mathbf{104.8}$

13. One possible combination is **rotation, reflection, translation.** Answers may vary. See student work.

14. (a) $A = Bh = 6(4) \ \text{m}^2 = \mathbf{24 \ m^2}$

(b) $A = \dfrac{1}{2}h(b_1 + b_2)$

$= \dfrac{1}{2}(5 \ \text{m})(4 + 7) \ \text{m}$

$= \mathbf{27.5 \ m^2}$

15. (a) $\dfrac{3}{1} = \dfrac{5}{x}$

$3x = 5$

$x = \dfrac{5}{3}$

(b) $z + 37° + 27° = 180°$

$z = 180° - 37° - 27°$

$z = \mathbf{116°}$

16. $am^2\left(2am + a + \dfrac{m}{b^2}\right)$

$= \mathbf{2a^2m^3 + a^2m^2 + \dfrac{am^2}{b^2}}$

17. $4a^3pm^2 + a^2pm^{-1} - 2apma^2m$

$= 4a^3pm^2 + a^2pm^{-1} - 2a^3pm^2$

$= \mathbf{2a^3pm^2 + a^2pm^{-1}}$

18.

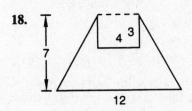

$V = A_{\text{Base}} \times \text{height}$

$= [A_{\text{Trapezoid}} - A_{\text{Rectangle}}] \times \text{height}$

$= \left[\dfrac{1}{2}(7)(4 + 12) - (4)(3)\right] \text{in.}^2 \times 12 \text{ in.}$

$= [56 - 12](12) \text{ in.}^3$

$= \mathbf{528 \text{ in.}^3}$

19. $P = (3 + 4 + 3 + 8.06 + 12 + 8.06) \text{ in.}$

$= \mathbf{38.12 \text{ in.}}$

20. $16x - 3 + 3x - 5 = -3x - 63$

$19x - 8 = -3x - 63$

$22x = -55$

$x = \mathbf{-\dfrac{5}{2}}$

21. $-2\dfrac{1}{2}x + 1\dfrac{1}{16} = \dfrac{1}{8}$

$-\dfrac{5}{2}x + \dfrac{17}{16} - \dfrac{17}{16} = \dfrac{1}{8} - \dfrac{17}{16}$

$-\dfrac{5}{2}x = -\dfrac{15}{16}$

$\left(-\dfrac{2}{5}\right)\left(-\dfrac{5}{2}\right)x = \left(-\dfrac{15}{16}\right)\left(-\dfrac{2}{5}\right)$

$x = \mathbf{\dfrac{3}{8}}$

22. $\dfrac{\frac{1}{5}}{-\frac{4}{9}} = \dfrac{\frac{3}{10}}{x}$

$\dfrac{1}{5}x = -\dfrac{2}{15}$

$x = \mathbf{-\dfrac{2}{3}}$

23. $4(7 - 2 \cdot 6) + \dfrac{1}{3^{-2}} - (-3^2) = 4(-5) + 9 + 9$

$= -20 + 9 + 9 = \mathbf{-2}$

24. $3mn^3pn^2p^3m = \mathbf{3m^2n^5p^4}$

25. $(-4)^3 + \sqrt[3]{-8} = -64 - 2 = \mathbf{-66}$

26. $\dfrac{-(-5^2) - 2^2(3 - 2 \cdot 3)}{-2(\sqrt[3]{-125})} = \dfrac{25 - 4(-3)}{-2(-5)}$

$= \dfrac{25 + 12}{10} = \mathbf{\dfrac{37}{10}}$

27. $4\dfrac{1}{2}\left(2\dfrac{1}{3} \cdot 1\dfrac{1}{2} - \dfrac{1}{4} \cdot \dfrac{4}{7}\right)$

$= \dfrac{9}{2}\left(\dfrac{7}{3} \cdot \dfrac{3}{2} - \dfrac{1}{4} \cdot \dfrac{4}{7}\right) = \dfrac{9}{2}\left(\dfrac{7}{2} - \dfrac{1}{7}\right)$

$= \dfrac{9}{2}\left(\dfrac{49}{14} - \dfrac{2}{14}\right) = \dfrac{9}{2}\left(\dfrac{47}{14}\right) = \mathbf{\dfrac{423}{28}}$

28. (a) $1^{-11} = \dfrac{1}{1^{11}} = \mathbf{1}$

(b) $\dfrac{2}{2^{-2}} = 2(2^2) = \mathbf{8}$

(c) $\dfrac{a^2}{a^{-3}} = a^2a^3 = \mathbf{a^5}$

29. $ab^2 + a^b = (2)(-3)^2 + (2^{-3})$

$= 18 + \dfrac{1}{8}$

$= \mathbf{18\dfrac{1}{8}}$

30. $\sqrt[a]{-b} - (ab)^{-1} = \left[\sqrt[3]{-(8)}\right] - [(3)(8)]^{-1}$

$= -2 - \dfrac{1}{24}$

$= \mathbf{-2\dfrac{1}{24}}$

PROBLEM SET 117

1. $0.7 \times 2000 = B$

$1400 = B$

Not bucolic = $2000 - 1400 = $ **600 paintings**

2. If 96% caviled, 4% did not cavil.

$0.04 \times N = 140$

$\dfrac{0.04N}{0.04} = \dfrac{140}{0.04}$

$N = $ **3500 newcomers**

3. $\dfrac{A}{S} = \dfrac{3}{14},\qquad \dfrac{A}{T} = \dfrac{3}{17},\qquad \dfrac{S}{T} = \dfrac{14}{17}$

$\dfrac{14}{17} = \dfrac{S}{1700}$

$23{,}800 = 17S$

$S = $ **1400 spectators**

4. $N + 40 = 6N - 5$

$45 = 5N$

$9 = N$

5. $P(H) = \dfrac{1}{2}$

6. $P(W, G) = \dfrac{2}{10} \cdot \dfrac{3}{10} = \dfrac{3}{50}$

7. Markdown $= 0.5 \times \$1.28 = \$0.64$
New price $= \$1.28 - \$0.64 = \$0.64$

Markup $= 0.5 \times \$0.64 = \$0.32$
New price $= \$0.64 + \$0.32 = \$0.96$

Markdown $= 0.5 \times \$0.96 = \$0.48$
New price $= \$0.96 - \$0.48 = \$0.48$

Markup $= 0.5 \times \$0.48 = \$0.24$
New price $= \$0.48 + \$0.24 = \$0.72$

The final price was **72¢ per lb.**

8. Interest 1st 6 mo $= 0.045 \times \$2000 = \$90$
Total after 6 mo $= \$2000 + \$90 = \$2090$

Interest 2nd 6 mo $= 0.045 \times \$2090 = \$94.05$
Total after 1 yr $= \$2090 + \$94.05 = \$2184.05$

Interest 3rd 6 mo $= 0.045 \times \$2184.05 = \$98.28$
Total after 18 mo $= \$2184.05 + \$98.28$
$\qquad\qquad\qquad = \$2282.33$

Interest 4th 6 mo $= 0.045 \times \$2282.33$
$\qquad\qquad\qquad = \$102.70$

Total after 2 yr $= \$2282.33 + \$102.70$
$\qquad\qquad\qquad = $ **\$2385.03**

9. Using the Pythagorean theorem:

$m^2 = 6^2 + 7^2$

$m^2 = 36 + 49$

$m^2 = 85$

$m = \pm\sqrt{85}$

$m = \sqrt{\mathbf{85}}$

10. Using the Pythagorean theorem:

$(\sqrt{5})^2 = m^2 + 1^2$

$5 = m^2 + 1$

$4 = m^2$

$\pm 2 = m$

$\mathbf{2} = m$

11. First, find the hypotenuse:

$c^2 = 6^2 + 8^2$

$c^2 = 36 + 64$

$c^2 = 100$

$c = \pm 10$

$c = 10$

$P = (6 + 8 + 8 + 8 + 10)\ \text{dm} = $ **40 dm**

12. $A = \dfrac{1}{2}\left(4\dfrac{1}{2}\right)(5 + 7)\ \text{in.}^2$

$A = \dfrac{1}{2}\left(\dfrac{9}{2}\right)(12)\ \text{in.}^2$

$A = $ **27 in.^{2}**

13. $0.6 \times 125 = B$

$75 = B$

$125 + 75 = $ **200**

14. $12\dfrac{1}{4}\% \times 200 = B$

$0.1225 \times 200 = B$

$24.5 = B$

$200 + 24.5 = $ **224.5**

15. **57 mm**

16. $3ac(ab + ac + bc) = \mathbf{3a^2bc + 3a^2c^2 + 3abc^2}$

17. $90° - 22\dfrac{1}{2}° = \mathbf{67\dfrac{1}{2}°}$

$180° - 22\dfrac{1}{2}° = \mathbf{157\dfrac{1}{2}°}$

18. $ab^2 + abb + 2a^2ab + 2abb - baa^2$
    $= ab^2 + ab^2 + 2a^3b + 2ab^2 - a^3b$
    $= \mathbf{4ab^2 + a^3b}$

19. $-xy^3 + x^2xyy^2 + x^3y^3 - 3xyy^2$
    $= -xy^3 + x^3y^3 + x^3y^3 - 3xy^3 = \mathbf{2x^3y^3 - 4xy^3}$

20. $3x + 4 - 2x - x = 4x - 5$
    $\qquad\qquad 4 = 4x - 5$
    $\qquad\qquad 9 = 4x$
    $\qquad\qquad \dfrac{9}{4} = x$

21. $\qquad 3\dfrac{1}{2}x - 1\dfrac{3}{4} = -\dfrac{3}{2}$
    $\qquad \dfrac{7}{2}x - \dfrac{7}{4} + \dfrac{7}{4} = -\dfrac{3}{2} + \dfrac{7}{4}$
    $\qquad\qquad \dfrac{7}{2}x = \dfrac{1}{4}$
    $\qquad \dfrac{2}{7} \cdot \dfrac{7}{2}x = \dfrac{1}{4} \cdot \dfrac{2}{7}$
    $\qquad\qquad x = \dfrac{1}{14}$

22. $6 + x - 2x + 3x - 4x = 3x + 4$
    $\qquad\qquad 6 - 2x = 3x + 4$
    $\qquad\qquad 2 = 5x$
    $\qquad\qquad \dfrac{2}{5} = x$

23. $\dfrac{-\dfrac{1}{4}}{\dfrac{1}{3}} = \dfrac{\dfrac{16}{9}}{x}$
    $\qquad -\dfrac{1}{4}x = \dfrac{16}{27}$
    $\qquad x = -\dfrac{64}{27}$

24. $x^2 = 1$
    $x = \pm\sqrt{1}$
    $x = \mathbf{\pm 1}$

25. $aba^2b^2a^3b = \mathbf{a^6b^4}$

26. $xy^3x^2xxy^3 = \mathbf{x^5y^6}$

27. $\dfrac{-(-3)^2 - 2^3(2^3 - 2 \cdot 6) + 3}{2^2(2^3 - 4)}$
    $= \dfrac{-9 - 8(-4) + 3}{4(4)} = \dfrac{13}{8}$

28. $\dfrac{1}{5}\left(2\dfrac{1}{3} \cdot \dfrac{1}{4} - \dfrac{7}{12}\right) = \dfrac{1}{5}\left(\dfrac{7}{3} \cdot \dfrac{1}{4} - \dfrac{7}{12}\right)$
    $= \dfrac{1}{5}\left(\dfrac{7}{12} - \dfrac{7}{12}\right) = \dfrac{1}{5}(0) = \mathbf{0}$

29. $-b^2 + a + b = -(-1)^2 + (-2) + (-1)$
    $\qquad\qquad\qquad = -1 - 2 - 1$
    $\qquad\qquad\qquad = \mathbf{-4}$

30. $-b^3 + 2ab = -(-2)^3 + 2(-1)(-2)$
    $\qquad\qquad\quad = 8 + 4$
    $\qquad\qquad\quad = \mathbf{12}$

PROBLEM SET 118

1. $P(5, 2) = \dfrac{1}{6} \cdot \dfrac{1}{6} = \dfrac{1}{36}$

2. $0.075 \times \$40 = T$
   $\qquad\quad \mathbf{\$3} = T$

3. $\dfrac{F}{L} = \dfrac{7}{2}, \qquad \dfrac{F}{T} = \dfrac{7}{9}, \qquad \dfrac{L}{T} = \dfrac{2}{9}$
   $\qquad \dfrac{7}{9} = \dfrac{F}{45,000}$
   $315,000 = 9F$
   $\qquad F = \mathbf{35,000 \ contestants}$

4. $-11N + 6 = 25N - 30$
   $\qquad\quad 36 = 36N$
   $\qquad\quad \mathbf{1} = N$

5. $3\dfrac{1}{8}$ **in.**

6. 1st yr: $0.035 \times \$2000 = \$70$
   2nd yr: $0.035 \times \$2070 = \$72.45$

   Total interest: $\$70 + \$72.45 = \mathbf{\$142.45}$

7. Markup: $0.5(\$61,000) = \$30,500$
   Intended price: $\$61,000 + \$30,500 = \$91,500$

   Selling price: $\dfrac{1}{2} \times \$91,500 = \$45,750$

   Profit = Selling price − Cost − Expenses
   $\qquad = \$45,750 - \$61,000 - \$12,000$
   $\qquad = \mathbf{-\$27,250}$

8. $y = 3x - 5$

x	y
0	-5
1	-2
2	1

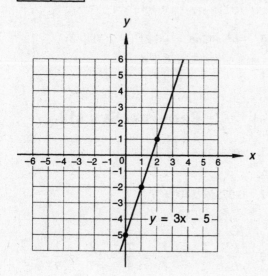

9. $20\frac{1}{2}\% \times 160 = B$

$0.205 \times 160 = B$

$32.8 = B$

$160 + 32.8 = \mathbf{192.8}$

10. $10 \cancel{mi}^3 \times \dfrac{5280 \cancel{ft}}{1 \cancel{mi}} \times \dfrac{5280 \cancel{ft}}{1 \cancel{mi}} \times \dfrac{5280 \cancel{ft}}{1 \cancel{mi}}$

$\times \dfrac{12 \text{ in.}}{1 \cancel{ft}} \times \dfrac{12 \text{ in.}}{1 \cancel{ft}} \times \dfrac{12 \text{ in.}}{1 \cancel{ft}}$

$= 10(5280)(5280)(5280)(12)(12)(12) \text{ in.}^3$

$\approx \mathbf{2.54 \times 10^{15} \text{ in.}^3}$

11. $1000 \cancel{mi}^3 \times \dfrac{5280 \cancel{ft}}{1 \cancel{mi}} \times \dfrac{5280 \cancel{ft}}{1 \cancel{mi}} \times \dfrac{5280 \cancel{ft}}{1 \cancel{mi}}$

$\times \dfrac{12 \text{ in.}}{1 \cancel{ft}} \times \dfrac{12 \text{ in.}}{1 \cancel{ft}} \times \dfrac{12 \text{ in.}}{1 \cancel{ft}}$

$= 1000(5280)(5280)(5280)(12)(12)(12) \text{ in.}^3$

$\approx \mathbf{2.54 \times 10^{17} \text{ in.}^3}$

12. One possible combination is **rotation, translation.** Answers may vary. See student work.

13. $9^2 = m^2 + 6^2$

$81 - 36 = m^2$

$45 = m^2$

$\sqrt{45} = m$

14. (a) $\dfrac{4}{8} = \dfrac{k}{9}$

$8k = 36$

$k = \dfrac{36}{8}$

$k = \mathbf{4\frac{1}{2}}$

(b) $Z + 100° + 23° = 180°$

$Z = 180° - 100° - 23°$

$Z = \mathbf{57°}$

15. $A = \dfrac{1}{2}h(b_1 + b_2) = \dfrac{b_1 h}{2} + \dfrac{b_2 h}{2}$

$= \left[\dfrac{3(3.9)}{2} + \dfrac{11(3.9)}{2}\right] \text{ft}^2$

$= \mathbf{27.3 \text{ ft}^2}$

16. $V = A_{\text{Base}} \times \text{height}$

$= [A_{\text{Semicircle}} + A_{\text{Rectangle}}] \times \text{height}$

$= \left[\dfrac{\pi(12)^2}{2} + (22)(12)\right] \text{cm}^2 \times 10 \text{ cm}$

$= [72\pi + 264](10) \text{ cm}^3$

$\approx \mathbf{4900.8 \text{ cm}^3}$

17. $L.A. = \text{Perimeter} \times \text{height}$

$= [(2)(22) + \pi(12) + (2)(12)] \text{ cm} \times 10 \text{ cm}$

$\approx [105.68](10) \text{ cm}^2$

$\approx \mathbf{1056.8 \text{ cm}^2}$

18. $x^2y^3(x + yx + 2xy^7)$

$= \mathbf{x^3y^3 + x^3y^4 + 2x^3y^{10}}$

19. $4zmn^3 - 2zm^2 + nzmn^2 - mzm$

$= 4zmn^3 - 2zm^2 + zmn^3 - zm^2$

$= \mathbf{5zmn^3 - 3zm^2}$

20. $4x - 9 + x = -3x + 31$

$8x = 40$

$x = \mathbf{5}$

21. $-1\frac{1}{2}x - \dfrac{1}{4} = 7\frac{1}{2}$

$-\dfrac{3}{2}x - \dfrac{1}{4} + \dfrac{1}{4} = \dfrac{15}{2} + \dfrac{1}{4}$

$-\dfrac{3}{2}x = \dfrac{31}{4}$

$x = -\dfrac{2}{3} \cdot \dfrac{31}{4}$

$x = \mathbf{-\dfrac{31}{6}}$

22.
$$\frac{-\dfrac{1}{6}}{\dfrac{4}{3}} = \frac{x}{-\dfrac{1}{7}}$$

$$\frac{1}{42} = \frac{4}{3}x$$

$$\frac{1}{56} = x$$

23. (a) $\dfrac{1}{7^{-2}} = 7^2 = \mathbf{49}$

(b) $(-3)^{-3} = \dfrac{1}{(-3)^3} = -\dfrac{1}{27}$

24. $-(-3^2) - 4(3 \cdot 2 - 2^2) = 9 - 4(2) = \mathbf{1}$

25. $7mn^4m^2bnb^4 = \mathbf{7m^3n^5b^5}$

26. $x \geq -2$

27. $12 = 2 \cdot 2 \cdot 3$
$20 = 2 \cdot 2 \cdot 5$
$24 = 2 \cdot 2 \cdot 2 \cdot 3$
LCM $(12, 20, 24) = 2 \cdot 2 \cdot 2 \cdot 3 \cdot 5 = \mathbf{120}$

28. GCF $(12, 20, 24) = 2 \cdot 2 = \mathbf{4}$

29. $ab^3 + (-ab)^3 = (-2)(3)^3 + [-(-2)(3)]^3$
$= -54 + 216$
$= \mathbf{162}$

30. $\sqrt{pz} + \sqrt[3]{z} + p^z$
$= \sqrt{(-2)(-8)} + \sqrt[3]{(-8)} + \left[(-2)^{-8}\right]$
$= 4 - 2 + \dfrac{1}{256}$
$= \mathbf{2\dfrac{1}{256}}$

PROBLEM SET 119

1. (a) There are five combinations out of a total of thirty-six that equal eight, thus
$$P(8) = \frac{5}{36}.$$

(b) There are four combinations out of a total of thirty-six that equal nine, thus
$$P(9) = \frac{4}{36} = \frac{1}{9}.$$

(c) There are three combinations out of a total of thirty-six that equal ten, thus
$$P(10) = \frac{3}{36} = \frac{1}{12}.$$

2. $P(S) = \dfrac{13 - 3}{52 - 3} = \dfrac{10}{49}$

3. If 20% were red, 80% were not red.
$$0.8 \times A = 1400$$
$$\frac{0.8A}{0.8} = \frac{1400}{0.8}$$
$$A = 1750 \text{ flowers total}$$
$1750 - 1400 = \mathbf{350 \text{ flowers}}$ were red

4. $\dfrac{E}{M} = \dfrac{3}{14}, \qquad \dfrac{E}{T} = \dfrac{3}{17}, \qquad \dfrac{M}{T} = \dfrac{14}{17}$

$$\frac{3}{17} = \frac{E}{3400}$$
$$10{,}200 = 17E$$
$$E = \mathbf{600 \text{ ballplayers}}$$

5. Markup $= 0.6 \times \$300 = \$180$
Regular price $= \$300 + \$180 = \$480$
Discount $= 0.2 \times \$480 = \$96$
Selling price $= \$480 - \$96 = \$384$
Profit $= \$384 - \$300 = \mathbf{\$84}$

6. $\dfrac{1}{4} \times 1744 = B$
$$436 = B$$
$1744 - 436 = \mathbf{1308 \text{ people}}$

7. $5N = 3N - 6$
$2N = -6$
$N = \mathbf{-3}$

8. $A^2 + 6^2 = 13^2$
$A^2 + 36 = 169$
$A^2 = 133$
$A = \pm\sqrt{133}$
$A = \mathbf{\sqrt{133}}$

9. $55 \text{ in.}^3 \times \dfrac{1 \text{ ft}}{12 \text{ in.}} \times \dfrac{1 \text{ ft}}{12 \text{ in.}} \times \dfrac{1 \text{ ft}}{12 \text{ in.}}$
$= \dfrac{55}{(12)(12)(12)} \text{ ft}^3 \approx \mathbf{0.032 \text{ ft}^3}$

10. $A = Bh$
$\quad\quad = (10 \text{ ft})(5 \text{ ft})$
$\quad\quad = \mathbf{50 \text{ ft}^2}$

11. $10,000 \text{ cm}^3 \times \dfrac{1 \text{ m}}{100 \text{ cm}} \times \dfrac{1 \text{ m}}{100 \text{ cm}} \times \dfrac{1 \text{ m}}{100 \text{ cm}}$

$\quad\quad \times \dfrac{1 \text{ km}}{1000 \text{ m}} \times \dfrac{1 \text{ km}}{1000 \text{ m}} \times \dfrac{1 \text{ km}}{1000 \text{ m}}$

$\quad = \dfrac{10,000}{(100)(100)(100)(1000)(1000)(1000)} \text{ km}^3$

$\quad = \mathbf{1 \times 10^{-11} \text{ km}^3}$

12. $12 \text{ m}^3 \times \dfrac{100 \text{ cm}}{1 \text{ m}} \times \dfrac{100 \text{ cm}}{1 \text{ m}} \times \dfrac{100 \text{ cm}}{1 \text{ m}}$

$\quad\quad \times \dfrac{1 \text{ L}}{1000 \text{ cm}^3} = \dfrac{12(100)(100)(100)}{1000} \text{ L}$

$\quad = \mathbf{12,000 \text{ L}}$

13. $P \cdot 70 = 112$

$\quad\quad P = \dfrac{112}{70}$

$\quad\quad P = 1.6$

$\quad\quad P = \mathbf{160\%}$

14. $y = \dfrac{1}{2}x + 2$

x	y
–2	1
0	2
2	3

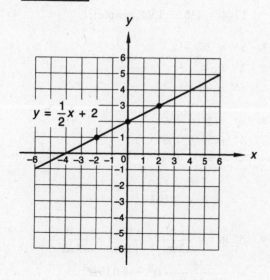

$y = \dfrac{1}{2}x + 2$

15. $1320 \text{ m}^3 \times \dfrac{100 \text{ cm}}{1 \text{ m}} \times \dfrac{100 \text{ cm}}{1 \text{ m}} \times \dfrac{100 \text{ cm}}{1 \text{ m}}$

$\quad\quad \times \dfrac{1 \text{ L}}{1000 \text{ cm}^3}$

$\quad = \dfrac{1320(100)(100)(100)}{1000} \text{ L} = \mathbf{1,320,000 \text{ L}}$

16. $15,321,156 \text{ mL} \times \dfrac{1 \text{ L}}{1000 \text{ mL}} = \mathbf{15,321.156 \text{ L}}$

17. $2ab(a + b - ab^2 + a^2b)$
$\quad = \mathbf{2a^2b + 2ab^2 - 2a^2b^3 + 2a^3b^2}$

18. $2xyy^2 + 3x^2xy - 6yxy^2 + 7xyy^2$
$\quad = 2xy^3 + 3x^3y - 6xy^3 + 7xy^3 = \mathbf{3xy^3 + 3x^3y}$

19.

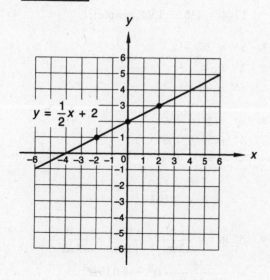

$V = A_{\text{Base}} \times \text{height}$
$\quad = [A_{\text{Rectangle}} + A_{\text{Triangle}}] \times \text{height}$
$\quad = \left[(40)(7) + \dfrac{(24)(12)}{2}\right]\left(\dfrac{(3)}{(100)(100)}\right) \text{ m}^3$
$\quad = \mathbf{0.1272 \text{ m}^3}$

20. $5x - 5 = -6x + 2x - 22$
$\quad\quad 9x = -17$
$\quad\quad x = -\dfrac{\mathbf{17}}{\mathbf{9}}$

21. $\quad\quad -\dfrac{2}{3}x - \dfrac{1}{5} = 1\dfrac{5}{6}$

$\quad -\dfrac{2}{3}x - \dfrac{1}{5} + \dfrac{1}{5} = \dfrac{11}{6} + \dfrac{1}{5}$

$\quad\quad\quad -\dfrac{2}{3}x = \dfrac{61}{30}$

$\quad -\dfrac{3}{2}\left(-\dfrac{2}{3}\right)x = -\dfrac{3}{2} \cdot \dfrac{61}{30}$

$\quad\quad\quad x = -\dfrac{\mathbf{61}}{\mathbf{20}}$

22. $m^2 = 25$
$\quad m = \pm\sqrt{25}$
$\quad m = \mathbf{\pm 5}$

23. $-[-(-2)^3] + (-2) = -8 - 2 = \mathbf{-10}$

24. $3a^2ba^3b^2b^3 = 3a^5b^6$

25. $\dfrac{-2^3 - 2^2(2^2 - 3) + 3}{3(2^3 - 1)}$

$= \dfrac{-8 - 4(1) + 3}{3(7)} = -\dfrac{9}{21} = -\dfrac{3}{7}$

26. $\dfrac{1}{3}\left(2\dfrac{1}{2} \cdot \dfrac{1}{3} - \dfrac{1}{6} \cdot \dfrac{1}{2}\right) = \dfrac{1}{3}\left(\dfrac{10}{12} - \dfrac{1}{12}\right)$

$= \dfrac{1}{3}\left(\dfrac{9}{12}\right) = \dfrac{1}{4}$

27. $xy^2 + 2x^2y = 1(-2)^2 + 2(1)^2(-2) = 4 - 4 = \mathbf{0}$

28. $400 = CD$

$40 = XL$

$4 = IV$

$444 = \mathbf{CDXLIV}$

29. **7 cm**

30. 621,321,131.2 is **six hundred twenty-one million, three hundred twenty-one thousand, one hundred thirty-one and two tenths**

PROBLEM SET 120

1. $P(T, T, T, T) = \dfrac{1}{2} \cdot \dfrac{1}{2} \cdot \dfrac{1}{2} \cdot \dfrac{1}{2} = \dfrac{1}{16}$

2. There are five combinations out of a total of thirty-six that equal eight, thus

$$P(8) = \dfrac{5}{36}.$$

3. $300\% \times A = 1200$

$3A = 1200$

$A = \mathbf{400\ people}$

4. $\dfrac{a}{t} = \dfrac{9}{7}, \qquad \dfrac{a}{T} = \dfrac{9}{16}, \qquad \dfrac{t}{T} = \dfrac{7}{16}$

$\dfrac{7}{16} = \dfrac{t}{1600}$

$16t = 7 \cdot 1600$

$t = \mathbf{700\ stones}$

5.

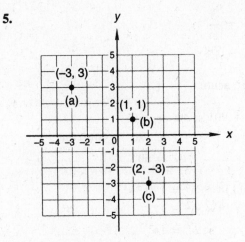

6. $S.A. = \pi r^2 + \pi r l$

$= \pi(2\text{ m})^2 + \pi(2\text{ m})(5\text{ m})$

$= 14\pi\text{ m}^2$

$\approx \mathbf{43.96\ m^2}$

7. $V_{\text{Cylinder}} = \pi r^2 h$

$= \pi(2\sqrt{2}\text{ m})^2(2\text{ m})$

$= \pi(8\text{ m}^2)(2\text{ m})$

$= 16\pi\text{ m}^3$

$\approx 50.24\text{ m}^3$

$V_{\text{Cone}} = \dfrac{1}{3}V_{\text{Cylinder}}$

$\approx \dfrac{1}{3}(50.24\text{ m}^3)$

$\approx \mathbf{16.75\ m^3}$

8. $V_{\text{Right Solid}} = (6\text{ m})^2(4\text{ m}) = (36)(4)\text{ m}^3$

$= 144\text{ m}^3$

$V_{\text{Pyramid}} = \dfrac{1}{3}V_{\text{Right solid}} = \dfrac{1}{3}(144\text{ m}^3) = \mathbf{48\ m^3}$

9. $S.A. = A_{\text{Base}} + 4A_{\text{Face}}$

$= (6\text{ m})^2 + 4\left[\dfrac{(6\text{ m})(5\text{ m})}{2}\right]$

$= 36\text{ m}^2 + 4(15)\text{ m}^2$

$= \mathbf{96\ m^2}$

10. $V_{\text{Cylinder}} = \pi r^2 h$

$= \pi(3\text{ m})^2(6\text{ m})$

$= 54\pi\text{ m}^3$

$\approx 169.56\text{ m}^3$

$V_{\text{Sphere}} = \dfrac{2}{3}V_{\text{Cylinder}}$

$\approx \dfrac{2}{3}(169.56)\text{ m}^3$

$\approx \mathbf{113.04\ m^3}$

11. $\dfrac{1}{3} = \dfrac{5}{m}$

 $m = \mathbf{15}$

12. **35°, acute**

13. $Z + 90° + 40° = 180°$

 $Z = 180° - 90° - 40°$

 $Z = \mathbf{50°}$

14. $(\sqrt{3})^2 + b^2 = (\sqrt{17})^2$

 $3 + b^2 = 17$

 $b^2 = 14$

 $b = \mathbf{\sqrt{14}}$

15. $A = \dfrac{1}{2}h(b_1 + b_2)$

 $= \dfrac{1}{2}(4\text{ ft})(10\text{ ft} + 6\text{ ft})$

 $= \dfrac{1}{2}(4)(16)\text{ ft}^2$

 $= \mathbf{32\text{ ft}^2}$

16. $B = 16\dfrac{1}{2}\% \times 1100$

 $B = 0.165 \times 1100$

 $B = \mathbf{181.5}$

17. $7^2 = 49$

 $8^2 = 64$

 So, $\mathbf{7 < \sqrt{57} < 8}$

18. $70{,}000{,}000 \ \cancel{\text{mL}} \times \dfrac{1 \ \cancel{\text{cm}^3}}{1 \ \cancel{\text{mL}}} \times \dfrac{1 \text{ m}}{100 \ \cancel{\text{cm}}} \times \dfrac{1 \text{ m}}{100 \ \cancel{\text{cm}}}$

 $\times \dfrac{1 \text{ m}}{100 \ \cancel{\text{cm}}}$

 $= \dfrac{70{,}000{,}000}{(100)(100)(100)} \text{ m}^3 = \mathbf{70 \text{ m}^3}$

19. $60 \ \cancel{\text{in.}^3} \times \dfrac{1 \ \cancel{\text{ft}}}{12 \ \cancel{\text{in.}}} \times \dfrac{1 \ \cancel{\text{ft}}}{12 \ \cancel{\text{in.}}} \times \dfrac{1 \ \cancel{\text{ft}}}{12 \ \cancel{\text{in.}}} \times \dfrac{1 \text{ mi}}{5280 \ \cancel{\text{ft}}}$

 $\times \dfrac{1 \text{ mi}}{5280 \ \cancel{\text{ft}}} \times \dfrac{1 \text{ mi}}{5280 \ \cancel{\text{ft}}}$

 $= \dfrac{60}{(12)(12)(12)(5280)(5280)(5280)} \text{ mi}^3$

 $\approx \mathbf{2.36 \times 10^{-13} \text{ mi}^3}$

20. $5x + 4 - x = 3x + 2$

 $4x + 4 = 3x + 2$

 $x = \mathbf{-2}$

21. $\qquad -4\dfrac{1}{3}x + \dfrac{1}{4} = -3\dfrac{1}{12}$

 $-\dfrac{13}{3}x + \dfrac{1}{4} - \dfrac{1}{4} = -\dfrac{37}{12} - \dfrac{1}{4}$

 $-\dfrac{13}{3}x = -\dfrac{40}{12}$

 $\left(-\dfrac{3}{13}\right)\left(-\dfrac{13}{3}\right)x = \left(-\dfrac{40}{12}\right)\left(-\dfrac{3}{13}\right)$

 $x = \mathbf{\dfrac{10}{13}}$

22. $\dfrac{-\dfrac{1}{6}}{\dfrac{2}{5}} = \dfrac{x}{\dfrac{1}{3}}$

 $-\dfrac{1}{18} = \dfrac{2}{5}x$

 $-\dfrac{5}{36} = x$

23. (a) $\dfrac{1}{9^{-2}} = 9^2 = \mathbf{81}$

 (b) $4^{-3} = \dfrac{1}{4^3} = \mathbf{\dfrac{1}{64}}$

24. $\dfrac{(-2)^3 - 3(-1^5 - 3 \cdot 2)}{2(3^3 - \sqrt[5]{-243})} = \dfrac{-8 - 3(-1 - 6)}{2[27 - (-3)]}$

 $= \dfrac{-8 - 3(-7)}{2(30)} = \mathbf{\dfrac{13}{60}}$

25. $6b^3ab^2a^5b = \mathbf{6a^6b^6}$

26. $\dfrac{1}{4}\left(2\dfrac{1}{2} \cdot \dfrac{1}{3} - \dfrac{1}{6} \cdot \dfrac{1}{5}\right) = \dfrac{1}{4}\left(\dfrac{5}{6} - \dfrac{1}{30}\right)$

 $= \dfrac{1}{4}\left(\dfrac{25}{30} - \dfrac{1}{30}\right) = \dfrac{1}{4}\left(\dfrac{24}{30}\right) = \mathbf{\dfrac{1}{5}}$

27. $ab^3 - b = (3)(-3)^3 - (-3) = -81 + 3 = \mathbf{-78}$

28. $cx^2 - c^5x = (-1)(-2)^2 - (-1)^5(-2)$

 $= -4 - 2$

 $= \mathbf{-6}$

29. Interest 1st yr = 0.1($8000) = $800

Total after 1 yr = $8000 + $800 = $8800

Interest 2nd yr = 0.1($8800) = $880

Total after 2 yr = $8800 + $880 = $9680

Interest 3rd yr = 0.1($9680) = $968

Total after 3 yr = $9680 + $968 = $10,648

Interest 4th yr = 0.1($10,648) = $1064.80

Total after 4 yr = $10,648 + $1064.80

= **$11,712.80**

30. $2abab - 4ba^2b + 7ab^2a + 3a^2b$

$= 2a^2b^2 - 4a^2b^2 + 7a^2b^2 + 3a^2b$

$= \mathbf{5a^2b^2 + 3a^2b}$

PROBLEM SET 121

1. $P(R, R) = \dfrac{9}{34} \cdot \dfrac{9}{34} = \mathbf{\dfrac{81}{1156}}$

2. $-17N - 5 = -9N + 3$

$-8N = 8$

$N = \mathbf{-1}$

3. If 30% could not produce, 70% could produce.

$\dfrac{70}{100} \times A = 1260$

$A = 1260 \cdot \dfrac{100}{70}$

$A = \mathbf{1800}$

4. $\dfrac{A}{N} = \dfrac{2}{3}, \qquad \dfrac{A}{T} = \dfrac{2}{5}, \qquad \dfrac{N}{T} = \dfrac{3}{5}$

$\dfrac{2}{5} = \dfrac{A}{325}$

$5A = 650$

$A = \mathbf{130 \text{ audiophiles}}$

5.

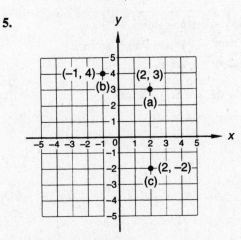

6. $A = \text{Base} \times \text{Height}$

$= 11 \text{ in.} \cdot 5 \text{ in.}$

$= \mathbf{55 \text{ in.}^2}$

7.

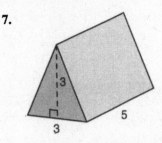

$V = A_{\text{Triangular End}} \times \text{length}$

$= \left(\dfrac{1}{2}\right)(3 \text{ in.})(3 \text{ in.}) \times 5 \text{ in.}$

$= \mathbf{22.5 \text{ in.}^3}$

8. $S.A. = A_{\text{Base}} + 4 \times A_{\text{Triangular Side}}$

$= (6 \cdot 6) \text{ ft}^2 + \left(4 \cdot \dfrac{1}{2} \cdot 6 \cdot 5\right) \text{ ft}^2$

$= 36 \text{ ft}^2 + 60 \text{ ft}^2$

$= \mathbf{96 \text{ ft}^2}$

9. $V_{\text{Cylinder}} = \pi r^2 h$

$= \pi(9 \text{ ft})^2(18 \text{ ft})$

$\approx 4578.12 \text{ ft}^3$

$V_{\text{Sphere}} = \dfrac{2}{3} V_{\text{Cylinder}}$

$\approx \dfrac{2}{3}(4578.12) \text{ ft}^3$

$\approx \mathbf{3052.08 \text{ ft}^3}$

10. $V = \dfrac{1}{3} \times A_{\text{Right Solid Base}} \times H_{\text{Right Solid}}$

$= \left(\dfrac{1}{3}\right)(\pi \cdot r^2)(\text{height})$

$\approx \left(\dfrac{1}{3}\right)(3.14 \cdot 10^2)(2) \text{ m}^3$

$\approx \mathbf{209.33 \text{ m}^3}$

11. (a) $z + 90° + 48° = 180°$

$z = 180° - 90° - 48°$

$z = \mathbf{42°}$

(b) $m^2 = 4^2 + 3^2$

$m^2 = 16 + 9$

$m^2 = 25$

$m = \mathbf{5}$

12.
$$\frac{m}{7} = \frac{2}{4}$$
$$4m = 14$$
$$m = \frac{14}{4}$$
$$m = \frac{7}{2}$$

13. **Figure (b) is symmetric about a point.**

(a)

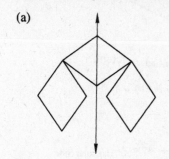

(b)

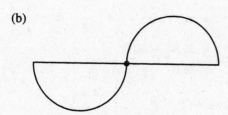

14. $A = 2[A_{\text{Large Circle}} - A_{\text{Small Circle}}]$
$= 2[\pi(4)^2 - \pi(2)^2]$ in.^{2}
$\approx$ **75.36 in.^{2}**

15. $C = 11\frac{3}{4}\% \times 4 \cdot \$2500$
$= 11.75\% \times \$10,000$
$= 0.1175 \times \$10,000$
$= \mathbf{\$1175}$

16. $y = -\frac{1}{2}x + 3$

x	y
-2	4
0	3
2	2

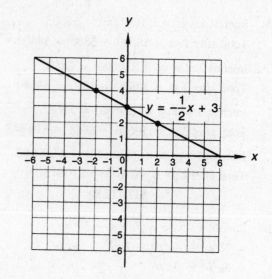

17. $9\frac{1}{2}\% \cdot A = 9500$
$0.095A = 9500$
$A = \dfrac{9500}{0.095}$
$A = \mathbf{100,000}$

18. $100,000 \text{ cm}^3 \times \dfrac{1 \text{ m}}{100 \text{ cm}} \times \dfrac{1 \text{ m}}{100 \text{ cm}} \times \dfrac{1 \text{ m}}{100 \text{ cm}}$
$\times \dfrac{1 \text{ km}}{1000 \text{ m}} \times \dfrac{1 \text{ km}}{1000 \text{ m}} \times \dfrac{1 \text{ km}}{1000 \text{ m}}$
$= \dfrac{100,000}{(100)(100)(100)(1000)(1000)(1000)} \text{ km}^3$
$= \mathbf{1 \times 10^{-10} \text{ km}^3}$

19. $20 \text{ mi}^3 \times \dfrac{5280 \text{ ft}}{1 \text{ mi}} \times \dfrac{5280 \text{ ft}}{1 \text{ mi}} \times \dfrac{5280 \text{ ft}}{1 \text{ mi}}$
$\times \dfrac{12 \text{ in.}}{1 \text{ ft}} \times \dfrac{12 \text{ in.}}{1 \text{ ft}} \times \dfrac{12 \text{ in.}}{1 \text{ ft}}$
$= 20(5280)(5280)(5280)(12)(12)(12) \text{ in.}^3$
$\approx \mathbf{5.09 \times 10^{15} \text{ in.}^3}$

20. $ab^2(ac + 3ab^2 - b) = \mathbf{a^2b^2c + 3a^2b^4 - ab^3}$

21. $2mp^2 - 3m^2p - 4pmp + 5mpm$
$= 2mp^2 - 3m^2p - 4mp^2 + 5m^2p$
$= \mathbf{-2mp^2 + 2m^2p}$

22. $4x - 6x - 6 = 3x + 14$
$-5x = 20$
$x = \mathbf{-4}$

23.

$$\dfrac{-\dfrac{1}{5}}{\dfrac{7}{8}} = \dfrac{-\dfrac{1}{20}}{x}$$

$$-\dfrac{1}{5}x = -\dfrac{7}{160}$$

$$x = \dfrac{7}{32}$$

3. Interest 1st yr = 0.095($10,000) = $950

Total after 1 yr = $10,000 + $950 = $10,950

Interest 2nd yr = 0.095($10,950) = $1040.25

Total after 2 yr = $10,950 + $1040.25

$$= \$11,990.25$$

Interest 3rd yr = 0.095($11,990.25) ≈ $1139.07

Total after 3 yr ≈ $11,990.25 + $1139.07

$$= \mathbf{\$13,129.32}$$

24. (a) $(-4)^{-3} = \dfrac{1}{(-4)^3} = -\dfrac{\mathbf{1}}{\mathbf{64}}$

(b) $\dfrac{1}{-4^{-2}} = -4^2 = \mathbf{-16}$

4. $\dfrac{P}{A} = \dfrac{2}{8}, \qquad \dfrac{P}{T} = \dfrac{2}{10}, \qquad \dfrac{A}{T} = \dfrac{8}{10}$

$$\dfrac{2}{10} = \dfrac{P}{1110}$$

$$10P = 2220$$

$$P = \mathbf{222 \text{ philatelists}}$$

$A = 1110 - 222 = \mathbf{888 \text{ archivists}}$

25. $-3^3 + (2)^3 - 3(2^3 - 3^2) = -27 + 8 - 3(-1)$
$$= \mathbf{-16}$$

26. $6m^3n^2bm^4n = \mathbf{6bm^7n^3}$

27. $\dfrac{1}{6}\left(2\dfrac{1}{2} \cdot \dfrac{2}{5} - \dfrac{1}{5} \cdot \dfrac{1}{4}\right) = \dfrac{1}{6}\left(1 - \dfrac{1}{20}\right)$
$$= \dfrac{\mathbf{19}}{\mathbf{120}}$$

5.

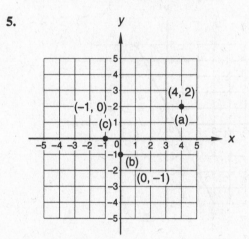

28. $x^y - y^x = \left[(-2)^{-3}\right] - \left[(-3)\right]^{-2}$
$$= \dfrac{1}{(-2)^3} - \dfrac{1}{(-3)^2}$$
$$= -\dfrac{1}{8} - \dfrac{1}{9}$$
$$= -\dfrac{9}{72} - \dfrac{8}{72}$$
$$= -\dfrac{\mathbf{17}}{\mathbf{72}}$$

6. There are 5! ways to order 5 objects in a row.

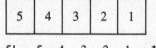

$5! = 5 \cdot 4 \cdot 3 \cdot 2 \cdot 1 = \mathbf{120}$

29. $\sqrt[q]{b} + \sqrt{b} + a^2 = \left(\sqrt[3]{64}\right) + \sqrt{(64)} + (3)^2$
$$= 4 + 8 + 9$$
$$= \mathbf{21}$$

7. (a) 7 objects:

7	6	5	4	3	2	1

$7! = 7 \cdot 6 \cdot 5 \cdot 4 \cdot 3 \cdot 2 \cdot 1$
$$= \mathbf{5040} \text{ different orders possible}$$

30. $x \nleq -4$

$x > -4$

(b) 3 of 7 objects:

There are $7 \cdot 6 \cdot 5 = \mathbf{210}$ different orders possible.

PROBLEM SET 122

1. $P\left(B_2, R_9\right) = \dfrac{2}{52} \cdot \dfrac{2}{52} = \dfrac{\mathbf{1}}{\mathbf{676}}$

2. $P(12) = \dfrac{1}{6} \cdot \dfrac{1}{6} = \dfrac{\mathbf{1}}{\mathbf{36}}$

8. $5 \text{ mi}^3 \times \dfrac{5280 \text{ ft}}{1 \text{ mi}} \times \dfrac{5280 \text{ ft}}{1 \text{ mi}} \times \dfrac{5280 \text{ ft}}{1 \text{ mi}}$

$= 5(5280)(5280)(5280) \text{ ft}^3 \approx 7.36 \times 10^{11} \text{ ft}^3$

9. $V_{\text{Pyramid}} = \dfrac{1}{3} V_{\text{Right Solid}} = \left(\dfrac{1}{3}\right)(A_{\text{Base}} \cdot \text{height})$

$= \left(\dfrac{1}{3}\right)(102 \cdot 30) \text{ m}^3$

$= 1000 \text{ m}^3$

10. $y = 2x + 1$

x	y
-2	-3
0	1
2	5

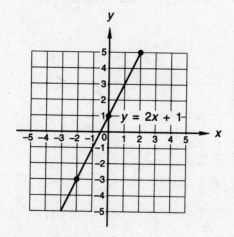

11. $P \cdot 60 = 84$

$P = \dfrac{84}{60}$

$P = 1.4$

$P = \mathbf{140\%}$

12. $70\% \cdot A = 651$

$\dfrac{70}{100} \cdot A = 651$

$\dfrac{100}{70} \cdot \dfrac{70}{100} \cdot A = 651 \cdot \dfrac{100}{70}$

$A = \mathbf{930}$

13. $1{,}000{,}000 \text{ cm}^3 \times \dfrac{1 \text{ m}}{100 \text{ cm}} \times \dfrac{1 \text{ m}}{100 \text{ cm}} \times \dfrac{1 \text{ m}}{100 \text{ cm}}$

$\times \dfrac{1 \text{ km}}{1000 \text{ m}} \times \dfrac{1 \text{ km}}{1000 \text{ m}} \times \dfrac{1 \text{ km}}{1000 \text{ m}}$

$= \dfrac{1{,}000{,}000}{(100)(100)(100)(1000)(1000)(1000)} \text{ km}^3$

$= 1 \times 10^{-9} \text{ km}^3$

14. $10{,}500 \text{ m}^3 \times \dfrac{100 \text{ cm}}{1 \text{ m}} \times \dfrac{100 \text{ cm}}{1 \text{ m}} \times \dfrac{100 \text{ cm}}{1 \text{ m}}$

$\times \dfrac{1 \text{ mL}}{1 \text{ cm}^3}$

$= 10{,}500(100)(100)(100)(1) \text{ mL}$

$= 1.05 \times 10^{10} \text{ mL}$

15.

To find the height of the pyramid, we must use the Pythagorean theorem.

$5^2 = h^2 + 3^2$

$25 = h^2 + 9$

$16 = h^2$

$4 \text{ ft} = h$

$V_{\text{Pyramid}} = \dfrac{1}{3} V_{\text{Solid}} = \dfrac{1}{3}[A_{\text{Base}} \times \text{height}]$

$= \dfrac{1}{3}[6(6) \times 4] \text{ ft}^3$

$= \mathbf{48 \text{ ft}^3}$

16. **Figure (a)** is symmetric about a point.

(a)

(b)

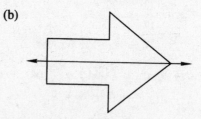

17. (a) $z + 90° + 60° = 180°$

$z = 180° - 90° - 60°$

$z = \mathbf{30°}$

(b) $10^2 = m^2 + 5^2$

$100 = m^2 + 25$

$75 = m^2$

$\sqrt{75} = m$

18. $\dfrac{m}{17} = \dfrac{5}{9}$

$9m = 85$

$m = \dfrac{\mathbf{85}}{\mathbf{9}}$

19. (a) $A_{\text{Trapezoid}}$

$$= \frac{b_1 h}{2} + \frac{b_2 h}{2}$$

$$= \left[\frac{11(2)}{2} + \frac{6.53(2)}{2}\right] \text{m}^2$$

$$= [11 + 6.53]\, \text{m}^2$$

$$= \mathbf{17.53\ m^2}$$

(b) $A_{\text{Parallelogram}} = Bh$

$$= (5)(2)\ \text{m}^2$$

$$= \mathbf{10\ m^2}$$

20. $7x + 3 + 2x = -4 - 2x + 3 - x$

$$9x + 3 = -1 - 3x$$

$$12x = -4$$

$$x = -\frac{1}{3}$$

21. $\quad -1\frac{4}{5}x + 2\frac{1}{2} = -\frac{1}{15}$

$$-\frac{9}{5}x + \frac{5}{2} - \frac{5}{2} = -\frac{1}{15} - \frac{5}{2}$$

$$-\frac{9}{5}x = -\frac{77}{30}$$

$$\left(-\frac{5}{9}\right)\left(-\frac{9}{5}\right)x = \left(-\frac{77}{30}\right)\left(-\frac{5}{9}\right)$$

$$x = \frac{77}{54}$$

22. $\quad \dfrac{\frac{1}{9}}{\frac{4}{7}} = \dfrac{x}{\frac{3}{14}}$

$$\frac{1}{42} = \frac{4}{7}x$$

$$\frac{1}{24} = x$$

23. (a) $\dfrac{1}{-2^{-6}} = -2^6 = \mathbf{-64}$

(b) $(-2)^{-6} = \dfrac{1}{(-2)^6} = \dfrac{\mathbf{1}}{\mathbf{64}}$

24. $\dfrac{(-1)^7 - 2(3^2 - 2 \cdot 4)}{3(\sqrt[3]{-216} + 2^3)} = \dfrac{-1 - 2(9 - 8)}{3(-6 + 8)}$

$$= \frac{-1 - 2(1)}{3(2)} = -\frac{1}{2}$$

25. $m^5 p^6 m^2 pmz = \mathbf{m^8 p^7 z}$

26. $\dfrac{2}{5}\left(3\dfrac{1}{4} \cdot \dfrac{1}{6} - \dfrac{1}{2} \cdot \dfrac{1}{4}\right) = \dfrac{2}{5}\left(\dfrac{13}{24} - \dfrac{3}{24}\right) = \dfrac{\mathbf{1}}{\mathbf{6}}$

27. $a^2z - z^2 = (-5)^2(-3) - (-3)^2$

$$= -75 - 9$$

$$= \mathbf{-84}$$

28. $\sqrt[p]{m} + p^z = \left(\sqrt[3]{-8}\right) + (3^{-2})$

$$= -2 + \frac{1}{9}$$

$$= -\frac{\mathbf{17}}{\mathbf{9}}$$

29. Tax = VIII = 8%

Cost = CXXV denarii = 125 denarii

Caesar paid 0.08×125 denarii or 10 denarii in taxes.

10 denarii = **X denarii**

30. $\dfrac{1}{2} - 0.2(\sqrt{36}) + 5^{-2} \div 1\dfrac{1}{5} + 1\dfrac{3}{4}$

$$= \frac{1}{2} - \frac{2}{10}(6) + \frac{1}{25} \div \frac{6}{5} + \frac{7}{4}$$

$$= \frac{1}{2} - \frac{12}{10} + \frac{1}{25} \cdot \frac{5}{6} + \frac{7}{4}$$

$$= \frac{1}{2} - \frac{12}{10} + \frac{1}{30} + \frac{7}{4}$$

$$= \frac{30}{60} - \frac{72}{60} + \frac{2}{60} + \frac{105}{60} = \frac{65}{60} = \frac{\mathbf{13}}{\mathbf{12}}$$

PROBLEM SET 123

1. $P(7T) = \dfrac{1}{2} \cdot \dfrac{1}{2} \cdot \dfrac{1}{2} \cdot \dfrac{1}{2} \cdot \dfrac{1}{2} \cdot \dfrac{1}{2} \cdot \dfrac{1}{2}$

$$= \frac{\mathbf{1}}{\mathbf{128}}$$

2. $110\% \cdot 840 = B$

$$1.1 \cdot 840 = B$$

$$924 = B$$

$$840 + 924 = \mathbf{1764}$$

3. $\dfrac{R}{N} = \dfrac{1}{29}, \quad \dfrac{R}{T} = \dfrac{1}{30}, \quad \dfrac{N}{T} = \dfrac{29}{30}$

$$\frac{1}{30} = \frac{R}{3930}$$

$$3930 = 30R$$

$$\mathbf{131} = R$$

4. $21N - 102 = 11N + 8$

$$10N = 110$$

$$N = \mathbf{11}$$

5. Initial rate $= \dfrac{90 \text{ mi}}{15 \text{ hr}} = \dfrac{6 \text{ mi}}{1 \text{ hr}}$

Doubled rate $= \dfrac{(2 \cdot 6) \text{ mi}}{1 \text{ hr}} = \dfrac{12 \text{ mi}}{1 \text{ hr}}$

$\dfrac{1 \text{ hr}}{12 \cancel{\text{ mi}}} \times 180 \cancel{\text{ mi}} = \mathbf{15 \text{ hr}}$

Total time $= (15 + 15) \text{ hr} = \mathbf{30 \text{ hr}}$

6. Avg $= \dfrac{14 + 17 + 23}{3} = \mathbf{18 \text{ assignments}}$

7. Markdown $= 25\% \cdot \$1500$

$= 0.25 \cdot \$1500$

$= \$375$

Sale price $= \$1500 - \$375 = \mathbf{\$1125}$

8. (a)

10	9	8	7	6	5	4	3	2	1

$10! = 10 \cdot 9 \cdot 8 \cdot 7 \cdot 6 \cdot 5 \cdot 4$

$\cdot\, 3 \cdot 2 \cdot 1$

$= \mathbf{3{,}628{,}800}$ different orders

(b)

10	9	8	7

There are $10 \cdot 9 \cdot 8 \cdot 7 = \mathbf{5040}$ different orders.

9. π belongs to both the **irrational numbers** and the **real numbers**.

10. (a) $\frac{2}{9}$ belongs to both the **rational numbers** and the **real numbers**.

(b) 10 belongs to the **whole numbers,** the **integers,** the **rational numbers,** and the **real numbers.**

11.

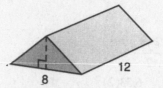

$V_{\text{Solid}} = A_{\text{Triangular End}} \times \text{length}$

To find the area of the triangular end, we must use the Pythagorean Theorem to determine the height of the triangle.

$4^2 + h^2 = 5^2$

$h^2 = 25 - 16$

$h = 3 \text{ m}$

$V_{\text{Solid}} = \left(\dfrac{1}{2}\right)(8)(3) \text{ m}^2 \times 12 \text{ m}$

$= \mathbf{144 \text{ m}^3}$

12.

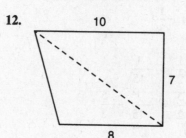

$A_{\text{Trapezoid}} = A_{\text{Triangle 1}} + A_{\text{Triangle 2}}$

$= \dfrac{1}{2}b_1 h_1 + \dfrac{1}{2}b_2 h_2$

$= \left(\dfrac{1}{2}\right)(8)(7) \text{ in.}^2 + \left(\dfrac{1}{2}\right)(10)(7) \text{ in.}^2$

$= 28 \text{ in.}^2 + 35 \text{ in.}^2$

$= \mathbf{63 \text{ in.}^2}$

13. (a) $z + 90° + 36° = 180°$

$z = 180° - 90° - 36°$

$z = \mathbf{54°}$

(b) $13^2 = 9^2 + m^2$

$169 = 81 + m^2$

$88 = m^2$

$\sqrt{\mathbf{88}} = m$

14. $\dfrac{m}{9} = \dfrac{2}{7}$

$7m = 18$

$m = \dfrac{\mathbf{18}}{\mathbf{7}}$

15. $y = 3x - 2$

x	y
–1	–5
0	–2
2	4

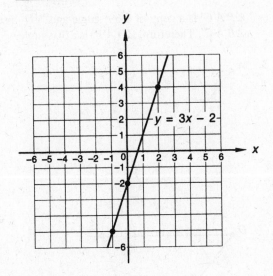

16. $11\frac{1}{4}\% = 11.25\% = 0.1125$

$0.1125 \cdot A = 562.5$

$A = \dfrac{562.5}{0.1125}$

$A = \mathbf{5000}$

17. $P \cdot 90 = 148.5$

$P = \dfrac{148.5}{90}$

$P = 1.65$

$P = \mathbf{165\%}$

18. $1000 \; \cancel{mL} \times \dfrac{1 \; \cancel{cm^3}}{1 \; \cancel{mL}} \times \dfrac{1 \; m}{100 \; \cancel{cm}} \times \dfrac{1 \; m}{100 \; \cancel{cm}}$

$\times \dfrac{1 \; m}{100 \; \cancel{cm}}$

$= \dfrac{\mathbf{1000}}{(100)(100)(100)} \; \mathbf{m^3} = \mathbf{0.001 \; m^3}$

19. $11{,}000{,}000 \; \cancel{mi^3} \times \dfrac{5280 \; \cancel{ft}}{1 \; \cancel{mi}} \times \dfrac{5280 \; \cancel{ft}}{1 \; \cancel{mi}} \times \dfrac{5280 \; \cancel{ft}}{1 \; \cancel{mi}}$

$\times \dfrac{12 \; in.}{1 \; \cancel{ft}} \times \dfrac{12 \; in.}{1 \; \cancel{ft}} \times \dfrac{12 \; in.}{1 \; \cancel{ft}}$

$= \mathbf{11{,}000{,}000(5280)(5280)(5280)(12)(12)(12) \; in.^3}$

$\approx \mathbf{2.8 \times 10^{21} \; in.^3}$

20. $m^2 b(m^3 + b^2 + cm) = \mathbf{m^5 b + m^2 b^3 + cm^3 b}$

21. $2abab - 4ba^2 b + 7ab^2 a + 3a^2 b$

$= 2a^2 b^2 - 4a^2 b^2 + 7a^2 b^2 + 3a^2 b$

$= \mathbf{5a^2 b^2 + 3a^2 b}$

22. $V_{\text{Cylinder}} = A_{\text{Base}} \times \text{height}$

$= \left[\pi(2\sqrt{2})^2 \times 2\right] \; m^3$

$\approx \mathbf{50.24 \; m^3}$

23. $V_{\text{Pyramid}} = \frac{1}{3} V_{\text{Solid}}$

$= \frac{1}{3}\left[A_{\text{Base}} \times \text{height}\right]$

$= \frac{1}{3}\left[(6)(6) \times 4\right] \; m^3$

$= \mathbf{48 \; m^3}$

24. $S.A. = A_{\text{Base}} + 4A_{\text{Face}}$

$= (6)(6) \; m^2 + 4\left[\dfrac{(6)(5)}{2}\right] \; m^2$

$= 36 \; m^2 + 60 \; m^2$

$= \mathbf{96 \; m^2}$

25. $5x - 4 + 2x - 1 = 3x + 6 + 2x + 1$

$7x - 5 = 5x + 7$

$2x = 12$

$x = \mathbf{6}$

26. $\dfrac{-\dfrac{1}{4}}{\dfrac{2}{7}} = \dfrac{x}{\dfrac{1}{3}}$

$-\dfrac{1}{12} = \dfrac{2}{7}x$

$-\dfrac{7}{24} = x$

27. $5! = 5 \cdot 4 \cdot 3 \cdot 2 \cdot 1 = \mathbf{120}$

28. $(-3)^{-2} = \dfrac{1}{(-3)^2} = \mathbf{\dfrac{1}{9}}$

29. $-a^{-2} + b^{-3} = -(3)^{-2} + (-5)^{-3}$

$= -\dfrac{1}{9} - \dfrac{1}{125}$

$= -\dfrac{125}{1125} - \dfrac{9}{1125}$

$= -\mathbf{\dfrac{134}{1125}}$

30. $x \geq 0$

PRACTICE A

1.

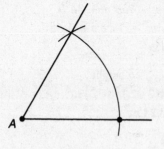

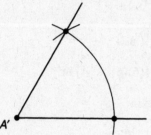

2.

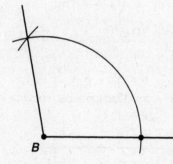

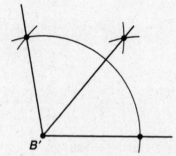

3.

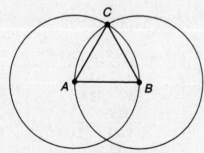

The triangle is **equilateral.**

4.

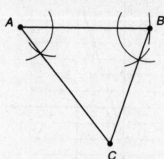

∠*B'A'C'* is a copy of a 60° angle and $\overline{A'D'}$ bisects ∠*B'A'C'*. Therefore, ∠*D'A'B'* is a 30° angle.

5.

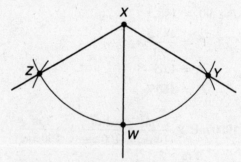

Δ*DEF* is a copy of Δ*ABC*.

6.

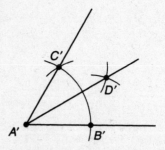

∠*YXW* is a copy of a 60° angle and so is ∠*ZXW*. Thus, ∠*YXZ* is a 120° angle.

7.

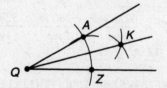

∠*AQZ* is a copy of a 30° angle and $\overline{KQ}$ bisects ∠*AQZ*. Therefore, ∠*KQZ* is a 15° angle.

8.

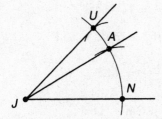

∠NJA is a copy of a 30° angle, and ∠AJU is a copy of a 15° angle. Thus, ∠NJU is a 45° angle.

9.

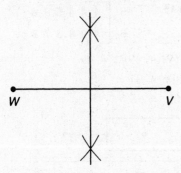

10.

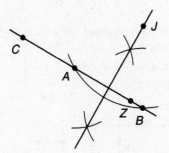

11.

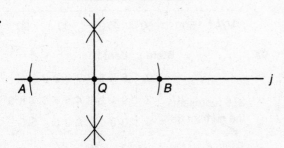

12.

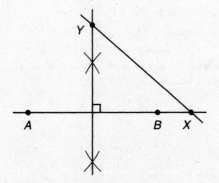

The right angle was produced by constructing the perpendicular bisector of $\overline{AB}$.

13.

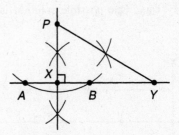

Construction D was used to make a right angle. Then a 60° angle was copied onto $\overline{PX}$. Since the sum of the angles in a triangle must be 180°, ∠XYP has to be a 30° angle. Therefore, △YPX is a 30°–60°–90° triangle.

14.

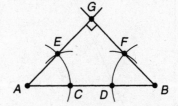

After drawing $\overline{AB}$, copies of 45° angles were drawn at each endpoint. Since the sum of the angles in a triangle must be 180°, ∠AGB has to be a 90° angle. Therefore, △ABG is a 45°–45°–90° triangle.

15. Answers may vary. One possible solution is the following:

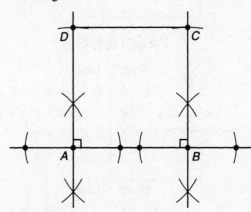

16. Answers may vary. One possible solution is the following:

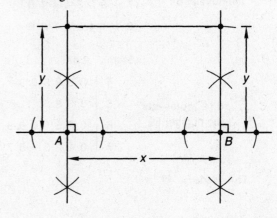

17. Answers may vary. One possible solution is the following:

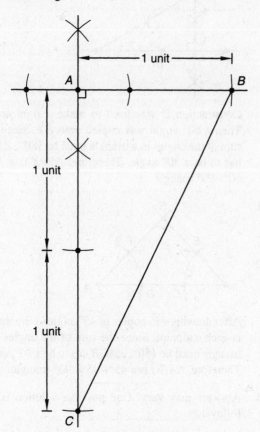

PRACTICE B

1.

Stem	Leaf
3	0 0 1 1 2 3 5 7 9
4	0 0 2 3 5 9 9 9 9
5	1 4 4 5 8 9
6	1 2 4 5 7 8

4 | 9 represents the number 49

2.

Stem	Leaf
5	0 3 4 5 6 8 9 9
6	1 3 3 3 4 4 6 9
7	0 2 3 4 4 7 8
8	1 4 5 5 7 8 9

6 | 4 represents the number 64

3. (a)

Stem	Leaf
4	2 8 9 9 9 9
5	3 6 8 9
6	0 2 5 5 6 6 8
7	0 1 2 3 4 5 7

5 | 9 represents the number 59

(b) Mode = **49**

(c) Range = 77 − 42 = **35**

(d) Median = (62 + 65)/2 = **63.5**

(e) Mean = $\frac{1476}{24}$ = **61.5**

4.

Stem	Leaf
3	0 0 1 1 2 3 5 7 9
4	0 0 2 3 5 9 9 9 9
5	1 4 4 5 8 9
6	1 2 4 5 7 8

4 | 9 represents the number 49

Median = (49 + 49)/2 = 49

Lower quartile = 37

Upper quartile = 58

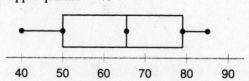

5. The data are already ordered, so a stem-and-leaf plot is unnecessary.

Median = (63 + 68)/2 = 65.5

Lower quartile = 50

Upper quartile = 79

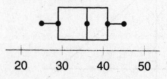

6.

Stem	Leaf
2	5 5 6 7 7 7 8 9
3	0 1 2 3 4 5 6 6 7 8 9 9
4	0 0 1 2 2 2 3 4 5 5

3 | 6 represents the number 36

Median = (36 + 36)/2 = 36

Lower quartile = 29

Upper quartile = 41

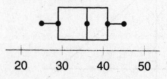

7.   **Distribution of Teams' Average Scores**

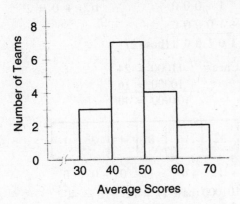

8.

Stem	Leaf
8	0 1 2 8 9
9	2 4 5 7 9
10	3 8
11	0 2 9

10 | 8 represents
the number 108

Distribution of Miles Traveled

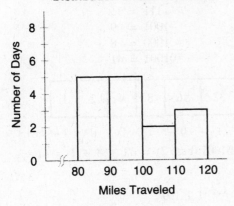

PRACTICE C

1.

64	32	16	8	4	2	1
1	1	0	0	1	0	0

1100100 (base 2) = 64 + 32 + 4 = **100**

2.

64	32	16	8	4	2	1
1	1	0	1	0	1	1

$$\begin{array}{r} 107 \\ -\ 64 \\ \hline 43 \\ -\ 32 \\ \hline 11 \end{array} \qquad \begin{array}{r} 11 \\ -\ 8 \\ \hline 3 \\ -\ 2 \\ \hline 1 \\ -\ 1 \\ \hline 0 \end{array}$$

107 (base 10) = **1101011 (base2)**

3.

32	16	8	4	2	1
1	0	1	0	0	1

101001 (base 2) = 32 + 8 + 1 = **41**

4.

16	8	4	2	1
1	1	1	1	1

11111 (base 2) = 16 + 8 + 4 + 2 + 1 = **31**

5.

64	32	16	8	4	2	1
1	0	0	1	0	1	0

$$\begin{array}{r} 74 \\ -\ 64 \\ \hline 10 \\ -\ 8 \\ \hline 2 \end{array} \qquad \begin{array}{r} 2 \\ -\ 2 \\ \hline 0 \end{array}$$

74 (base 10) = **1001010 (base 2)**

6.

16	8	4	2	1
1	0	1	1	1

10111 (base 2) = 16 + 4 + 2 + 1 = **23**

7.

64	32	16	8	4	2	1
1	1	0	0	0	0	0

$$\begin{array}{r} 96 \\ -\ 64 \\ \hline 32 \\ -\ 32 \\ \hline 0 \end{array}$$

96 (base 10) = **1100000 (base 2)**

8.

32	16	8	4	2	1
1	1	1	1	1	0

111110 (base 2) = 32 + 16 + 8 + 4 + 2 = **62**

9.

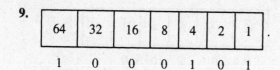

64	32	16	8	4	2	1
1	0	0	0	1	0	1

1000101 (base 2) = 64 + 4 + 1 = **69**

10.

16	8	4	2	1
1	0	0	0	1

10001 (base 2) = 16 + 1 = **17**

11.

```
  1 1 1
  1 1 0 1          1(2) + 0 = 2
+ 1 0 1 1          1(2) + 1 = 3
1 1 0 0 0 (base 2)
```

Check: 1101 = 13
        + 1011 = 11
          11000 = 24

16	8	4	2	1
1	1	0	0	0

11000 (base 2) = 16 + 8 = 24

12.

```
    1 1 1 1
    1 1 1 1             1(2) + 0 = 2
+ 1 0 0 0 0 1
  1 1 0 0 0 0 (base 2)
```

Check: 1111 = 15
        + 100001 = 33
           110000 = 48

32	16	8	4	2	1
1	1	0	0	0	0

110000 (base 2) = 32 + 16 = 48

13.

```
1   1 1 1 1
  1 0 1 1 0 1          1(2) + 0 = 2
+ 1 0 0 0 1 1
1 0 1 0 0 0 0 (base 2)
```

Check: 101101 = 45
        + 100011 = 35
         1010000 = 80

64	32	16	8	4	2	1
1	0	1	0	0	0	0

1010000 (base 2) = 64 + 16 = 80

14.

```
1
  1 1 0 0 0          1(2) + 0 = 2
+ 1 0 0 0 0
1 0 1 0 0 0 (base 2)
```

Check: 11000 = 24
        + 10000 = 16
         101000 = 40

32	16	8	4	2	1
1	0	1	0	0	0

101000 (base 2) = 32 + 8 = 40

15.

```
2 1 1 1
  1 0 0 1          1(2) + 1 = 3
  1 1 1 1          1(2) + 0 = 2
  1 0 0 1          2(2) + 1 = 5
+ 1 0 0 0
1 0 1 0 0 1 (base 2)
```

Check: 1001 = 9
            1111 = 15
            1001 = 9
        +   1000 = 8
          101001 = 41

32	16	8	4	2	1
1	0	1	0	0	1

101001 (base 2) = 32 + 8 + 1 = 41

16.

```
1 1 1 1
  1 1 1 0
+ 1 1 1 1 0 1
1 0 0 1 0 1 1 (base 2)
```

Check: 1110 = 14
        + 111101 = 61
         1001011 = 75

64	32	16	8	4	2	1
1	0	0	1	0	1	1

1001011 (base 2) = 64 + 8 + 2 + 1 = 75

17.
$$\overset{1\ \ \overset{1}{} \ \overset{1}{}\overset{1}{}}{1\ 0.1\ 1}$$
$+\ 1\ 0\ 1.0\ 1\ 1$

$1(2) + 0 = 2$

$\mathbf{1\ 0\ 0\ 0.0\ 0\ 1}$ **(base 2)**

Check:

10.11 (base 2)

$= 1(2^1) + 1(2^{-1}) + 1(2^{-2})$

$= 2 + \dfrac{1}{2} + \dfrac{1}{4}$

$= 2.75$

101.011 (base 2)

$= 1(2^2) + 1(2^0) + 1(2^{-2}) + 1(2^{-3})$

$= 4 + 1 + \dfrac{1}{4} + \dfrac{1}{8}$

$= 5.375$

$2.75 + 5.375 = 8.125$

1000.001 (base 2)

$= (2^3) + 1(2^{-3}) = 8 + \dfrac{1}{8} = 8.125$

18.
$$\overset{1\ \ \overset{2}{}\ \overset{2}{}\ \overset{1}{}\ \overset{1}{}\overset{1}{}\ }{1\ 1\ 1.0\ 1\ 1}$$
$1\ 1\ 0.1$
$+\ \ \ 1\ 1.0\ 1\ 1$

$1(2) + 0 = 2$
$1(2) + 1 = 3$
$2(2) + 0 = 4$

$\mathbf{1\ 0\ 0\ 0\ 1.0\ 1\ 0}$ **(base 2)**

Check:

111.011 (base 2)

$= 1(2^2) + 1(2^1) + 1(2^0) + 1(2^{-2}) + 1(2^{-3})$

$= 4 + 2 + 1 + \dfrac{1}{4} + \dfrac{1}{8}$

$= 7.375$

110.1 (base 2)

$= 1(2^2) + 1(2^1) + 1(2^{-1})$

$= 4 + 2 + \dfrac{1}{2}$

$= 6.5$

11.011 (base 2)

$= 1(2^1) + 1(2^0) + 1(2^{-2}) + 1(2^{-3})$

$= 2 + 1 + \dfrac{1}{4} + \dfrac{1}{8} = 3.375$

$7.375 + 6.5 + 3.375 = 17.25$

10001.01 (base 2)

$= 1(2^4) + 1(2^0) + 1(2^{-2})$

$= 16 + 1 + \dfrac{1}{4} = 17.25$

19.
$$\overset{1\ \ \overset{1}{}\ \overset{1}{}\ \ \overset{1}{}\ \overset{1}{}\ \overset{1}{}}{1\ 1\ 1\ 0\ 1\ 1.0\ 1}$$
$+\ \ \ \ 1\ 0\ 1\ 1.1\ 1$

$1(2) + 0 = 2$
$1(2) + 1 = 3$

$\mathbf{1\ 0\ 0\ 0\ 1\ 1\ 1.0\ 0}$ **(base 2)**

Check:

111011.01 (base 2)

$= 1(2^5) + 1(2^4) + 1(2^3) + 1(2^1)$
$\quad + 1(2^0) + 1(2^{-2})$

$= 32 + 16 + 8 + 2 + 1 + \dfrac{1}{4} = 59.25$

1011.11 (base 2)

$= 1(2^3) + 1(2^1) + 1(2^0) + 1(2^{-1}) + 1(2^{-2})$

$= 8 + 2 + 1 + \dfrac{1}{2} + \dfrac{1}{4} = 11.75$

$59.25 + 11.75 = 71$

1000111 (base 2)

$= 1(2^6) + 1(2^2) + 1(2^1) + 1(2^0)$

$= 64 + 4 + 2 + 1 = 71$

20.
$$\overset{1\ \ \overset{1}{}\ \overset{1}{}\ \overset{1}{}\overset{1}{}}{1\ 1\ 1.0\ 1}$$
$1\ 1\ 1.1\ 0\ 1$
$+\ 1\ 0\ 0\ 0\ 0.0\ 1$

$\mathbf{1\ 1\ 0\ 1\ 1.0\ 0\ 1}$ **(base 2)**

Check:

11.01 (base 2)

$= 1(2^1) + 1(2^0) + 1(2^{-2})$

$= 2 + 1 + \dfrac{1}{4} = 3.25$

111.101 (base 2)

$= 1(2^2) + 1(2^1) + 1(2^0) + 1(2^{-1}) + 1(2^{-3})$

$= 4 + 2 + 1 + \dfrac{1}{2} + \dfrac{1}{8} = 7.625$

10000.01 (base 2)

$= 1(2^4) + 1(2^{-2}) = 16 + \dfrac{1}{4} = 16.25$

$3.25 + 7.625 + 16.25 = 27.125$

11011.001 (base 2)

$= 1(2^4) + 1(2^3) + 1(2^1) + 1(2^0) + 1(2^{-3})$

$= 16 + 8 + 2 + 1 + \dfrac{1}{8} = 27.125$

21.

```
    1 1 0 1
  ×     1 1
    1 1 0 1
    1 1 0 1
```
1 0 0 1 1 1 (base 2)

Check:

1101 (base 2)

$= 1(2^3) + 1(2^2) + 1(2^0)$

$= 8 + 4 + 1 = 13$

11 (base 2)

$= 1(2^1) + 1(2^0) = 2 + 1 = 3$

$3 \cdot 13 = 39$

100111 (base 2)

$= 1(2^5) + 1(2^2) + 1(2^1) + 1(2^0)$

$= 32 + 4 + 2 + 1 = 39$

22.

```
    1 0 1 0 1
  ×   1 0 1 1
    1 0 1 0 1
    1 0 1 0 1
  1 0 1 0 1
```
1 1 1 0 0 1 1 1 (base 2)

Check:

10101 (base 2)

$= 1(2^4) + 1(2^2) + 1(2^0)$

$= 16 + 4 + 1 = 21$

1011 (base 2)

$= 1(2^3) + 1(2^1) + 1(2^0)$

$= 8 + 2 + 1 = 11$

$21 \cdot 11 = 231$

11100111 (base 2)

$= 1(2^7) + 1(2^6) + 1(2^5)$

$\quad + 1(2^2) + 1(2^1) + 1(2^0)$

$= 128 + 64 + 32 + 4 + 2 + 1 = 231$

23.

```
      1 1
  ×   1 1
      1 1
    1 1
```
1 0 0 1 (base 2)

Check:

11 (base 2)

$= 1(2^1) + 1(2^0) = 3$

$3 \cdot 3 = 9$

1001 (base 2)

$= 1(2^3) + 1(2^0) = 8 + 1 = 9$

24.

```
    1 1 0 1 0
  ×     1 1.1
    1 1 0 1 0
    1 1 0 1 0
  1 1 0 1 0
```
1 0 1 1 0 1 1.0 (base 2)

Check:

11010 (base 2)

$= 1(2^4) + 1(2^3) + 1(2^1)$

$= 16 + 8 + 2 = 26$

11.1 (base 2)

$= 1(2^1) + 1(2^0) + 1(2^{-1})$

$= 2 + 1 + \dfrac{1}{2} = 3.5$

$26 \cdot 3.5 = 91$

1011011 (base 2)

$= 1(2^6) + 1(2^4) + 1(2^3) + 1(2^1) + 1(2^0)$

$= 64 + 16 + 8 + 2 + 1 = 91$

25.

```
    1.0 0 1
  ×   0.1 1
    1 0 0 1
   1 0 0 1
```
0.1 1 0 1 1 (base 2)

Check:

1.001 (base 2)

$= 1(2^0) + 1(2^{-3}) = 1 + \dfrac{1}{8} = 1.125$

0.11 (base 2)

$= 1(2^{-1}) + (2^{-2}) = \dfrac{1}{2} + \dfrac{1}{4} = 0.75$

$0.75 \times 1.125 = 0.84375$

0.11011 (base 2)

$= 1(2^{-1}) + 1(2^{-2}) + 1(2^{-4}) + 1(2^{-5})$

$= \dfrac{1}{2} + \dfrac{1}{4} + \dfrac{1}{16} + \dfrac{1}{32} = \dfrac{27}{32} = 0.84375$

26.

$$\begin{array}{r} 1.0\,1 \\ \times\,1.0\,1 \\ \hline 1\,0\,1 \\ \end{array}$$

$$\begin{array}{r} 1\,0\,1 \\ \hline \mathbf{1.1\,0\,0\,1}\ \textbf{(base 2)} \end{array}$$

Check:

1.01 (base 2)

$= 1(2^0) + 1(2^{-2}) = 1 + \dfrac{1}{4} = 1.25$

$1.25 \times 1.25 = 1.5625$

1.1001 (base 2)

$= 1(2^0) + 1(2^{-1}) + 1(2^{-4})$

$= 1 + \dfrac{1}{2} + \dfrac{1}{16} = 1\dfrac{9}{16} = 1.5625$

27.

$$\begin{array}{r} 1\,1.0\,0\,1 \\ \times\,1\,0\,1.0\,1 \\ \hline 1\,1\,0\,0\,1 \\ 1\,1\,0\,0\,1 \\ \end{array}$$

$$\begin{array}{r} 1\,1\,0\,0\,1 \\ \hline \mathbf{1\,0\,0\,0\,0.0\,1\,1\,0\,1}\ \textbf{(base 2)} \end{array}$$

Check:

11.001 (base 2)

$= 1(2^1) + 1(2^0) + 1(2^{-3})$

$= 2 + 1 + \dfrac{1}{8} = 3.125$

101.01 (base 2)

$= 1(2^2) + 1(2^0) + 1(2^{-2})$

$= 4 + 1 + \dfrac{1}{4} = 5.25$

$5.25 \cdot 3.125 = 16.40625$

10000.01101 (base 2)

$= 1(2^4) + 1(2^{-2}) + 1(2^{-3}) + 1(2^{-5})$

$= 16 + \dfrac{1}{4} + \dfrac{1}{8} + \dfrac{1}{32} = 16\dfrac{13}{32} = 16.40625$

PRACTICE D

1. $b = d = f = \mathbf{100°}$ corresponding and
                   alternate interior angles

      $b + a = 180°$ supplementary angles

   $100° + a = 180°$

          $a = 80°$

   $a = c = e = g = \mathbf{80°}$ corresponding and
                          alternate interior
                          angles

2. $y = \mathbf{90°}$ alternate interior angles

   $x + y = 180°$ supplementary angles

   $x + 90° = 180°$

       $x = \mathbf{90°}$

3. $3m + 36° = 180°$ supplementary angles

       $3m = 144°$

         $m = \dfrac{144°}{3}$

         $m = \mathbf{48°}$

      $3m = n - m$ corresponding and

   $3(48°) = n - 48°$ alternate interior

  $144° + 48° = n$ angles

     $192° = n$

4. $p + 60° = 180°$ supplementary angles

        $p = \mathbf{120°}$

        $q = \mathbf{40°}$ corresponding angles

5. $3a + 90° = 180°$ supplementary angles

       $3a = 90°$

        $a = \dfrac{90°}{3}$

        $a = 30°$

   $b + 68° = 180°$ supplementary angles

        $b = \mathbf{112°}$

6. $2c + 94° = 180°$ supplementary angles

       $2c = 86°$

        $c = \dfrac{86°}{2}$

        $c = \mathbf{43°}$

   $3d + 43° = 180°$ supplementary angles

       $3d = 137°$

        $d = \dfrac{137°}{3}$

        $d = \mathbf{45\dfrac{2°}{3}}$

7. $x = \mathbf{40°}$

   $x + y = 180°$

   $40° + y = 180°$

        $y = \mathbf{140°}$

8. $x = \dfrac{40°}{2} = \mathbf{20°}$

$x + y = \dfrac{140°}{2}$

$20° + y = 70°$

$y = \mathbf{50°}$

9. $\text{m}\overset{\frown}{BC} = \mathbf{50°}$

$\text{m}\overset{\frown}{CD} = \mathbf{60°}$

$\text{m}\overset{\frown}{BD} = \text{m}\overset{\frown}{BC} + \text{m}\overset{\frown}{CD} = 50° + 60° = \mathbf{110°}$

10. $\text{m}\overset{\frown}{WZ} = \mathbf{60°}$

$\text{m}\angle XYW + 60° = 180°$

$\text{m}\angle XYW = \mathbf{120°}$

$\text{m}\overset{\frown}{XW} = \text{m}\angle XYW = \mathbf{120°}$

11. $\text{m}\overset{\frown}{AC} = 2\text{m}\angle ABC = 2 \cdot 50° = \mathbf{100°}$

$\text{m}\angle ECB = \dfrac{1}{2}\text{m}\overset{\frown}{BE} = \dfrac{70°}{2} = \mathbf{35°}$

12. $\angle BDC$ must equal $\angle BAC$ because both angles subtend the same arc, $\overset{\frown}{BC}$, and both angles have their vertex on the circle. Thus, $\text{m}\angle BDC = \text{m}\angle BAC = \mathbf{23°}$.

13. $x = \dfrac{50°}{2} = \mathbf{25°}$

$x + y + 90° = 180°$

$25° + y + 90° = 180°$

$y = \mathbf{65°}$

14. $\text{m}\angle BAC = \dfrac{1}{2}\text{m}\overset{\frown}{BC} = \dfrac{72°}{2} = 36°$

$36° + 2\text{m}\angle ABC = 180°$

$2\text{m}\angle ABC = 144°$

$\text{m}\angle ABC = \mathbf{72°}$

$\text{m}\angle BCA = \text{m}\angle ABC = \mathbf{72°}$

PRACTICE E

1. $\sqrt{2}$: Initial estimate $= x_1 = 1$

$x_2 = \dfrac{(1)^2 + 2}{2(1)} = \dfrac{1 + 2}{2} = \dfrac{3}{2} = 1.5$

$x_3 = \dfrac{(1.5)^2 + 2}{2(1.5)} = 1.41\overline{6}$

$x_4 = \dfrac{(1.41\overline{6})^2 + 2}{2(1.41\overline{6})} = 1.41421...$

$x_5 = \dfrac{(1.41421)^2 + 2}{2(1.41421)} = 1.41421...$

Since the first 3 places to the right of the decimal point did not change, we may round to 2 places: $\sqrt{2} \approx \mathbf{1.41}$.

2. $\sqrt{3}$: Initial estimate $= x_1 = 2$

$x_2 = \dfrac{(2)^2 + 3}{2(2)} = \dfrac{7}{4} = 1.75$

$x_3 = \dfrac{(1.75)^2 + 3}{2(1.75)} = 1.73214...$

$x_4 = \dfrac{(1.73214)^2 + 3}{2(1.73214)} = 1.73205...$

Since the first 3 places to the right of the decimal point did not change, we may round to 2 places: $\sqrt{3} \approx \mathbf{1.73}$.

3. $\sqrt{5}$: Initial estimate $= x_1 = 2$

$x_2 = \dfrac{(2)^2 + 5}{2(2)} = \dfrac{9}{4} = 2.25$

$x_3 = \dfrac{(2.25)^2 + 5}{2(2.25)} = 2.23611...$

$x_4 = \dfrac{(2.23611)^2 + 5}{2(2.23611)} = 2.23606...$

Since the first 3 places to the right of the decimal point did not change, we may round to 2 places: $\sqrt{5} \approx \mathbf{2.24}$.

4. $\sqrt{10}$: Initial estimate $= x_1 = 3$

$x_2 = \dfrac{(3)^2 + 10}{2(3)} = \dfrac{19}{6} = 3.1\overline{6}$

$x_3 = \dfrac{(3.1\overline{6})^2 + 10}{2(3.1\overline{6})} = 3.16228...$

$x_4 = \dfrac{(3.16228)^2 + 10}{2(3.16228)} = 3.16227...$

Since the first 3 places to the right of the decimal point did not change, we may round to 2 places: $\sqrt{10} \approx \mathbf{3.16}$.

5. $\sqrt{90}$: Initial estimate $= x_1 = 9$

$$x_2 = \frac{(9)^2 + 90}{2(9)} = \frac{171}{18} = 9.5$$

$$x_3 = \frac{(9.5)^2 + 90}{2(9.5)} = 9.48684\ldots$$

$$x_4 = \frac{(9.48684)^2 + 90}{2(9.48684)} = 9.48683\ldots$$

Since the first 3 places to the right of the decimal point did not change, we may round to 2 places: $\sqrt{90} \approx \mathbf{9.49.}$

6. $\sqrt{27}$: Initial estimate $= x_1 = 5$

$$x_2 = \frac{(5)^2 + 27}{2(5)} = \frac{52}{10} = 5.2$$

$$x_3 = \frac{(5.2)^2 + 27}{2(5.2)} = 5.19615\ldots$$

$$x_4 = \frac{(5.19615)^2 + 27}{2(5.19615)} = 5.19615\ldots$$

Since the first 3 places to the right of the decimal point did not change, we may round to 2 places: $\sqrt{27} \approx \mathbf{5.20.}$

7. $\sqrt{28}$: Initial estimate $= x_1 = 5$

$$x_2 = \frac{(5)^2 + 28}{2(5)} = \frac{53}{10} = 5.3$$

$$x_3 = \frac{(5.3)^2 + 28}{2(5.3)} = 5.29150\ldots$$

$$x_4 = \frac{(5.29150)^2 + 28}{2(5.29150)} = 5.29150\ldots$$

Since the first 3 places to the right of the decimal point did not change, we may round to 2 places: $\sqrt{28} \approx \mathbf{5.29.}$

8. $\sqrt{105}$: Initial estimate $= x_1 = 10$

$$x_2 = \frac{(10)^2 + 105}{2(10)} = \frac{205}{20} = 10.25$$

$$x_3 = \frac{(10.25)^2 + 105}{2(10.25)} = 10.24695\ldots$$

$$x_4 = \frac{(10.24695)^2 + 105}{2(10.24695)} = 10.24695\ldots$$

Since the first 3 places to the right of the decimal point did not change, we may round to 2 places: $\sqrt{105} \approx \mathbf{10.25.}$

9. $\sqrt{6}$: Initial estimate $= x_1 = 2$

$$x_2 = \frac{(2)^2 + 6}{2(2)} = \frac{10}{4} = 2.5$$

$$x_3 = \frac{(2.5)^2 + 6}{2(2.5)} = 2.45$$

$$x_4 = \frac{(2.45)^2 + 6}{2(2.45)} = 2.4494897\ldots$$

$$x_5 = \frac{(2.4494897)^2 + 6}{2(2.4494897)} = 2.4494897\ldots$$

Since the first 5 places to the right of the decimal point did not change, we may round to 4 places: $\sqrt{6} \approx \mathbf{2.4495.}$

10. $\sqrt{7}$: Initial estimate $= x_1 = 3$

$$x_2 = \frac{(3)^2 + 7}{2(3)} = \frac{16}{6} = 2.\overline{6}$$

$$x_3 = \frac{(2.\overline{6})^2 + 7}{2(2.\overline{6})} = 2.6458\overline{3}$$

$$x_4 = \frac{(2.6458\overline{3})^2 + 7}{2(2.6458\overline{3})} = 2.6457513\ldots$$

$$x_5 = \frac{(2.6457513)^2 + 7}{2(2.6457513)} = 2.6457513\ldots$$

Since the first 5 places to the right of the decimal point did not change, we may round to 4 places: $\sqrt{7} \approx \mathbf{2.6458.}$

11. $\sqrt{11}$: Initial estimate $= x_1 = 3$

$$x_2 = \frac{(3)^2 + 11}{2(3)} = \frac{20}{6} = 3.\overline{3}$$

$$x_3 = \frac{(3.\overline{3})^2 + 11}{2(3.\overline{3})} = 3.31\overline{6}$$

$$x_4 = \frac{(3.31\overline{6})^2 + 11}{2(3.31\overline{6})} = 3.3166247\ldots$$

$$x_5 = \frac{(3.3166247)^2 + 11}{2(3.3166247)} = 3.3166247\ldots$$

Since the first 5 places to the right of the decimal point did not change, we may round to 4 places: $\sqrt{11} \approx \mathbf{3.3166.}$

12. $\sqrt{12}$: Initial estimate $= x_1 = 3$

$$x_2 = \frac{(3)^2 + 12}{2(3)} = \frac{21}{6} = 3.5$$

$$x_3 = \frac{(3.5)^2 + 12}{2(3.5)} = 3.4642857...$$

$$x_4 = \frac{(3.4642857)^2 + 12}{2(3.4642857)} = 3.4641016...$$

$$x_5 = \frac{(3.4641016)^2 + 12}{2(3.4641016)} = 3.4641016...$$

Since the first 5 places to the right of the decimal point did not change, we may round to 4 places:
$\sqrt{12} \approx$ **3.4641.**

13. $\sqrt{24}$: Initial estimate $= x_1 = 5$

$$x_2 = \frac{(5)^2 + 24}{2(5)} = \frac{49}{10} = 4.9$$

$$x_3 = \frac{(4.9)^2 + 24}{2(4.9)} = 4.8989795...$$

$$x_4 = \frac{(4.8989795)^2 + 24}{2(4.8989795)} = 4.8989795...$$

Since the first 5 places to the right of the decimal point did not change, we may round to 4 places:
$\sqrt{24} \approx$ **4.8990.**

14. $\sqrt{29}$: Initial estimate $= x_1 = 5$

$$x_2 = \frac{(5)^2 + 29}{2(5)} = \frac{54}{10} = 5.4$$

$$x_3 = \frac{(5.4)^2 + 29}{2(5.4)} = 5.3851851...$$

$$x_4 = \frac{(5.3851851)^2 + 29}{2(5.3851851)} = 5.3851648...$$

$$x_5 = \frac{(5.3851648)^2 + 29}{2(5.3851648)} = 5.3851648...$$

Since the first 5 places to the right of the decimal point did not change, we may round to 4 places:
$\sqrt{29} \approx$ **5.3852.**

15. $\sqrt[3]{12}$: Initial estimate $= x_1 = 2$

$x_1 \cdot x_1 \cdot a_1 = 12$
$2 \cdot 2 \cdot a_1 = 12$
$\qquad a_1 = 3$

$$x_2 = \frac{2(x_1) + a_1}{3} = \frac{2(2) + 3}{3} = \frac{7}{3} = 2.\overline{3}$$

$x_2 \cdot x_2 \cdot a_2 = 12$
$(2.\overline{3})(2.\overline{3})a_2 = 12$
$\qquad a_2 = 2.20408...$

$$x_3 = \frac{2(x_2) + a_2}{3} = \frac{2(2.\overline{3}) + 2.20408}{3}$$
$$= 2.29024...$$

$x_3 \cdot x_3 \cdot a_3 = 12$
$(2.29024)(2.29024)a_3 = 12$
$\qquad a_3 = 2.28780...$

$$x_4 = \frac{2(x_3) + a_3}{3} = \frac{2(2.29024) + 2.28780}{3}$$
$$= 2.28942...$$

$x_4 \cdot x_4 \cdot a_4 = 12$
$(2.28942)(2.28942)a_4 = 12$
$\qquad a_4 = 2.28944...$

$$x_5 = \frac{2(x_4) + a_4}{3} = \frac{2(2.28942) + 2.28944}{3}$$
$$= 2.28944...$$

The first 3 places to the right of the decimal point were stable, so we may round to 2 places:
$\sqrt[3]{12} \approx$ **2.29.**

16. $\sqrt[3]{5}$: Initial estimate $= x_1 = 2$

$x_1 \cdot x_1 \cdot a_1 = 5$
$2 \cdot 2 \cdot a_1 = 5$
$\qquad a_1 = \frac{5}{4} = 1.25$

$$x_2 = \frac{2(x_1) + a_1}{3} = \frac{2(2) + 1.25}{3} = 1.75$$

$x_2 \cdot x_2 \cdot a_2 = 5$
$(1.75)(1.75)a_2 = 5$
$\qquad a_2 = 1.63265...$

$$x_3 = \frac{2(x_2) + a_2}{3} = \frac{2(1.75) + 1.63265}{3}$$
$$= 1.71088...$$

$x_3 \cdot x_3 \cdot a_3 = 5$
$(1.71088)(1.71088)a_3 = 5$
$\qquad a_3 = 1.70816...$

$$x_4 = \frac{2(x_3) + a_3}{3} = \frac{2(1.71088) + 1.70816}{3}$$
$$= 1.70997...$$

$x_4 \cdot x_4 \cdot a_4 = 5$
$(1.70997)(1.70997)a_4 = 5$
$\qquad a_4 = 1.70998...$

$$x_5 = \frac{2(x_4) + a_4}{3} = \frac{2(1.70997) + 1.70998}{3}$$
$$= 1.70997...$$

The first 3 places to the right of the decimal point were stable, so we may round to 2 places:
$\sqrt[3]{5} \approx$ **1.71.**

17. $\sqrt[3]{100}$: Initial estimate $= x_1 = 5$

$x_1 \cdot x_1 \cdot a_1 = 100$

$5 \cdot 5 \cdot a_1 = 100$

$a_1 = \dfrac{100}{25} = 4$

$x_2 = \dfrac{2(x_1) + a_1}{3} = \dfrac{2(5) + 4}{3} = \dfrac{14}{3} = 4.\overline{6}$

$x_2 \cdot x_2 \cdot a_2 = 100$

$(4.\overline{6})(4.\overline{6})a_2 = 100$

$a_2 = 4.59183\ldots$

$x_3 = \dfrac{2(x_2) + a_2}{3} = \dfrac{2(4.\overline{6}) + 4.59183}{3}$

$= 4.64172\ldots$

$x_3 \cdot x_3 \cdot a_3 = 100$

$(4.64172)^2 \cdot a_3 = 100$

$a_3 = 4.64132\ldots$

$x_4 = \dfrac{2(x_3) + a_3}{3} = \dfrac{2(4.64172) + 4.64132}{3}$

$= 4.64158\ldots$

The first 3 places to the right of the decimal point were stable, so we may round to 2 places: $\sqrt[3]{100} \approx \mathbf{4.64.}$

18. $\sqrt[3]{120}$: Initial estimate $= x_1 = 5$

$x_1 \cdot x_1 \cdot a_1 = 120$

$5 \cdot 5 \cdot a_1 = 120$

$a_1 = \dfrac{120}{25} = 4.8$

$x_2 = \dfrac{2(x_1) + a_1}{3} = \dfrac{2(5) + 4.8}{3} = 4.9\overline{3}$

$x_2 \cdot x_2 \cdot a_2 = 120$

$(4.9\overline{3})^2 \cdot a_2 = 120$

$a_2 = 4.930606\ldots$

$x_3 = \dfrac{2(x_2) + a_2}{3} = \dfrac{2(4.9\overline{3}) + 4.930606}{3}$

$= 4.932424\ldots$

$x_3 \cdot x_3 \cdot a_3 = 120$

$(4.932424)^2 \cdot a_3 = 120$

$a_3 = 4.932424\ldots$

$x_4 = \dfrac{2(x_3) + a_3}{3}$

$= \dfrac{2(4.932424) + (4.932424)}{3}$

$= 4.932424$

The first 3 places to the right of the decimal point were stable, so we may round to 2 places: $\sqrt[3]{120} \approx \mathbf{4.93.}$

19. $\sqrt[4]{6}$: Initial estimate $= x_1 = 2$

$x_1 \cdot x_1 \cdot x_1 \cdot a_1 = 6$

$2 \cdot 2 \cdot 2 \cdot a_1 = 6$

$a_1 = \dfrac{6}{8} = 0.75$

$x_2 = \dfrac{3(x_1) + a_1}{4} = \dfrac{3(2) + 0.75}{4}$

$= \dfrac{6.75}{4} = 1.6875$

$x_2 \cdot x_2 \cdot x_2 \cdot a_2 = 6$

$(1.6875)^3 \cdot a_2 = 6$

$a_2 = 1.24859\ldots$

$x_3 = \dfrac{3(x_2) + a_2}{4} = \dfrac{3(1.6875) + 1.24859}{4}$

$= 1.57777\ldots$

$x_3 \cdot x_3 \cdot x_3 \cdot a_3 = 6$

$(1.57777)^3 \cdot a_3 = 6$

$a_3 = 1.52763\ldots$

$x_4 = \dfrac{3(x_3) + a_3}{4} = \dfrac{3(1.57777) + 1.52763}{4}$

$= 1.565235$

$x_4 \cdot x_4 \cdot x_4 \cdot a_4 = 6$

$(1.565235)^3 \cdot a_4 = 6$

$a_4 = 1.56463\ldots$

$x_5 = \dfrac{3(x_4) + a_4}{4} = \dfrac{3(1.565235) + 1.56463}{4}$

$= 1.56508$

The first 3 places to the right of the decimal point were stable, so we may round to 2 places: $\sqrt[4]{6} \approx \mathbf{1.57.}$

20. $\sqrt[4]{25}$: Initial estimate $= x_1 = 2$

$x_1 \cdot x_1 \cdot x_1 \cdot a_1 = 25$

$(2)^3 \cdot a_1 = 25$

$a_1 = 3.125$

$x_2 = \dfrac{3(x_1) + a_1}{4} = \dfrac{3(2) + 3.125}{4}$

$= 2.28125$

$x_2 \cdot x_2 \cdot x_2 \cdot a_2 = 25$

$(2.28125)^3 \cdot a_2 = 25$

$a_2 = 2.10582...$

$x_3 = \dfrac{3(x_2) + a_2}{4} = \dfrac{3(2.28125) + 2.10582}{4}$

$= 2.23739...$

$x_3 \cdot x_3 \cdot x_3 \cdot a_3 = 25$

$(2.23739)^3 \cdot a_3 = 25$

$a_3 = 2.23210...$

$x_4 = \dfrac{3(x_3) + a_3}{4} = \dfrac{3(2.23739) + 2.23210}{4}$

$= 2.23606...$

$x_4 \cdot x_4 \cdot x_4 \cdot a_4 = 25$

$(2.23606)^3 \cdot a_4 = 25$

$a_4 = 2.23609$

$x_5 = \dfrac{3(x_4) + a_4}{4} = \dfrac{3(2.23606) + 2.23609}{4}$

$= 2.23606...$

The first 3 places to the right of the decimal point were stable, so we may round to 2 places:
$\sqrt[4]{25} \approx$ **2.24.**

21. $\sqrt[4]{109}$: Initial estimate $= x_1 = 3$

$x_1 \cdot x_1 \cdot x_1 \cdot a_1 = 109$

$(3)^3 \cdot a_1 = 109$

$a_1 = 4.\overline{037}$

$x_2 = \dfrac{3(x_1) + a_1}{4} = \dfrac{3(3) + 4.\overline{037}}{4} = 3.\overline{259}$

$x_2 \cdot x_2 \cdot x_2 \cdot a_2 = 109$

$(3.\overline{259})^3 \cdot a_2 = 109$

$a_2 = 3.148254...$

$x_3 = \dfrac{3(x_2) + a_2}{4} = \dfrac{3(3.\overline{259}) + 3.148254}{4}$

$= 3.231507...$

$x_3 \cdot x_3 \cdot x_3 \cdot a_3 = 109$

$(3.231508)^3 \cdot a_3 = 109$

$a_3 = 3.230064...$

$x_4 = \dfrac{3(x_3) + a_3}{4}$

$= \dfrac{3(3.231508) + (3.230064)}{4}$

$= 3.231146...$

The first 3 places to the right of the decimal point were stable, so we may round to 2 places:
$\sqrt[4]{109} \approx$ **3.23.**

22. $\sqrt[5]{15}$: Initial estimate $= x_1 = 2$

$x_1 \cdot x_1 \cdot x_1 \cdot x_1 \cdot a_1 = 15$

$(2)^4 \cdot a_1 = 15$

$a_1 = 0.9375$

$x_2 = \dfrac{4(x_1) + a_1}{5} = \dfrac{4(2) + 0.9375}{5}$

$= 1.7875$

$x_2 \cdot x_2 \cdot x_2 \cdot x_2 \cdot a_2 = 15$

$(1.7875)^4 \cdot a_2 = 15$

$a_2 = 1.46928...$

$x_3 = \dfrac{4(x_2) + a_2}{5} = \dfrac{4(1.7875) + 1.46928}{5}$

$= 1.72385...$

$x_3 \cdot x_3 \cdot x_3 \cdot x_3 \cdot a_3 = 15$

$(1.72385)^4 \cdot a_3 = 15$

$a_3 = 1.69860...$

$x_4 = \dfrac{4(x_3) + a_3}{5} = \dfrac{4(1.72385) + 1.69860}{5}$

$= 1.71880$

$x_4 \cdot x_4 \cdot x_4 \cdot x_4 \cdot a_4 = 15$

$(1.71880)^4 \cdot a_4 = 15$

$a_4 = 1.71865...$

$x_5 = \dfrac{4(x_4) + a_4}{5} = \dfrac{4(1.71880) + 1.71865}{5}$

$= 1.71877$

The first 3 places to the right of the decimal point were stable, so we may round to 2 places:
$\sqrt[5]{15} \approx$ **1.72.**

23. $\sqrt[5]{155}$: Initial estimate $= x_1 = 3$

$x_1 \cdot x_1 \cdot x_1 \cdot x_1 \cdot a_1 = 155$

$(3)^4 \cdot a_1 = 155$

$a_1 = 1.913580...$

$x_2 = \dfrac{4(x_1) + a_1}{5} = \dfrac{4(3) + 1.913580}{5}$

$= 2.782716$

$$x_2 \cdot x_2 \cdot x_2 \cdot x_2 \cdot a_2 = 155$$
$$(2.782716)^4 \cdot a_2 = 155$$
$$a_2 = 2.584973\ldots$$

$$x_3 = \frac{4(x_2) + a_2}{5}$$
$$= \frac{4(2.782716) + 2.584973}{5}$$
$$= 2.743167\ldots$$

$$x_3 \cdot x_3 \cdot x_3 \cdot x_3 \cdot a_3 = 155$$
$$(2.743167)^4 \cdot a_3 = 155$$
$$a_3 = 2.737301\ldots$$

$$x_4 = \frac{4(x_3) + a_3}{5}$$
$$= \frac{4(2.743167) + 2.737301}{5}$$
$$= 2.741993\ldots$$

$$x_4 \cdot x_4 \cdot x_4 \cdot x_4 \cdot a_4 = 155$$
$$(2.741993)^4 \cdot a_4 = 155$$
$$a_4 = 2.741992\ldots$$

$$x_5 = \frac{4(x_4) + a_4}{5}$$
$$= \frac{4(2.741993) + 2.741992}{5}$$
$$= 2.741992\ldots$$

The first 3 places to the right of the decimal point were stable, so we may round to 2 places: $\sqrt[5]{155} \approx \textbf{2.74.}$

24. $\sqrt[5]{200}$: Initial estimate $= x_1 = 3$

$$x_1 \cdot x_1 \cdot x_1 \cdot x_1 \cdot a_1 = 200$$
$$(3)^4 \cdot a_1 = 200$$
$$a_1 = 2.469135\ldots$$

$$x_2 = \frac{4(x_1) + a_1}{5} = \frac{4(3) + 2.469135}{5}$$
$$= 2.893827$$

$$x_2 \cdot x_2 \cdot x_2 \cdot x_2 \cdot a_2 = 200$$
$$(2.893827)^4 \cdot a_2 = 200$$
$$a_2 = 2.851935\ldots$$

$$x_3 = \frac{4(x_2) + a_2}{5}$$
$$= \frac{4(2.893827) + 2.851935}{5}$$
$$= 2.8854486\ldots$$

$$x_3 \cdot x_3 \cdot x_3 \cdot x_3 \cdot a_3 = 200$$
$$(2.8854486)^4 \cdot a_3 = 200$$
$$a_3 = 2.885204\ldots$$

$$x_4 = \frac{4(x_3) + a_3}{5}$$
$$= \frac{4(2.8854486) + 2.885204}{5}$$
$$= 2.885399\ldots$$

The first 3 places to the right of the decimal point were stable, so we may round to 2 places: $\sqrt[5]{200} \approx \textbf{2.89.}$

25. $\sqrt[6]{85}$: Initial estimate $= x_1 = 2$

$$x_1 \cdot x_1 \cdot x_1 \cdot x_1 \cdot x_1 \cdot a_1 = 85$$
$$(2)^5 \cdot a_1 = 85$$
$$a_1 = 2.65625$$

$$x_2 = \frac{5(x_1) + a_1}{6} = \frac{5(2) + 2.65625}{6}$$
$$= 2.109375$$

$$x_2 \cdot x_2 \cdot x_2 \cdot x_2 \cdot x_2 \cdot a_2 = 85$$
$$(2.109375)^5 \cdot a_2 = 85$$
$$a_2 = 2.035400\ldots$$

$$x_3 = \frac{5(x_2) + a_2}{6}$$
$$= \frac{5(2.109375) + 2.035400}{6}$$
$$= 2.097045\ldots$$

$$x_3 \cdot x_3 \cdot x_3 \cdot x_3 \cdot x_3 \cdot a_3 = 85$$
$$(2.097045)^5 \cdot a_3 = 85$$
$$a_3 = 2.095946\ldots$$

$$x_4 = \frac{5(x_3) + a_3}{6}$$
$$= \frac{5(2.097045) + 2.095946}{6}$$
$$= 2.096861$$

$$x_4 \cdot x_4 \cdot x_4 \cdot x_4 \cdot x_4 \cdot a_4 = 85$$
$$(2.096861)^5 \cdot a_4 = 85$$
$$a_4 = 2.096866\ldots$$

$$x_5 = \frac{5(x_4) + a_4}{6}$$
$$= \frac{5(2.096861) + 2.096866}{6}$$
$$= 2.096861\ldots$$

The first 3 places to the right of the decimal point were stable, so we may round to 2 places: $\sqrt[6]{85} \approx \textbf{2.10.}$

PRACTICE F

1. $(3x^3 + 2x^2 - x + 4) - (x^2 - 7x - 5)$
$= 3x^3 + 2x^2 - x + 4 - x^2 + 7x + 5$
$= \mathbf{3x^3 + x^2 + 6x + 9}$

2. $-(3x^5 + 6x^4 - 7x^3 - 5)$
$\quad + (x^5 - x^4 + 3x^2 - 2x - 8)$
$= -3x^5 - 6x^4 + 7x^3 + 5 + x^5 - x^4$
$\quad + 3x^2 - 2x - 8$
$= \mathbf{-2x^5 - 7x^4 + 7x^3 + 3x^2 - 2x - 3}$

3. $(3x^2 + x^4 - 6x + 2)$
$\quad - (15x^4 + 2x^3 - 6x^2 + 5x - 3)$
$= x^4 + 3x^2 - 6x + 2 - 15x^4 - 2x^3 + 6x^2$
$\quad - 5x + 3$
$= \mathbf{-14x^4 - 2x^3 + 9x^2 - 11x + 5}$

4. $(6k^7 - 5k^5 + 3k^3 - k)$
$\quad + (5k^6 + 4k^4 + 2k^3 + 3k^2)$
$= \mathbf{6k^7 + 5k^6 - 5k^5 + 4k^4 + 5k^3 + 3k^2 - k}$

5. $(5k^4 + 3k^3 - 2k^2 - 1)$
$\quad - (4k^4 + 2k^3 - 3k^2 - k - 2)$
$= 5k^4 + 3k^3 - 2k^2 - 1 - 4k^4 - 2k^3 + 3k^2$
$\quad + k + 2$
$= \mathbf{k^4 + k^3 + k^2 + k + 1}$

6. $(2x^3 + 5x^2 - 7) + (2x^2 - 4x + 1)$
$\quad - (3x^4 + 4x^2 - 2x)$
$= 2x^3 + 5x^2 - 7 + 2x^2 - 4x + 1 - 3x^4$
$\quad - 4x^2 + 2x$
$= \mathbf{-3x^4 + 2x^3 + 3x^2 - 2x - 6}$

7. $(x^4 - 3x^3 + x^2 + 2) - (3x^2 + 7x - 2)$
$\quad + (-x^3 + 2x^2 - 4)$
$= x^4 - 3x^3 + x^2 + 2 - 3x^2 - 7x + 2 - x^3$
$\quad + 2x^2 - 4$
$= \mathbf{x^4 - 4x^3 - 7x}$

8. $(3x^3y - 4x^2y + 2xy - y)$
$\quad + (-x^3y + 5x^2y - 6xy - 3y)$
$= \mathbf{2x^3y + x^2y - 4xy - 4y}$

9. $(7xyz)(3wy^2z) = \mathbf{21wxy^3z^2}$

10. $(2x^2y^2)(4xy^5z^3)(-3x^5y^{10}z^{19})$
$= 8x^3y^7z^3(-3x^5y^{10}z^{19}) = \mathbf{-24x^8y^{17}z^{22}}$

11. $(4x + 2)(x - 5) = 4x(x - 5) + 2(x - 5)$
$= 4x^2 - 20x + 2x - 10 = \mathbf{4x^2 - 18x - 10}$

12. $(4x + 2)(3x - 5) = 4x(3x - 5) + 2(3x - 5)$
$= 12x^2 - 20x + 6x - 10 = \mathbf{12x^2 - 14x - 10}$

13. $(3x + 2)^2 = (3x + 2)(3x + 2)$
$= 3x(3x + 2) + 2(3x + 2)$
$= 9x^2 + 6x + 6x + 4 = \mathbf{9x^2 + 12x + 4}$

14. $(x - 27y)(3x - 3y)$
$= x(3x - 3y) - 27y(3x - 3y)$
$= 3x^2 - 3xy - 81xy + 81y^2$
$= \mathbf{3x^2 - 84xy + 81y^2}$

15. $4x^2(3x - y + 5x^2y) = 12x^3 - 4x^2y + 20x^4y$
$= \mathbf{20x^4y + 12x^3 - 4x^2y}$

16. $(5x)(3y)(5x - 3y) = 15xy(5x - 3y)$
$= \mathbf{75x^2y - 45xy^2}$

17. $abc(abc - def) = \mathbf{a^2b^2c^2 - abcdef}$

18. $12n^2(2n - 3n^2) = 24n^3 - 36n^4$
$= \mathbf{-36n^4 + 24n^3}$

19. $\dfrac{140x^2}{7x} = \mathbf{20x}$

20. $\dfrac{88x^{10}y}{11xy} = \mathbf{8x^9}$

21. $\dfrac{90m^4n^3}{6mn^3} = \mathbf{15m^3}$

22. $\dfrac{200m^3n^{10}q}{10m^2n} = \mathbf{20mn^9q}$

23. $\dfrac{200m^3 - 10m}{5m} = \dfrac{200m^3}{5m} - \dfrac{10m}{5m}$
$= \mathbf{40m^2 - 2}$

24. $\dfrac{6x^3 - 3x^2 + 9x}{3x} = \dfrac{6x^3}{3x} - \dfrac{3x^2}{3x} + \dfrac{9x}{3x}$
$= \mathbf{2x^2 - x + 3}$

25. $\dfrac{80x^5 - 40x^2y + 8x}{8x}$
$= \dfrac{80x^5}{8x} - \dfrac{40x^2yz}{8x} + \dfrac{8x}{8x} = \mathbf{10x^4 - 5xyz + 1}$

26. $\dfrac{40a - 4a^3 + 8a^2b}{4a} = \dfrac{40a}{4a} - \dfrac{4a^3}{4a} + \dfrac{8a^2b}{4a}$
$= 10 - a^2 + 2ab = \mathbf{-a^2 + 2ab + 10}$

27. $\dfrac{12c^3 + 20c^2d - 400cd}{4c}$

$= \dfrac{12c^3}{4c} + \dfrac{20c^2d}{4c} - \dfrac{400cd}{4c}$

$= 3c^2 + 5cd - 100d$

28. $\dfrac{90y^3 + 18xy^2 - 45y}{9y}$

$= \dfrac{90y^3}{9y} + \dfrac{18xy^2}{9y} - \dfrac{45y}{9y}$

$= 10y^2 + 2xy - 5$

PRACTICE G

1. $T: (x, y) \longrightarrow (x + 3, y + 2)$

$T: (1, 1) \longrightarrow (4, 3) = A'$

$T: (0, 2) \longrightarrow (3, 4) = B'$

$T: (-2, -3) \longrightarrow (1, -1) = C'$

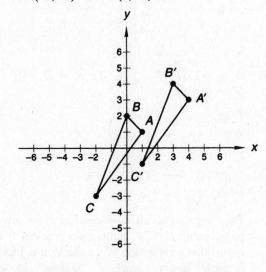

2. $T: (x, y) \longrightarrow (x - 1, y + 4)$

$T: (-2, 0) \longrightarrow (-3, 4) = A'$

$T: (-2, -2) \longrightarrow (-3, 2) = B'$

$T: (0, -2) \longrightarrow (-1, 2) = C'$

$T: (0, 0) \longrightarrow (-1, 4) = D'$

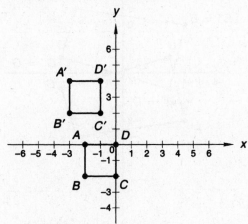

3. $T: (x, y) \longrightarrow (x + 4, y)$

$T: (-4, -2) \longrightarrow (0, -2) = A'$

$T: (-2, 1) \longrightarrow (2, 1) = B'$

$T: (0, 0) \longrightarrow (4, 0) = C'$

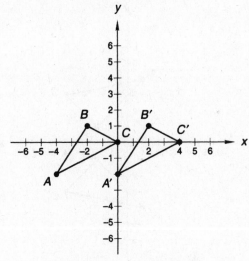

4. $R: (x, y) \longrightarrow (x + 8, y - 1)$

$R: (-5, -3) \longrightarrow (3, -4) = A'$

$R: (-4, 1) \longrightarrow (4, 0) = B'$

$R: (-1, -1) \longrightarrow (7, -2) = C'$

$S: (x, y) \longrightarrow (x - 1, y + 4)$

$S: (3, -4) \longrightarrow (2, 0) = A''$

$S: (4, 0) \longrightarrow (3, 4) = B''$

$S: (7, -2) \longrightarrow (6, 2) = C''$

The transformation T is a combination of the transformations R and S. We apply S to the result of R.

$T: (x, y) \longrightarrow ((x + 8) - 1, (y - 1) + 4)$
$\qquad\qquad = (x + 7, y + 3)$

$T: (x, y) \longrightarrow (x + 7, y + 3)$

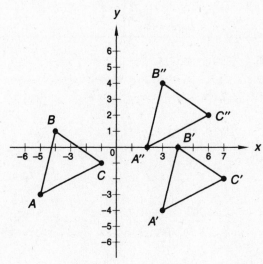

5. $R: (x, y) \rightarrow (x, -y)$

$R: (-1, 4) \rightarrow (-1, -4) = A'$

$R: (2, 1) \rightarrow (2, -1) = B'$

$R: (4, 3) \rightarrow (4, -3) = C'$

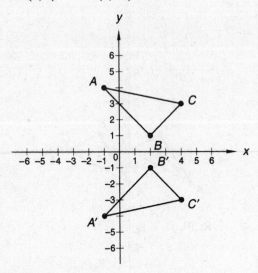

6. $S: (x, y) \rightarrow (-x, y)$

$S: (-1, 4) \rightarrow (1, 4) = A'$

$S: (2, 1) \rightarrow (-2, 1) = B'$

$S: (4, 3) \rightarrow (-4, 3) = C'$

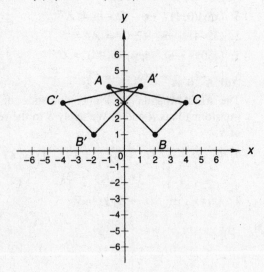

7. $T: (x, y) \rightarrow (x, -y)$

$T: (-3, -1) \rightarrow (-3, 1) = D'$

$T: (-4, 3) \rightarrow (-4, -3) = E'$

$T: (3, 4) \rightarrow (3, -4) = F'$

$T: (4, 1) \rightarrow (4, -1) = G'$

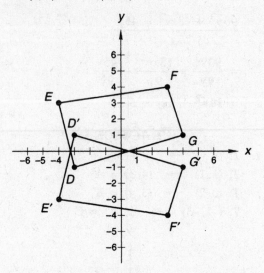

8. $D' = (-3, 1)$; $E' = (-4, -3)$; $F' = (3, -4)$; $G' = (4, -1)$

$T: (x, y) \rightarrow (x, -y)$

$T: (-3, 1) \rightarrow (-3, -1) = D''$

$T: (-4, -3) \rightarrow (-4, 3) = E''$

$T: (3, -4) \rightarrow (3, 4) = F''$

$T: (4, -1) \rightarrow (4, 1) = G''$

The image points are now identical to the original image points D, E, F, and G from problem 7. This means that **a reflection of a reflection is just the original image.** Since T "undoes" its own operation, it is its own inverse operation.

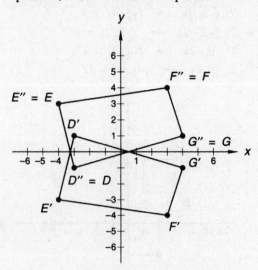

9. $R: (x, y) \rightarrow (-y, x)$

$R: (0, 0) \rightarrow (0, 0) = A'$

$R: (-1, -3) \rightarrow (3, -1) = B'$

$R: (1, -2) \rightarrow (2, 1) = C'$

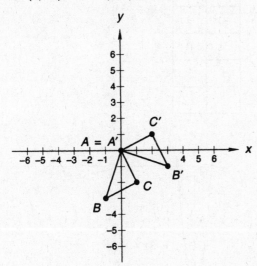

10. $S: (x, y) \rightarrow (-x, -y)$

$S: (-1, 0) \rightarrow (1, 0) = D'$

$S: (-4, 1) \rightarrow (4, -1) = E'$

$S: (-3, 3) \rightarrow (3, -3) = F'$

$S: (0, 4) \rightarrow (0, -4) = G'$

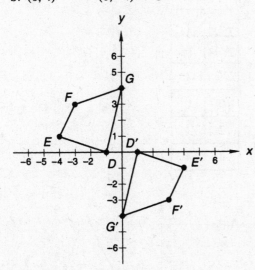

11. $T: (x, y) \rightarrow (y, -x)$

$T: (-1, 0) \rightarrow (0, 1) = H'$

$T: (-3, 0) \rightarrow (0, 3) = I'$

$T: (-2, 3) \rightarrow (3, 2) = J'$

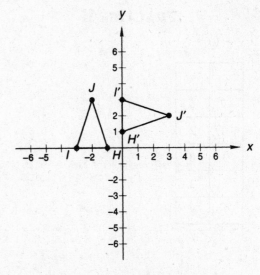

12. (a) $R: (x, y) \rightarrow (-y, x)$

$R: (0, 1) \rightarrow (-1, 0) = A'$

$R: (-1, -3) \rightarrow (3, -1) = B'$

$R: (-3, 2) \rightarrow (-2, -3) = C'$

(b) $R: (-1, 0) \rightarrow (0, -1) = A''$

$R: (3, -1) \rightarrow (1, 3) = B''$

$R: (-2, -3) \rightarrow (3, -2) = C''$

(c) $S: (x, y) \rightarrow (-x, -y)$

$S: (0, 1) \rightarrow (0, -1) = A'''$

$S: (-1, -3) \rightarrow (1, 3) = B'''$

$S: (-3, 2) \rightarrow (3, -2) = C'''$

(d) $A'' = A'''$, $B'' = B'''$, $C'' = C'''$
They are the same points.

13. (a) $T: (x, y) \rightarrow (-x, y)$

$T: (-1, 0) \rightarrow (1, 0) = D'$

$T: (-4, 1) \rightarrow (4, 1) = E'$

$T: (-3, 3) \rightarrow (3, 3) = F'$

$T: (0, 4) \rightarrow (0, 4) = G'$

(b) $W: (x, y) \rightarrow (x, -y)$

$W: (1, 0) \rightarrow (1, 0) = D''$

$W: (4, 1) \rightarrow (4, -1) = E''$

$W: (3, 3) \rightarrow (3, -3) = F''$

$W: (0, 4) \rightarrow (0, -4) = G''$

(c) $S: (x, y) \rightarrow (-x, -y)$

$S: (-1, 0) \rightarrow (1, 0) = D'''$

$S: (-4, 1) \rightarrow (4, -1) = E'''$

$S: (-3, 3) \rightarrow (3, -3) = F'''$

$S: (0, 4) \rightarrow (0, -4) = G'''$

(d) $D'' = D'''$, $E'' = E'''$, $F'' = F'''$, $G'' = G'''$
They are the same points.

PRACTICE H

1. $y = 2x^2$

x	y
-2	8
-1	2
0	0
1	2
2	8

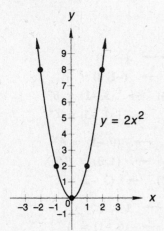

$y = 2x^2$

2. $y = \frac{1}{2}x^2$

x	y
-4	8
-2	2
0	0
2	2
4	8

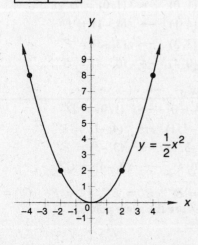

$y = \frac{1}{2}x^2$

3. $y = 3x^2$

x	y
-2	12
-1	3
0	0
1	3
2	12

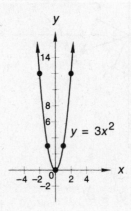

$y = 3x^2$

4. $y = \frac{1}{4}x^2$

x	y
-4	4
-2	1
0	0
2	1
4	4

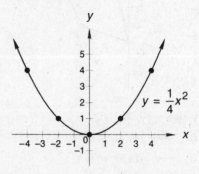

$y = \frac{1}{4}x^2$

5. $y = 5x^2$

x	y
-2	20
-1	5
0	0
1	5
2	20

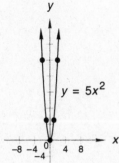

8. $y = -3x^2$

x	y
-2	-12
-1	-3
0	0
1	-3
2	-12

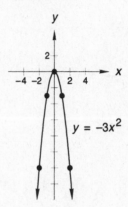

6. The higher the value of the coefficient, the steeper the curve. The smaller the coefficient, the flatter or shallower the curve. Alternatively, the larger the coefficient, the thinner the opening of the curve. The smaller the coefficient, the wider the opening of the curve.

7. $y = -x^2$

x	y
-3	-9
-2	-4
0	0
2	-4
3	-9

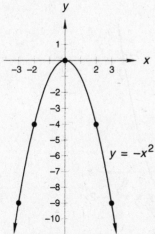

9. $y = 4x^2$

x	y
-2	16
-1	4
0	0
$\frac{1}{2}$	1
1	4
$\frac{3}{2}$	9
2	16

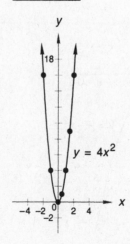

10. $y = -2x^2$

x	y
-2	-8
-1	-2
0	0
1	-2
2	-8

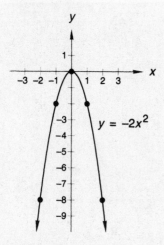

11. $y = -\frac{1}{2}x^2$

x	y
-4	-8
-2	-2
0	0
2	-2
4	-8

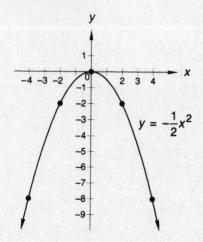

12. When the quadratic term (x^2) has a negative coefficient, the graph is a parabola that opens downward.

13. $y = x^2 + 1$

x	y
-2	5
-1	2
0	1
1	2
2	5

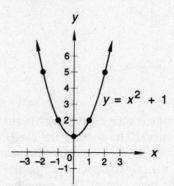

14. $y = x^2 - 2$

x	y
-3	7
-2	2
0	-2
2	2
3	7

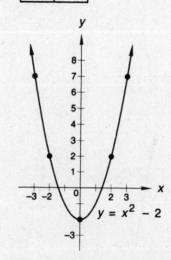

15. $y = x^2 + 4$

x	y
–2	8
–1	5
0	4
1	5
2	8

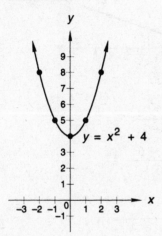

$y = x^2 + 4$

16. $y = x^2 - 5$

x	y
–3	4
–2	–1
0	–5
2	–1
3	4

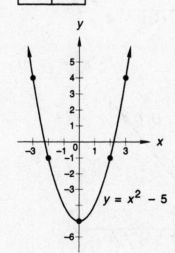

$y = x^2 - 5$

17. Adding a number to the x^2 term shifts (or slides or translates) the parabola up that many units, and subtracting a number from the x^2 term shifts (or slides or translates) the parabola down that many units.

18. $y = 2x^2 - 6$

x	y
–2	2
–1	–4
0	–6
1	–4
2	2

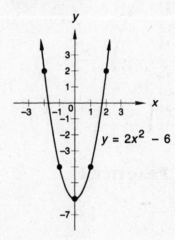

$y = 2x^2 - 6$

19. $y = 4x^2 - 10$

x	y
–2	6
–1	–6
0	–10
1	–6
2	6

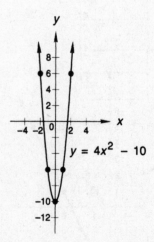

$y = 4x^2 - 10$

20. $y = -\frac{1}{2}x^2 + 5$

x	y
-4	-3
-2	3
0	5
2	3
4	-3

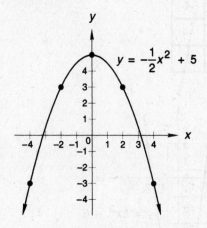

$y = -\frac{1}{2}x^2 + 5$

PRACTICE I

1. slope $= \dfrac{y_2 - y_1}{x_2 - x_1}$

$\qquad = \dfrac{5 - 1}{5 - 3}$

$\qquad = \dfrac{4}{2}$

$\qquad = \mathbf{2}$

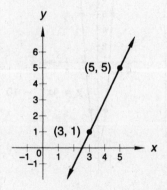

(5, 5)

(3, 1)

2. slope $= \dfrac{y_2 - y_1}{x_2 - x_1}$

$\qquad = \dfrac{2 - 0}{4 - (-2)}$

$\qquad = \dfrac{2}{6}$

$\qquad = \dfrac{\mathbf{1}}{\mathbf{3}}$

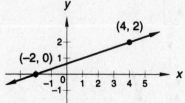

(4, 2)

(-2, 0)

3. slope $= \dfrac{y_2 - y_1}{x_2 - x_1}$

$\qquad = \dfrac{6 - 6}{(-10) - 5}$

$\qquad = \dfrac{0}{-15}$

$\qquad = \mathbf{0}$

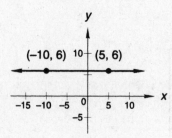

(-10, 6)   (5, 6)

4. slope $= \dfrac{y_2 - y_1}{x_2 - x_1}$

$\qquad = \dfrac{7 - 1}{4 - (-1)}$

$\qquad = \dfrac{\mathbf{6}}{\mathbf{5}}$

(4, 7)

(-1, 1)

5. slope $= \dfrac{y_2 - y_1}{x_2 - x_1}$

$= \dfrac{(-1) - 3}{4 - 2}$

$= \dfrac{-4}{2}$

$= \mathbf{-2}$

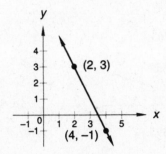

6. slope $= \dfrac{1 - (-2)}{2 - (-3)} = \dfrac{3}{5}$

7. slope $= \dfrac{2 - 0}{5 - 2} = \dfrac{2}{3}$

8. Two estimated points are (0, 1) and (1, 3).

slope $\approx \dfrac{3 - 1}{1 - 0} = \dfrac{2}{1} = 2$

The slope is **about 2.**

9. Two estimated points are (−10, 2) and (−4, −2).

slope $\approx \dfrac{-2 - 2}{-4 - (-10)} = \dfrac{-4}{6} = -\dfrac{2}{3}$

The slope is **about** $-\dfrac{2}{3}$.

10. $y = 2x + 1$

x	y
0	1
2	5

slope $= \dfrac{1 - 5}{0 - 2} = \dfrac{-4}{-2} = \mathbf{2}$

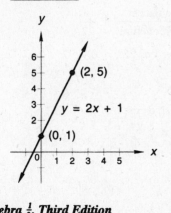

11. $y = -5x + 6$

x	y
0	6
1	1

slope $= \dfrac{6 - 1}{0 - 1} = \dfrac{5}{-1} = \mathbf{-5}$

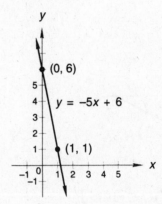

12. $y = 4x$

x	y
0	0
1	4

slope $= \dfrac{4 - 0}{1 - 0} = \dfrac{4}{1} = \mathbf{4}$

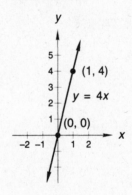

13. $y = -3x + 2$

x	y
0	2
1	-1

slope $= \dfrac{2 - (-1)}{0 - 1} = \dfrac{3}{-1}$

$= \mathbf{-3}$

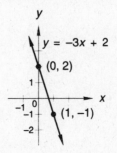

14. $y = \frac{1}{2}x - 3$

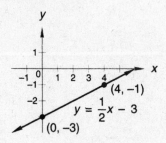

x	y
0	-3
4	-1

slope $= \dfrac{-3 - (-1)}{0 - 4} = \dfrac{-2}{-4}$

$= \dfrac{1}{2}$

PRACTICE J

1. $\sin A = \dfrac{10}{26} = \dfrac{5}{13}$

2. $\cos B = \dfrac{10}{26} = \dfrac{5}{13}$

3. $\tan D = \dfrac{8}{15}$

4. $\sin B = \dfrac{24}{26} = \dfrac{12}{13}$

5. $\cos A = \dfrac{24}{26} = \dfrac{12}{13}$

6. $\tan E = \dfrac{15}{8}$

7. $\sin D = \dfrac{8}{17}$

8. $\tan A = \dfrac{10}{24} = \dfrac{5}{12}$

9. $\cos E = \dfrac{8}{17}$

10. $\tan B = \dfrac{24}{10} = \dfrac{12}{5}$

11. $\cos D = \dfrac{15}{17}$

12. $\sin E = \dfrac{15}{17}$

13. $\tan 15° \approx \mathbf{0.2679}$

14. $\cos 30° \approx \mathbf{0.8660}$

15. $\tan 35° \approx \mathbf{0.7002}$

16. $\sin 10° \approx \mathbf{0.1736}$

17. $\sin 25° \approx \mathbf{0.4226}$

18. $\tan 10° \approx \mathbf{0.1763}$

19. $\cos 15° \approx \mathbf{0.9659}$

20. $\sin 45° \approx \mathbf{0.7071}$

21. $\cos 20° \approx \mathbf{0.9397}$

22. $\tan 45° = \mathbf{1}$

23. $\cos 10° \approx \mathbf{0.9848}$

24. $\sin 30° = \mathbf{0.5}$

25. The tangent of **25°** corresponds to 0.4663.

26. The sine of **15°** corresponds to 0.2588.

27. The cosine of **35°** corresponds to 0.8192.

28. The sine and cosine of **45°** correspond to the same value, 0.7071.

29. $\tan 40° = \dfrac{\text{opposite}}{\text{adjacent}} = \dfrac{12}{a}$

From the table, $\tan 40° \approx 0.8391$.

$0.8391 \approx \dfrac{12}{a}$

$0.8391a \approx 12$

$a \approx \dfrac{12}{0.8391}$

$a \approx \mathbf{14.3}$

30. $\cos 22° = \dfrac{\text{adjacent}}{\text{hypotenuse}} = \dfrac{y}{22}$

From the table, $\cos 22° \approx 0.9272$.

$0.9272 \approx \dfrac{y}{22}$

$y \approx 22(0.9272)$

$y \approx \mathbf{20.4}$

31. The length of the loading ramp corresponds to the length of the hypotenuse of the right triangle.

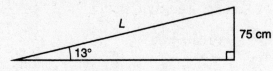

$$\sin 13° = \frac{\text{opposite}}{\text{hypotenuse}} = \frac{75 \text{ cm}}{L}$$

From the table, $\sin 13° \approx 0.2250$.

$$0.2250 \approx \frac{75 \text{ cm}}{L}$$
$$0.2250\, L \approx 75 \text{ cm}$$
$$L \approx \frac{75 \text{ cm}}{0.2250}$$
$$L \approx \textbf{333 cm}$$